ALTERNATIVES TO THE INTERNAL COMBUSTION ENGINE

IMPACTS ON ENVIRONMENTAL QUALITY

ALTERNATIVES TO THE INTERNAL COMBUSTION ENGINE

IMPACTS ON ENVIRONMENTAL QUALITY

by

Robert U. Ayres

and

Richard P. McKenna

Published for Resources for the Future, Inc.
by The Johns Hopkins University Press, Baltimore and London

Resources for the Future is a nonprofit corporation for research and education in the development, conservation, and use of natural resources and the improvement of the quality of the environment. It was established in 1952 with the cooperation of the Ford Foundation. Part of the work of Resources for the Future is carried out by its resident staff; part is supported by grants to universities and other nonprofit organizations. Unless otherwise stated, interpretations and conclusions in RFF publications are those of the authors; the organization takes responsibility for the selection of significant subjects for study, the competence of the researchers, and their freedom of inquiry.

This book is one of RFF's studies in quality of the environment, which are directed by Allen V. Kneese and Blair T. Bower. Robert U. Ayres, a physicist who engaged in research on this subject while he was a visiting scholar at RFF in 1967, is now vice-president of International Research and Technology, Inc. Richard P. McKenna, an engineer, is also associated with IR&T.

RFF editors: Henry Jarrett, Vera W. Dodds, Nora E. Roots, Tadd Fisher.

Copyright © 1972 by The Johns Hopkins University Press
All rights reserved
Manufactured in the United States of America

The Johns Hopkins University Press, Baltimore, Maryland 21218
The Johns Hopkins University Press Ltd., London

Library of Congress Catalog Card Number 74-181555
ISBN-0-8018-1369-7

Preface

This book provides a comprehensive analysis of existing and potential power plants for automotive vehicles. While technological and economic features of these devices are treated in detail the whole discussion revolves around the environmental effects of the alternatives considered. This book is highly important and timely since, in terms of poundage, the presently omnipresent internal combustion engine is by far the largest source of gaseous emissions to the atmosphere. In many of our urban areas it overwhelms all other sources combined by a huge margin. Moreover, it is a rapidly growing source, since the population of automobiles seems to rise inexorably. While air pollution is the central problem currently associated with internal combustion engines the ICE is also a major source of noise in our urban areas.

Until very recently only small attention was given to these environmental effects and, accordingly, few resources were expended to control them. Under pressure of rapidly deteriorating air quality in our urban areas the national government has now moved vigorously to restrict automotive emissions—and some progress is evident in the engines now being produced. Emissions of carbon monoxide and unburned hydrocarbons have been curtailed substantially by the use of relatively simple methods. But in the face of a growing population of automobiles it is hard to achieve much improvement in air quality without drastic reductions in emissions. Furthermore, virtually no progress has been made in curbing the emissions of one of the most serious components of ICE exhausts, the oxides of nitrogen. Indeed, the amount of this material has been somewhat increased by the technology used to curb the others. To successfully reduce this gas while still controlling the others is difficult and will require significantly more complex engine designs to achieve. The simplicity and low cost of the ICE, which have been its greatest virtues, are being continuously eroded as tighter control of emissions is required.

Nevertheless, the Congress has served notice on the automobile industry that it expects development of a very low emission vehicle within the next five

v

years—a feat that some automobile company executives claim to be impossible. This book is an extremely timely contribution since it gathers together in one place what is known about the technology and economics of reducing automobile emissions, as well as presenting some original analyses by the authors. It should help the reader assess what is technologically feasible, and provide a basis for consideration of policies for reducing automotive externalities.

Some parts of the book are moderately technical but the authors have made a special effort to present their material clearly. Some knowledge of basic physics and chemistry plus a feeling for how mechanical things work should carry the reader comfortably through the bulk of the book. In Part I (chapters 1 and 2) the authors establish the context of the study and describe the internal combustion engine as a source of environmental problems. In Part II they examine energy conversion and energy storage and analyze the various technological options, including further modification of the internal combustion engine. These subjects, which are examined in considerable detail in chapters 3–12, are summarized in chapter 13 (primarily for readers who are not interested in so much detail), and the various options are compared in terms of performance and cost in chapter 14. In Part III (chapter 15) the authors conclude their study with a brief comment on some policy aspects of controlling automotive externalities.

Allen V. Kneese, Director
Quality of the Environment Program
August 1971   Resources for the Future, Inc.

Contents

PART I. INTRODUCTION

CHAPTER 1.—The Public Interest in Automotive Transportation 3
  Institutional Considerations 5
  Rationale of the Analysis 7

CHAPTER 2.—Internal Combustion Engines and External Effects 12
  Emissions from Internal Combustion Engines 12
  Emissions from Urban Vehicle Populations 15
  Emissions and Air Quality 21
  Concluding Comment 31

PART II. ANALYSIS OF THE INTERNAL
COMBUSTION ENGINE AND ITS ALTERNATIVES

CHAPTER 3.—Energy Conversion Requirements for Automotive
  Purposes 35
  Patterns of Vehicle Utilization 35
  Automotive Vehicle Performance 40

CHAPTER 4.—Energy Conversion 51
  Internal Combustion E-Engines 52
  Heat (Q) Engines 68
  Electrochemical Conversion 75
  Other Direct Conversion Schemes 84

CHAPTER 5.—Energy Storage 86
  Electrochemical Energy Storage 86
    Commercially Available Storage Cells 86

The Lead-Acid Cell   86
The Nickel-Iron Cell   88
The Nickel-Cadmium Cell   90
The Silver-Zinc Cell   91
The Silver-Cadmium Cell   93
The Nickel-Zinc (Drumm) Cell   93
High-Energy Cells Under Development   93
The Metal-Air Cells   93
Alkali-Metal Anodes with Organic Electrolytes   98
Liquid Metal Cells/Molten Salt Electrolytes   100
General Comments on Batteries   105
Thermal Energy Storage   108
Mechanical Energy Storage   109

CHAPTER 6.–New Developments in Internal Combustion Engines   113
Rotary Internal Combustion Engines   114
The Constant Pressure Reciprocating Internal Combustion Engine   121
Control of Emissions Due to Incomplete Combustion in Conventional
ICEs   122
Fuel Injection   123
Stratified Charge Systems   123
Exhaust Control Systems   126
Evaporative Losses   128
Future Prospects for Emissions Control   129
Alternate Fuels   133

CHAPTER 7.–Applications of Gas Turbines to Automotive Vehicles   137

**CHAPTER 8.–Applications of Rankine-Cycle Engines to Automotive
Vehicles**   148
Development History   149
Besler   149
Sentinel Truck   149
Henschel Bus   150
McCulloch "Paxton" Engine   150
Yuba Steam Tractor   151
The Keen Steam Car   152
Thermo Electron (TECo)   153
Williams Brothers   154
Pritchard   156
Other Developers   156
Efficiency and "Water Rate"   159
Operational Characteristics   161

The Fluid Circulation System 169
Rankine-Cycle Turbines 175

**CHAPTER 9.–Application of Noncondensing Gas Cycle–External
Combustion Engines to Auto Use** 180
The Stirling Cycle 180
A Modified Stirling Engine 185
The Brayton Cycle Hot-Air Engine 186

CHAPTER 10.–Electric Propulsion Systems for Automobiles 190
Torque-Producing (Rotary) Motors 190
Electronic Controls 199
Weight and Cost 201

CHAPTER 11.–The Electric Vehicle 206
General Principles 206
Electric Vehicle Development Efforts 219
Alden Self-Transit Systems Corporation 219
American Motors–Gulton Industries 219
Army Engineer Research and Development Laboratories 222
Chrysler Corporation 222
ESB-Battronic 222
Electric Fuel Propulsion, Inc. 223
Ford Motor Company 223
General Electric 224
General Motors Corporation 224
Gould–National Batteries, Inc. 225
Linear-Alpha, Inc. 225
National Union Electric Corporation 225
Rowan Controller Company 225
Stelber Industries 226
West Penn Power Company 226
Westinghouse 227
Gar Wood 227
Yardney Electric 227
Miscellaneous Electric Vehicles 228
Foreign Developments 228
Fuel Cells 229

CHAPTER 12.–Hybrid Power Systems 230

CHAPTER 13.–Technological Review 245
Energy Requirements for Automotive Purposes 245

Energy Conversion 247
Energy Storage 249
New Developments in Internal Combustion Engines 252
The Gas Turbine 255
Rankine-Cycle External Combustion Engines 256
Noncondensing External Combustion Engines 257
Electric Propulsion 258
Electric Vehicles 260
Hybrid Vehicles 262

CHAPTER 14.—Comparisons of Performance and Cost 265
Performance 265
Cost Comparisons 268

PART III. CONCLUSION

CHAPTER 15.—Comments on Effects and Policies 283
Impacts 283
Policies for Controlling Automotive Externalities 286
Public Utilities 287
Emission Standards 288
Subsidies and/or Tax Incentives 288
Sponsorship of New Technology 289
A Perspective on the Present R&D Effort 292

APPENDIX

Fuel Cells 295
Hydrogen-Oxygen Fuel Cells 295
Hydrocarbon Fuel Cells 298
Soluble Fuel Cells 299
Present Status 302

Index 312

List of Tables

2-1. Typical automobile exhaust gas characteristics, Otto-cycle engine with no exhaust controls, urban driving conditions 16

2-2. Emissions coefficients, uncontrolled engines 16

2-3. Emissions per thousand vehicle miles (urban driving conditions, average speed of 24 mph) 17

2-4. Emissions variation with traffic speed (auto) 17

2-5. Emissions produced by various urban transportation systems . . . 20

2-6. Important fuel composition characteristics 22

2-7. Concentrations of emissions in Los Angeles for 1947, 1955, and 1965, and projections for 1980 and 2000 28

2-8. Emission requirements for United States and California 31

3-1. LA-4 synthetic driving cycle (XC-15 dynamometer sequence) . . . 36

3-2. Numerical values of power coefficients 47

4-1. Standard electrode potentials, 25°C 77

4-2. Molar free energy values of various reactions 78

4-3. Energetics of selected primary cell systems 82

5-1. Theoretical energy densities of possible battery couples 94

5-2. Various lithium organic systems 99

5-3. Ratings for various batteries 107

5-4. Battery materials costs 107

5-5. Single molten salts suitable for thermal energy storage 109

5-6. Applications of flywheels for energy storage 111

6-1. Emission control methods and estimated cost per vehicle 130

6-2. New Jersey ACID system: Cost and effectiveness per car 132

7-1. Vehicular gas-turbine parameters 142

7-2. Emissions from the Rover 2S/140 gas turbine 147

8-1. Summary of characteristics for three TECo steam engines 153

8-2. Modern reciprocating steam or vapor engines 155

8-3. Effect of various approaches to increasing steam engine efficiency 161

xi

8-4.	Performance trade-offs for reciprocating steam engine	164
8-5.	Comparison of four steam generators	167
8-6.	Summary table for Rankine-cycle systems with 90°F, 200°F, and 400°F condensers	172
8-7.	Possible organic working fluids for vapor turbines	176
9-1.	Data on four experimental Philips Stirling-cycle engines	185
10-1.	Morphological summary of torque-producing devices	194
11-1.	Parameters for the function $E^{-1}(P_w)$	212
11-2.	Total battery energy available per pound	217
11-3.	Profile of possible electric car development	218
11-4.	Comparisons of three types of electric cars with conventional counterparts.	220
12-1.	Hybrid combinations	237
12-2.	The Minicar hybrid propulsion system	238
12-3.	Alternative hybrid configuration, 2½-ton army truck	243
12-4.	Overall efficiencies of a hybrid system	244
14-1.	Specific fuel consumption	266
14-2.	Comparison of alternative prime movers providing a maximum of 100 hp at wheels	269
14-3.	Comparative levels of automobile exhaust emissions	272
14-4.	Calculation of power-system first cost, relative to ICE	274
14-5.	Estimated first cost of power systems	275
14-6.	Estimated ten-year costs for fuel or energy	276
14-7.	Estimated ten-year maintenance costs for power systems	277
14-8.	Ten-year net cost differences from baseline ICE	278
15-1.	Relations of automobile manufacturing to other industries	286
A-1.	General fuel cell ratings	303
A-2.	Some complete fuel cell systems	306
A-3.	Costs per pound per kwh produced of some common fuel cell reactants	308
A-4.	Cost per kwh of energy produced by H_2 from various sources, based on the costs of the required fuels used	308

List of Figures

1-1. Trip-making behavior in cities 8
1-2. Distribution of trip modes versus length 9
1-3. Trip length frequency via automobile for various cities 10
2-1. Internal combustion engine exhaust composition as a function of
air-fuel ratio 13
2-2. NO_x production as a function of air-fuel ratio 14
2-3. Carbon monoxide emissions per mile and gross hydrocarbon emissions per mile versus vehicle speed 18
2-4. Processes resulting in smog production in air 23
2-5. Photochemical reactions produced experimentally 24
2-6. Frequency-time distribution of gaseous pollutants 26
2-7. Projections of hydrocarbon and NO_x concentrations in Los
Angeles 29
2-8. Emission estimates for the United States based on legislative
standards for 1970 and for 1973 and 1975 30
3-1. Influence of stop-and-go driving on average driving speed 37
3-2. Limiting cruise and average speeds in stop-and-go driving 38
3-3. Marginal distributions for acceleration, gradient, and logarithm of
speed, Pittsburgh cycle. 39
3-4. Rolling resistance versus speed as a function of tire pressure 42
3-5. Rolling resistance versus speed for different tire materials 43
3-6. Power required to cruise, for typical vehicle weights 44
3-7. Power available to the driver at various speeds (1960 compact car) 45
3-8. Typical acceleration characteristics (1960 compact car) 46
3-9. Power required at rear wheels to accelerate a 2,000-lb vehicle at
various rates 47
3-10. Distribution of power during the Pittsburgh cycle 50
4-1. The air standard Otto cycle 53

4-2. Sequence of events in 4-stroke Otto cycle and 2-stroke Otto cycle   54

4-3. Thermal efficiency and mean effective pressure for the Otto-cycle internal combustion engine   56

4-4. Variation of specific fuel consumption with power for Otto-cycle engine   57

4-5. Internal combustion engine horsepower characteristics   58

4-6. The air standard diesel cycle   59

4-7. Sequence of events in 4-stroke diesel engine   61

4-8. Thermal efficiencies of the diesel cycle with constant pressure combustion   62

4-9. Mean effective pressure for the diesel cycle.   63

4-10. Variation of specific fuel consumption with power for the diesel-cycle engine   64

4-11. The air standard Brayton cycle   64

4-12. Schematic diagram of simple and compound gas turbines   65

4-13. Optimum compression ratio for peak efficiency for various inlet temperatures and heat exchanger efficiencies   66

4-14. Variation of specific fuel consumption with power for the gas turbine.   67

4-15. The Carnot cycle   70

4-16. The Stirling cycle   71

4-17. The Ericsson cycle   72

4-18. The Rankine cycle   73

4-19. Rankine-cycle engine efficiencies for various conditions and working fluids   74

4-20. Enthalpy efficiency variation with temperature of O_2-H_2 reaction   79

4-21. Fuel cell schematic   80

5-1. Energy storage capabilities of various lead-acid cells   89

5-2. Lead-acid cell trends   89

5-3. Improvements in specific energy as percentage of maximum for nickel-cadmium batteries   92

5-4. Schematic of General Atomic's zinc-air battery   96

5-5. Performance characteristics of rechargeable zinc-air cell   97

5-6. Schematic of the GM lithium-chlorine cell   102

5-7. Schematic of the Ford sodium-sulfur cell.   103

5-8. Energy storage capacity versus specific power for various battery types   106

6-1. The Wankel rotary engine   115

6-2. Trochoidal rotor configurations   117

6-3. The Kauertz and Tschudi rotary engine configurations   118

6-4. The Kal-Pac engine   119

6-5. The Mallory rotary engine   120

6-6. Reciprocating internal combustion engine with separate constant pressure combustion chamber 121

6-7. Schematic of the stratified charge principle 124

6-8. Estimated costs of controlling automotive emissions 131

7-1. Schematic of split-shaft "free" turbine 143

7-2. Torque characteristics of simple and free turbines 144

7-3. Schematic of differential turbine 145

8-1. Torque-speed characteristics of Rankine-cycle (steam) and diesel engines using the same engine block 157

8-2. Gibbs and Hosick elliptocline steam expander 158

8-3. Single-acting uniflow engine 162

8-4. Fluid cycle schematic 169

8-5. Efficiency of steam Rankine cycle 175

8-6. Efficiency of various turbine systems 178

9-1. Schematic of Philips Stirling engine 182

9-2. Efficiency versus specific power for various temperature-pressure conditions 183

9-3. Efficiency versus specific power for various working fluids 184

9-4. The modified Stirling engine of Vannevar Bush in schematic form 186

9-5. Idealized air engine cycle 187

9-6. Typical power and torque curves for rotary hot-air engines 188

10-1. DC motor types 192

10-2. Polyphase AC motor types 193

10-3. Torque characteristics of various DC electric motors and ICE with automatic transmission 195

10-4. Operation of a "chopper" 199

10-5. Basic electronic control circuit schematics 200

10-6. Stall torque versus weight for various electric motor types 203

11-1. Energy requirements per ton-mile for various battery types 214

11-2. Projected lead-acid battery cycle life versus depth of discharge (1972) 216

11-3. Withdrawable energy factor for lead-acid battery 217

12-1. Distribution of battery energy levels in the Pittsburgh cycle for various maximum energies and generator power levels assuming no regeneration 234

12-2. Distribution of battery energy levels in the Pittsburgh cycle for various maximum energies and generator power levels assuming 80% regeneration 235

12-3. Discharge fraction (E_{min}/E_{max}) for various maximum energies versus generator power, during the Pittsburgh cycle 236

12-4. Efficiency versus power for a battery-fuel cell hybrid system ... 244

14-1. Energy and fuel consumption for various engines 267

xvi   *Contents*

A-1.   Specific energy versus discharge time for several batteries and a
        typical fuel cell 303
A-2.   Fuel-cell power trends 309
A-3.   Allis-Chalmers fuel-cell performance trends 310

PART I

Introduction

Chapter One

The Public Interest in Automotive Transportation

In the United States, transportation is overwhelmingly automotive. Most of the population use private automobiles en route to and from work, and the majority of the rest travel on buses. Intercity travel is even more heavily dominated by motor vehicles: in 1968 about 87% of intercity passenger miles was accounted for by private cars, some 2½% by motor coaches, and a little over 9% by aircraft.[1] The movement of freight is still largely by railroad (mainly diesel power), but trucks have become the second most important mode, accounting for about one-quarter of all freight ton-miles in 1967, for example, and having a virtual monopoly of local distribution of goods within urban areas. Since almost all forms of self-propelled vehicles currently use some form of internal combustion engine, and since so many of them (almost 100 million in the United States) use the familiar spark-ignition reciprocating-piston Otto-cycle engine, it is clear why the latter plays such a central role in the transportation picture.

The Otto-cycle engine is, of course, not the only internal combustion engine in general use. About 400,000 trucks and 65,000 buses, not counting off-highway vehicles, are powered by diesel engines,[2] but the percentage of vehicles powered by diesels is only a small fraction of the overall total. The field is dominated by the Otto-cycle engine.[3]

One of the outstanding difficulties associated with a transportation system based on such an engine is its tremendous output of material and energy

---

[1] *Automobile Facts and Figures* (American Automobile Manufacturers Association, 1969 ed.).

[2] Ibid.

[3] Nevertheless, in most large central business districts, diesels contribute a comparatively large percentage of certain types of emissions because city buses and large trucks are powered by them. Surprisingly, however, emissions from the diesel engine, while more obnoxious, are probably less harmful to health than emissions from the Otto-cycle gasoline engine in that they contain significantly less carbon monoxide and fewer unburned hydrocarbons.

residuals. The disposal of residuals—solid, liquid, gaseous, and thermal wastes, plus noise—imposes large direct costs on the community, and these costs must be paid by its citizens. To the extent that residuals are not "disposed of," but merely dispersed in the environment, a further indirect cost is imposed on the community in the form of degradation of important and valuable public assets—the landscape, the atmosphere, and the biosphere.

Private property and private exchange in markets are the central elements of our traditional system for valuing goods and services and allocating them among alternative uses. But they are not applicable to such inherently "common property" or "collective" resources as the landscape or the atmosphere. As economic development and population growth proceed, these common property resources become congested, and when conflicts in their use occur, they are not mediated by market exchange. These resources are unpriced, and therefore tend to be treated as though they were superabundant, even though in reality they are scarce and valuable. To some extent the cost of internal combustion engines has been artificially lowered because their heavy use of common property resources has been unpriced.

→ The consumption of fuel for propulsion power accounts for about one-quarter of the total primary energy produced in the United States (and for the heavy emphasis on energy conversion technology in later chapters of this book). Almost 100% of this power is derived from petroleum products. In 1966 the automotive transportation sector (94 million cars, trucks, and buses) consumed 224 million tons of petroleum products, of which 200 million tons were gasoline and 24 million tons were diesel fuel.[4] This accounts for close to 70% of all fuel used for transportation purposes.

The impact of automotive transportation as a *system* exceeds that of transportation as a *service* just as the latter dwarfs the significance of the automobile as an *artifact*. The system involves not only automobile manufacturing, which is a colossal industry in its own right providing markets for energy and equipment and employment for people, but also the petroleum industry and highway construction.

In toto, these industries involved nearly 819,000 business establishments in 1966. They accounted for between 10% and 15% of the gross national product (GNP) and of total private investment, employment, and revenues.[5] Direct sales of autos, trucks, buses, and parts were over $40 billion in 1965. Consumer expenditures for automotive transportation (depreciation, maintenance, gasoline, oil, and insurance) were $1,000 per household in 1965, adding up to $60 billion of "final" expenditures, or close to 10% of all personal income. Moreover, automotive transportation purchased by business constitutes an

---

[4] *Petroleum Facts and Figures* (American Petroleum Institute, 1967 ed.).

[5] *The Automobile and Air Pollution—A Program for Progress*, Report of the Panel on Electrically Powered Vehicles to the Secretary of Commerce, Part II, December 1967.

additional indirect component of other goods and services purchased by the final consumer. The patterns of interrelationships thus created reach through the whole of our economy. Large fractions of the output of the steel, cement, sand and gravel, lead, glass, and rubber industries, among others, are utilized in turn by the automotive transportation system, including the petroleum and highway components.

On the other side of the coin, automotive transportation is a fundamental economic input to many—if not all—modern industries. Workers are increasingly dependent on it to bring them to their jobs. Raw materials must be gathered from increasingly remote places, collected for processing (sometimes in several stages), then fabricated into producer or consumer goods and distributed for "final" use. Even this is not the end, however, because trash and waste products must be collected and disposed of. A change in any aspect of this complex will have profound effects on the functioning of the economy as a whole.

The influence of automotive transportation extends still further. It touches the quality of our social and cultural life in almost every conceivable way. The building of trolley lines and railroads to bring people downtown in the late nineteenth century led to the first wave of urban decentralization. This was greatly reinforced by the advent of private automobiles. The spread of automobiles has since created its own type of low-density residential area—a type that no other system will adequately serve—and it has promoted the appearance of a host of new service industries and facilities oriented to serve the driver and his machine.

INSTITUTIONAL CONSIDERATIONS

Transportation services have been developed through a combination of public and private efforts—public planning, construction, and operation of facilities and private provision of facilities, equipment, and services, sometimes under public regulations. Thus public roads and highways normally are planned, constructed, and maintained by agencies of general government, whereas automobiles are designed and built (within certain general government specifications such as maximum width) by private enterprises for profit. Oil producers operate under public output limitations and quotas—not to mention a considerable incentive in the form of a depletion allowance and other tax benefits—but otherwise without restriction. Only in the area of traffic control does the public authority enforce significant limitations on the behavior of automotive vehicle users.

It can be argued that this combination of private and public enterprise has produced a system that renders transportation services *as such* in a reasonably effective and efficient manner. It is clear, however, that many of the broader social effects associated with transportation systems have received little or no consideration from either public agencies or private enterprises. In the case of private enterprises the reasons are clear and rather well explained by the economists' concept of external costs. Because the dispersal of gaseous and energy residuals calls upon common property resources, there is no market for noise or air pollution abatement. Furthermore, abatement of these environmental impacts is a "public good." No one can purchase improved air quality via air pollution abatement for himself alone without automatically buying it for everyone else at the same time. Because no individual can obtain a benefit in proportion to his expenditure (unless the cost of controlling pollution is extremely low), it is not worth his while to go it alone. The individual polluter—for example, a single auto driver—ordinarily contributes so slightly to his own damages from overall emissions to the atmosphere that he does not find it worth his while to curb his emissions in his own interest. The result of this situation is that there is no market for silent, odorless trucks, buses, or automobiles even though the total willingness of large numbers of people to pay for such vehicles might be large. Thus the external costs associated with automotive transportation may be high compared with the costs of control but the private market does not provide any effective means for controlling them.

In our society, when the market cannot satisfy the demands of large numbers of people, the remedy is found in the political arena. Government agencies play a large role in the automotive transportation system, but until quite recently they have done little to protect the interests of third parties who are affected by the rendering of such services.[6] An example of legislation that was meant to control, and has indeed reduced, one externality associated with the use of internal combustion engines for transportation is the requirement that such engines be equipped with mufflers. This was made easier, perhaps, by the fact that an unmuffled engine is a source of annoyance to most drivers. As other unwanted by-products from our present transportation system have begun to exceed the threshold of toleration in an increasingly crowded, prosperous, and aware society, legislation has begun to appear on driver education, compulsory insurance, auto safety, highway beautification, junk car disposal, and air pollution control.

However, important as these measures are, they must be further adapted as economic and population growth continue and new technological opportunities arise. It appears that they may serve to buy a few years—perhaps two decades at the most—before more fundamental and far-reaching changes in the automotive transportation system itself will be required. If the present system, which is based largely on the private automobile, is ultimately replaced by something else, it will be found that the substitution brings new problems with it. We can be sure, however, that social efficiency will require—and society will demand—

---

[6] Although the historic role of government in the transportation system is also a response to market failures in the rendering of transportation services—especially its provisions of transportation routes. In economists' terms the provision of transportation routes is a rather extreme case of a "decreasing cost industry." It can be demonstrated that market exchange will not allocate investment in capacity in such industries up to the point where the cost of the last unit provided just equals the willingness of a user to pay for it—one of the criteria for optimal resources allocation.

that future technological decisions be made with a greater recognition of the broader effects on society than has heretofore been the case.

Before society embarks on an ambitious scheme to design and mold the future transportation system of the country not only from the point of view of rendering transportation services but also of controlling external costs (and optimizing external benefits, wherever they can be found), some progress must be made in understanding the range of choices available and the direct and indirect costs associated with each of them.

RATIONALE OF THE ANALYSIS

One may well ask why take so narrow a view of the alternatives to current automotive technology. As a matter of fact, in the early drafts of this book, an attempt was made to draw the line more broadly—to include alternative public transportation systems, for instance. Obviously mass transit is a conceivable alternative to the private automobile: so, for that matter, is the taxi. Bicycles, rail rapid transit, VSTOLs (vertical and short takeoff and landing aircraft), or air-cushion vehicles might substitute for some uses of autos, buses, or trucks. Pipelines, conveyor belts, and barges can also compete with automotive vehicles in some applications; telephones and teletypes provide viable alternatives in certain instances; and advanced forms of electronic communications may even displace some categories of automobile (and other vehicular) trips in the future.

Controlling investment in freeways and altering the shape of the city are still further alternatives. The demand for trips is a function of the accessibility of destinations and the cost, time, and stress involved in travel. Often new freeways, supposedly built to relieve congestion over existing urban routes, stimulate enough new traffic from additional trips to become overburdened and congested themselves in a few years. This phenomenon is simply a consequence of the fact that each new route makes a great number of destinations more accessible to a great number of potential users. Thus, investment in freeways (or any other type of transportation facility) tends to increase the utilization of—and the apparent demand for—that particular mode.

Attention to the possibilities of comprehensive urban planning as a tool for minimizing the demand for automotive transportation is based on the empirical fact that trips by auto decrease as density increases. The reason is that in high-density residential or business areas, many destinations are accessible by a combination of public transport (including elevators and escalators) and walking. In the future, moving sidewalks or "people-movers" will be added to the repertory. Vehicular trip-making behavior, mode choice, and trip length as a function of population density and/or distance from downtown can be seen in figures 1-1 to 1-3.

In light of the foregoing data, it is evident that vertical and horizontal transportation are, to some extent, interchangeable. Thus, actions (such as zoning

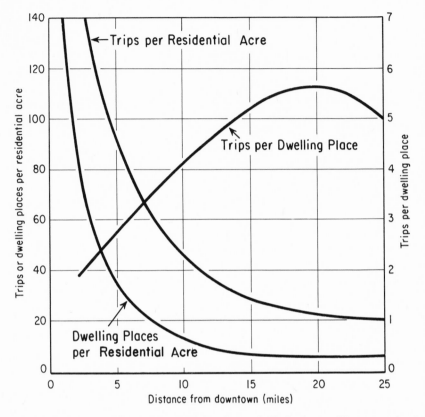

Figure 1-1. Trip-making behavior in cities. (From J. W. Dyckman, "Transportation in Cities," *Scientific American*, September 1965.)

changes) that stimulate vertical development will tend to reduce both the number of automotive trips taken, and their length. On the other side of the coin, however, the *density* of such trips (i.e., vehicular congestion) tends to increase in direct proportion to the intensity of general land use. Moreover, external effects, such as air pollution and noise, broadly speaking, become more serious in direct proportion to the number of persons affected—also a function of the intensity of land use. Thus, an extremely decentralized society would probably be more tolerant of air pollution and noise in an absolute sense than a highly compact one.

The technological and institutional alternatives that could provide trade-offs either for automotive transportation as such, or for associated externalities, cover an enormous range. One cannot hope to do justice to all of them in a book such as this.

Even more pertinent, however, is the fact that if automotive transportation were to be toppled from its present position of overwhelming economic impor-

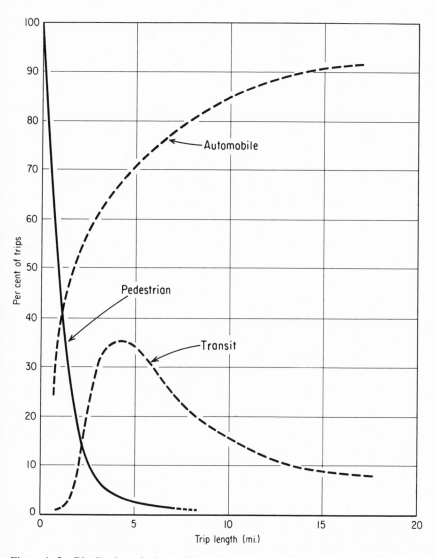

Figure 1-2. Distribution of trip modes versus length. (Adapted from Wilfred Owen, *The Metropolitan Transportation Problem*, The Brookings Institution, Washington, D.C., 1956; rev. 1966, App. table 20.)

tance in the U.S. economy—by any combination of mass transportation, improved communications technology, and changes in the urban structure—a *minimum* of 10% to 15% of the U.S. economy would have to be scrapped and replaced. Realistically, we cannot believe that this is a viable alternative in the short run, though, of course, it could occur as a result of evolutionary changes over a span of three or four decades. For this reason, primarily, we have elected

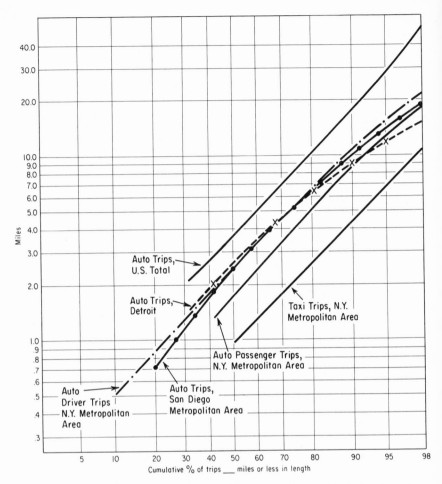

Figure 1-3.   Trip length frequency via automobile for various cities. (Unpublished data from Tri-State Transportation Agency, 100 Church Street, New York, N.Y.)

to restrict the scope of the study to alternative power sources for the vehicles themselves. This limitation does not reflect a lack of imagination or unwillingness on the part of the authors to consider the possibility of broader, more far-reaching strategies to improve the quality of our urban environment. Many such alternatives deserve attention. However, we feel that the longer-term issues should not be dealt with in the same context as the more immediate technological alternatives to which this book is addressed.

When we discuss the emissions from automotive vehicles—as we do in the next chapter—we restrict ourselves to a survey of the sources and quantities of pollutants, and do not attempt a detailed review of the many aspects of air pollution. In a book such as this, it seems reasonable to skip over the environmental health aspects, especially as these have been studied intensively by many others.[7] Economic aspects are omitted for another reason. If there were a well-defined body of knowledge with respect to the losses of value that people suffer from their own or other people's ill-health, sensory insults, and psychic stress, a summary would certainly be appropriate.[8] At this stage, however, there is little to report on measuring these damages or determining "how much it would be worth" to avoid the ill effects of air pollution. We can only point out that public concern about air pollution problems suggests that the economic damages associated with the physical and biological effects of air pollution are widespread and large.

---

[7]See *Air Pollution*, World Health Organization, Geneva, 1961; *Motor Vehicles, Air Pollution, and Health*, Report of the Surgeon-General to the United States Congress, June 1962; A. C. Stern, ed., *Air Pollution* (New York: Academic Press, 1962); and J. R. Goldsmith and S. A. Landaw, "Carbon Monoxide and Human Health," *Science*, Vol. 162 (1968), pp. 1352–59.

[8]Two exploratory studies are: R. G. Ridker, *Economic Costs of Air Pollution: Studies in Measurement* (New York: Frederick A. Praeger, 1967); R. U. Ayres and Allen V. Kneese, "Production, Consumption, and Externalities," *American Economic Review*, June 1969.

Chapter Two

Internal Combustion Engines and External Effects

EMISSIONS FROM INTERNAL COMBUSTION ENGINES

Theoretically, about 14.7 pounds of air are needed to burn one pound of gasoline completely; any deviation from this ideal will result in a change in the balance of exhaust products. For example, too little air results in the production of carbon monoxide, unburned hydrocarbons, and possibly hydrogen, although the latter will not be observed unless the air deficiency is drastic. With too much air, the mixture is too lean, oxygen is present in the exhaust, ignition becomes progressively more difficult, and engine power is reduced. The effects on exhaust composition of variations in the air-fuel ratio are shown graphically in figure 2-1.

The stoichiometric, or theoretically ideal, air-fuel ratio for complete combustion (14.7 : 1) is quite close to the observed ratio for maximum fuel economy (15 : 1); however, maximum power is achieved with a richer mixture of about 12 : 1, which results in carbon monoxide and hydrogen being present in the exhaust. Moreover, combustion is a complex process, and even in two "identical" cylinders, the mixing and distribution are never exactly homogeneous. Thus, even with "ideal" mixtures, some residual carbon monoxide and hydrogen are often found in the exhaust gases.

Furthermore, because the cylinder wall is relatively cool, the combustion process is somewhat quenched near the outer edges, particularly in the space between the cylinder wall and the piston itself, adjacent to the piston rings. The result is that a layer of air-fuel mixture adjacent to the cylinder wall is not completely burned. Thus, some unburned or partially burned hydrocarbons are always slushed out at the end of the power stroke, while some escape (blow by) past the piston rings into the crankcase from which they were simply "vented" until the 1968 model cars.[1] In cars currently manufactured this blow-by is returned to the intake manifold.

---

[1] Except in California, where crankcase blow-by has been controlled by law on new cars sold since 1962.

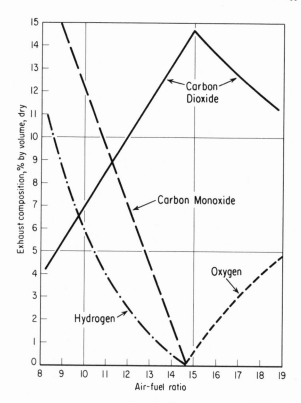

Figure 2-1. Internal combustion engine exhaust composition as a function of air-fuel ratio. (From A. H. Rose, Jr., "Automotive Exhaust Emissions," in Arthur Stern, ed., *Air Pollution*, Academic Press, New York, 1962.)

To further complicate the picture, under conditions of high temperature and high pressure, oxygen and nitrogen tend to combine to form various oxides of nitrogen (NO_x), particularly in lean mixtures, where excess oxygen is present. The production of NO_x as a function of the air-fuel ratio is shown in figure 2-2.

Sulfur oxides—principally sulfur dioxide—occur in direct proportion to the sulfur content of the fuel. In addition, finely divided particulate matter is found in the average gasoline exhaust—at an average emission rate of 0.78 mg/gm of gasoline. Most of the particulate matter is in the form of lead compounds, which result from the widespread use of tetraethyl and tetramethyl lead as a gasoline additive to prevent "knocking" in engines.

The engine exhaust, although the principal contributor, is by no means the sole source of emissions. Large amounts of hydrocarbons are also evaporated directly from the gas tank and the carburetor. Gasoline vapor escapes from the tank of an automobile, service station, or tank-truck through the air inlet vents, which are needed to prevent the tank from collapsing as fuel is consumed. Under

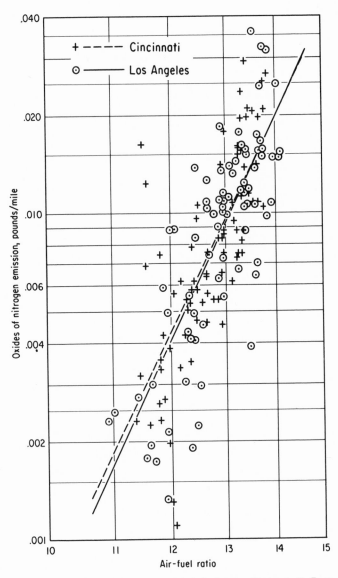

Figure 2–2. NO$_x$ production as a function of air-fuel ratio. (From A. H. Rose, Jr., et al., "Comparison of Auto Exhaust Emissions from Two Major Cities," *Journal of the Air Pollution Control Association*, August 1965.)

certain conditions—if the tank is being filled or if it is being heated by the sun, for example—saturated vapor will be pushed out through the vent. It is estimated that vapor losses from tank filling account for losses of 0.45% of the gasoline throughput (by weight), while other evaporation losses from the gas tank and carburetor account for an additional 1.5%.[2] Carburetor losses are essentially of two types: carburetor vent losses that occur when the engine is operating; and the so-called hot soak losses that occur when the engine is turned off. The latter occur because the cooling system ceases to function when the engine is turned off, and residual engine heat then causes a temperature rise in auxiliary equipment, which, in turn, results in evaporation of gasoline left in the carburetor bowl. The relative contribution of hydrocarbons (HC) from these three sources has been quoted as follows for a car without emission controls: exhaust 46%, crankcase 30%, and evaporation 24%.[3]

In principle, at least, crankcase blow-by emissions can be completely eliminated by returning them, via the suction force of the intake manifold vacuum, to the combustion cylinder for reburning; or they can be burned outside the engine. Most cars today utilize a so-called PCV (positive crankcase ventilation) valve which recycles these gases to the manifold. Blow-by gases consist of approximately 80% unburned air-fuel mixture and 20% cylinder combustion products.

The volumetric concentration of carbon monoxide (CO) and hydrocarbons in the exhaust is influenced by the engine operating condition, which in turn depends on the vehicle driving mode. It is customary to identify four basic modes: acceleration, deceleration, cruise, and idle. The variation in emission concentrations with driving mode is shown in table 2-1. Average emissions from uncontrolled automobiles in urban driving conditions, on a volumetric basis, are now generally taken to be 3.5% carbon monoxide, 900 parts per million (ppm) hydrocarbons, and 1,500 ppm nitrogen oxides.[4]

EMISSIONS FROM URBAN VEHICLE POPULATIONS

It has been estimated that in 1965 the average daily gasoline consumption in Los Angeles County was 7.4 million gallons, of which 5.75 million gallons were used by cars and small pickup-type trucks. Assuming hexane (C_6H_{14}) as a typical constituent, an idealized combustion process would be:

$$C_6H_{14} + 9.5O_2 = 6CO_2 + 7H_2O.$$

Thus the 7.4 million gallons, or 22,600 tons, of gasoline (assuming an average specific gravity of 0.73) burned in Los Angeles would theoretically combine with 80,000 tons of atmospheric oxygen to yield a total of 102,600 tons of

---

[2] R. L. Duprey, "Compilation of Air Pollutant Emissions Factors," U.S. Public Health Service Publication No. 999–AP–42 (1968).

[3] K. Steinhagen, "Air Pollution Control Methods for the Gasoline-Powered Internal Combustion Engine," Briefing before the Panel on Electrically Powered Vehicles, Office of the Secretary, U.S. Department of Commerce, 1967.

[4] U.S. Department of Commerce, "The Automobile and Air Pollution: A Program for Progress," Subpanel Report to the Panel on Electrically Powered Vehicles, October 1967.

Table 2-1. Typical Automobile Exhaust Gas Characteristics, Otto-Cycle Engine with No Exhaust Controls, Urban Driving Conditions

Mode of driving	Driving pattern (California cycle)		Carbon monoxide (CO) concentration		Hydrocarbons (HC) concentration		Oxides of nitrogen (NO_x) concentration	
	% of time	% of volume	% in exhaust	% of total	ppm in exhaust	% of total	ppm in exhaust	% of total
Idle	15	4.2	5.2	5.0	750	4.4	30	0.05
Cruise	16	16.8	0.8	3.1	300	7.0	1,500	10.6
Acceleration	36.7	69.9	5.2	83.1	400	38.5	3,000	89.3
Deceleration	32.3	9.1	4.2	8.8	4,000	50.2	60	0.1
Average concentration	–	–	3.5	–	900	–	1,500	–

Source: National Air Pollution Control Administration.

reaction products, of which about 69,500 tons would be carbon dioxide and the rest water vapor.

In reality, of course, sizable quantities of residuals other than carbon dioxide and water vapor are produced. By combining bench-test results, such as in table 2-1, with driving cycle data (discussed in chapter 3), it is possible to compute average emissions in pounds or kilograms per thousand miles and to estimate totals for a city or an area. Alternatively, of course, emissions can be measured directly by instrumented test cars in city traffic. The National Air Pollution Control Administration—now the Office of Air Programs (OAP) of the Environmental Protection Agency—has followed both approaches.

Table 2-2 shows emissions coefficients for Otto-cycle and diesel engines in pounds of residuals per pound of fuel, and table 2-3 shows emissions in kilograms per thousand miles for urban driving. In each case, average route speed is

Table 2-2. Emissions Coefficients, Uncontrolled Engines

(lb of residuals per lb of fuel)

	CO	HC	NO_x	SO_2	Particulates
Otto[a]	0.377	0.082[b]	0.0185	0.0015	0.002
Diesel	0.0085	0.0194	0.0316	0.0057	0.0157

Source: Adapted from R. L. Duprey, "Compilation of Air Pollutant Emissions Factors," U.S. Public Health Service Publication 999-AP-42 (1968).

[a] Based on an average route speed of 24 mph.
[b] Assuming 40% of HC emissions are accounted for by the exhaust.

assumed to be 24 miles per hour. A more accurate treatment would allow for the fact that specific fuel consumption tends to vary with average traffic speed. Emissions coefficients are somewhat lower in rural areas, where there is a greater proportion of steady cruising at moderate speeds and less stop-start driving. Emissions in areas where average driving speeds are lower than 24 mph are correspondingly higher, as shown in table 2-4. Basic data for two cities

Table 2-3. Emissions per Thousand Vehicle Miles (Urban Driving Conditions, Average Speed of 24 mph)

	Specific fuel consumption (miles per gal.)	Fuel per thousand miles (kg)	Carbon monoxide (CO) (kg)	HC (kg)	NO_x (kg)	SO_2 (kg)	Particulates (kg)
Auto	14.4	177	75	17.0	3.86	0.27	0.364
Trucks (gasoline engine)	7.85	324	137	31.2	7.1	0.5	0.68
Bus (diesel)	4.67	636	5.4	12.4	20.14	3.64	10

Source: Average fuel consumption from "Automobile Facts and Figures," Automobile Manufacturers Association, 1966 edition; other data derived from table 2-2.

Table 2-4. Emissions Variation with Traffic Speed (Auto)

Location	Average traffic speed (mph)	Index (24 mph = 1)	
		CO emissions	HC emissions
Central business district	10	2.06	1.77
Residential area	18	1.24	1.15
Arterial road	24	1.0	1.0
Cruise	45	0.59	0.65

Source: R. L. Duprey, "Compilation of Air Pollutant Emissions Factors," U.S. Public Health Service Publication 999-AP-42 (1968).

(Cincinnati and Los Angeles) are displayed in figure 2-3. Note that emissions per vehicle mile are sharply higher in congested areas than the averages shown in tables 2-2 and 2-3. This means that densely populated cities such as New York—especially Manhattan—suffer much more air pollution than a simple calculation based on the number of vehicle miles traveled would suggest. The average speed of traffic, *ceteris paribus*, is inversely related to the degree of congestion, which is in turn roughly proportional to the density of vehicles in use, or the number of vehicle miles traveled per square mile. Thus the air pollution produced by a vehicle population, measured in emissions per square mile per hour, increases approximately as the square of the traffic density measured in vehicle-miles traveled per square mile per hour.[5]

---

[5] Actually, the emissions of CO seem to vary as $\sim ds^{-0.9}$ while HC varies as $\sim ds^{-0.7}$, where d is traffic density and s is speed. For a considerable range of speeds, both might be expressed with sufficient accuracy by the rule:

$$\text{Emissions} \approx ds^{-0.8}$$

Congestion models suggest various relationships between the speed and the traffic density. A typical form is:

$$s/s_{max} = \left[1 - K(d/d_{max})\right]^n.$$

(*continued on p. 20*)

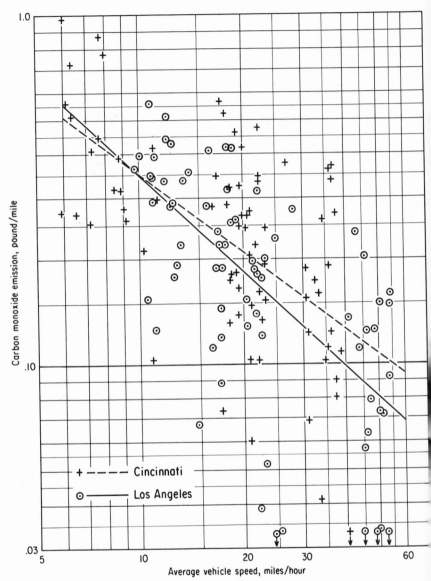

Figure 2-3. Carbon monoxide emissions per mile (left) and gross hydrocarbon emissions per mile (right) versus vehicle speed. (From A. H. Rose, Jr., et al., "Comparison of Auto

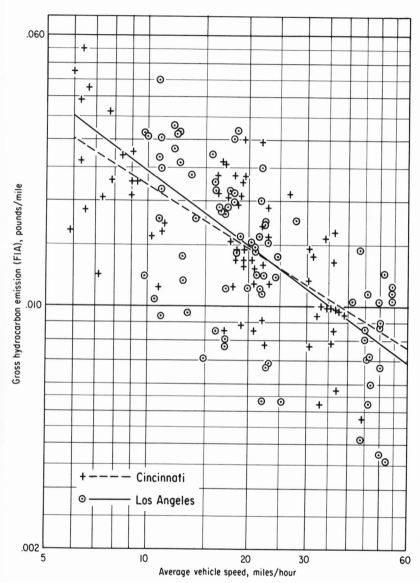

Exhaust Emissions from Two Major Cities," *Journal of the Air Pollution Control Association*, August 1965.)

It is interesting from the standpoint of comparison with other transportation systems to calculate coefficients for residuals in terms of passenger miles. We can include rail rapid transit in such a comparison by assuming that the electric utility power for propulsion is originally generated in plants burning fossil fuels. In cities it is reasonable to assume something like 1.25 riders per automobile, 20 passengers per bus, and 20 riders per transit car. In the New York City subway system where energy consumption is 5.5 kilowatt-hours (kwh) per car mile, 275 kwh of electricity are required for each thousand passenger miles traversed.[6] Emission values per thousand passenger miles for automobiles, diesel buses, and rapid transit systems are shown in table 2-5.

Table 2-5. Emissions Produced by Various Urban Transportation Systems

	Passengers per vehicle (no.)	Emissions per thousand passenger miles (kg)				
		Carbon monoxide	Hydro-carbons	Oxides of nitrogen	Sulfur dioxide	Particulates
Auto (uncontrolled), 25-mph average speed	1.25	60	13.64	3.1	0.227	0.295
Diesel bus	20	0.273	0.62	1	0.182	0.5
Rapid transit Case A[a]	20	0.025	0.011	1.023	4.86	1.273
Case B[b]	20	0.00036	0.028	0.91	3.45	0.0123

Source: Compiled by the authors, using data from R. L. Duprey, "A Compilation of Air Pollutant Emissions Factors," U.S. Public Health Service Publication 999-AP-42 (1968).

[a]Case A: Utility coal-fired thermal power plant (2.5% sulfur, 10% ash), heat rate 10,582 Btu/kwh; particle collection efficiency assumed to be 86%.
[b]Case B: Residual (No. 6) oil-fired thermal power plant (2.5% sulfur), heat rate 10,582 Btu/kwh; particle collection efficiency assumed to be 86%.

Emissions by a vehicle population clearly vary considerably from one city to another due to differences in climate, topography, traffic density, vehicle mix, driving habits, and average speed of automobile traffic. The variation of CO and HC emissions with average speed was discussed above. NO_x is not dependent on route speed, but all pollutants seem to be strongly dependent on altitude. According to OAP, emissions of CO and HC are, respectively, 60% and 30% greater in Denver (altitude 5,280 feet) than in low-altitude cities such as Los

---

Thus one might derive a more complex relationship of the form:

$$\text{Emissions} \cong S_{max}^{-0.8} \left[1 - K(d/d_{max})\right]^{-0.8n} d.$$

[6]D. S. Berry, G. W. Blomme, O. W. Shuldiner, and J. H. Jones, *The Technology of Urban Transportation* (Evanston, Ill.: Northwestern University Press, 1963), pp. 74-78.

Angeles and Cincinnati, while emissions of NO_x are 50% less.[7] With information of this sort, supplemented by additional data, it should be possible to extrapolate total emissions for cities that have not yet been studied.

EMISSIONS AND AIR QUALITY

Total emissions generated by automobiles are related roughly to total gasoline consumption or total vehicle miles traveled. However, *air quality* is a somewhat subtler concept related to ambient concentrations, as a function of time. Air quality is a function not only of total emissions per square mile but also of local meteorological conditions (such as wind speed and atmospheric stability), topography, and other factors.

One point of importance in areas like southern California and Arizona is that photochemical reactions in the atmosphere, stimulated by strong sunlight, result in peak concentrations of various derivative substances occurring at different times of the day. Starting with volatile hydrocarbons (such as isobutene) and nitrogen dioxide (NO_2) produced by motor vehicle exhausts, and ultraviolet light from the sun, a number of highly reactive, unstable compounds are formed which are usually lumped together as "oxidants."

Petroleum fuels vary considerably in their photochemical reactivity, or propensity to form smog. The relative activities of saturates (including paraffins), aromatics, and olefins are roughly in the ratios of $1 : 3 : 10$. As compared to olefins, gasoline vapor ranges from 20% to 30% reactive, while automotive exhaust vapor is about 36% reactive. Other hydrocarbon fuels tend to be correspondingly less active due to a smaller content of olefins and aromatics, as shown in table 2-6.

A schematic version of what happens in the air is shown in figure 2-4. A simplified laboratory simulation of the process, carried out in an irradiation chamber at Stanford Research Institute, led to altered composition as a function of time as shown in figure 2-5. Actual measurements in Los Angeles showed quite distinct peaks for various pollutants during the course of a day.

It has been shown that the concentrations of all major pollutants are very nearly log-normally distributed as a function of time for all cities where extensive air sampling measurements have been made and for all averaging times from one minute to three years.[8] Log-normal distributions of concentration for the jth pollutant in the kth city, being of the form

$$N(C^{jk}) = \frac{1}{\sqrt{2\pi}} \frac{1}{\sigma_{jk} C^{jk}} \exp\left[-\frac{1}{2\sigma_{jk}^2} \left\{ \ln C^{jk} - \mu_{jk} \right\}^2 \right],$$

---

[7] Duprey, "Compilation of Air Pollutant Emissions Factors."

[8] C. E. Zimmer and R. I. Larsen, "Calculating Air Quality and Its Control," *Journal of the Air Pollution Control Association*, Vol. 15 (1965), pp. 565-72.

Table 2-6. Important Fuel Composition Characteristics

Fuel	Saturates (%)	Olefins (%)	Aromatics (%)	Average reactivity (calculated)	Average volatility vapor pressures @ 100°F (psi)
Gasoline					
(typical)	60	19	21	0.31	7-9
Diesel No. 1	80.9	2.6	16.5	0.16	∿0.1
Diesel No. 2	71.9	3.5	24.6	0.18 ± 0.10	∿0.1
	(59-92)[a]	(1-8.5)	(14-38)		
Kerosene No. 1	71-86	1.5-4.0	12.0-26.5	0.16 ± 0.04	<1
JP-4	85	1-2	15	0.15 ± 0.025	3
(jet fuel)	(75-90)	(<5 max)	(10-25)		
			(<25 max)		
Commercial					
propane	95-100	0-5	0	0.12 ± 0.02	192
LNG					
(liquefied					
natural					
gas)	100	0	0	0.10	high
Reactivity	0.1	1.0	0.3		

Source: R. U. Ayres and Roy A. Renner, "Automotive Emission Control: Alternatives to the Internal Combustion Engine." Paper presented to the Fifth Technical Meeting, West Coast Section, Air Pollution Control Association, San Francisco, October 8-9, 1970.

[a]Figures in parentheses are ranges of values.

are completely characterized by a mean μ_{jk} and a standard deviation σ_{jk}. These parameters are based on studies carried out in six cities, over the three-year period 1962-64.[9]

Some graphic representations of these distributions are shown as cumulative-probability curves in figure 2-6. The concentration that is exceeded by a given fraction of measurement increases steadily as shorter and shorter averaging times are considered.

For each city, the mean concentration of each pollutant is presumably related to the total emission of that pollutant in that city (or "airshed"), the size of the area, and the mean rate of atmospheric mixing. Thus the percentage reduction in emission source needed to achieve a given percentage reduction in mean concentration is presumably proportional to the mean rate of atmospheric mixing.[10] Moreover if the reduction is reasonably uniform over an area, the effect would probably be to shift the whole frequency-time distribution down. (By the same token, in the normal course of events, with increased population density and

[9]R. I. Larsen, C. E. Zimmer, D. A. Lynn, and K. G. Blemel, "Analyzing Air Pollutant Concentration and Dosage Data," *Journal of the Air Pollution Control Association*, Vol. 1 (February 1967).

[10]R. I. Larsen, "A Method for Determining Source Reduction Required to Meet Air Quality Standards," *Journal of the Air Pollution Control Association*, Vol. 2 (1961) pp. 7-76.

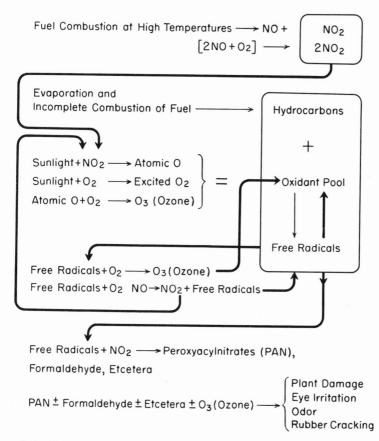

Figure 2-4. Processes resulting in smog production in air. (From S. Tilson, "Air Pollution," *International Science and Technology*, Stamford, Conn., June 1965.)

vehicle-use, the entire distribution tends to shift upward.) The detailed relation between total emissions and mean concentration in an area can be approached theoretically by means of diffusion-type calculations.[11] It is interesting to assume, for purposes of argument, that the diffusion and dispersion mechanisms for a given city remain relatively constant over the years, so that the form of the log-normal distribution function remains unaffected by time. Then only the mean concentration may change, and it is reasonable to suppose that this is simply proportional to total emissions of a given pollutant in the airshed. Thus we can project not only absolute annual quantities of effluents in the future under various circumstances, but also the probabilities of local concentrations

---

[11] F. Pooler, Jr., "A Prediction Model of Mean Urban Pollution for Use with Standard Wind Roses," *Journal, Air and Water Pollution*, Vol. 4 (1960), pp. 199–211. See also "Air Over Cities," Symposium, SEC Technical Report A62–65, 1961.

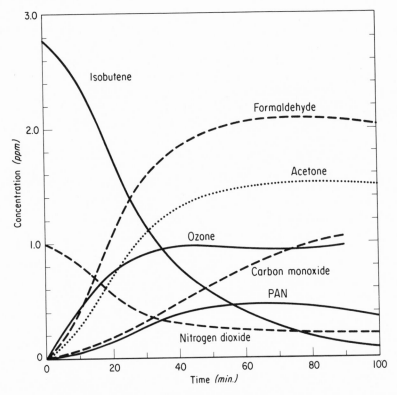

Figure 2-5. Photochemical reactions produced experimentally. (From S. Tilson, "Air Pollution," *International Science and Technology*, Stamford, Conn., June 1965.)

exceeding a given threshold level over an arbitrary period (such as 8 hours). In terms of physiological effects or aesthetic impact, the latter is probably the more important measure.

Some implications of projecting the frequency-time distributions, as described above, are shown in table 2-7. The significance of these projections is, in effect, that during a period when absolute emissions are increasing slowly, the frequency of days with severe smog increases *much more rapidly* than the frequency of smog days (see figure 2-7). For instance, the average eight-hour concentration of hydrocarbons in the atmosphere in Los Angeles exceeded five parts per million (ppm) on only one day in twenty in 1950 but on six days out of twenty in 1965. Yet gasoline consumption only doubled in the same fifteen-year period. The frequency of occurrence of high concentrations of NO_x also rose dramatically: from only one day in twenty for eight-hour NO_x concentrations above 2 ppm in 1950 to one day in four (5 in 20) by 1965.

The effect of various percentage reductions in the original levels of emissions by various dates can also be shown. This exercise can be carried out much more elaborately to show the projected effects of alternative control policies or "standards," and estimates of emissions for the United States are shown in figure 2-8. These are OAP projections based on legislative standards for 1970 and for 1973 and 1975; they should not be regarded as predictions, since, like all projections, they are based on a series of assumptions that may not come true. The actual emission requirements of pollutant emissions per mile for the United States and California during this time are given in table 2-8.

The medical consequences of air pollution have been discussed enough so that it seems reasonable to forgo any summary of the literature in this book. However, the physiological and psychological effects of living in a noisy environment are not so widely known or understood.

Quantitatively, the most important effect of exposure to high levels of noise over long periods of time is hearing damage (presbycusis) beyond the normal effects of aging. Deafness induced by job-connected noise is a recognized cause for receiving workmen's compensation in thirty-two states out of fifty. It was estimated in 1968 that 4.9 million people in the United States were in jobs where continued employment would cause them to suffer hearing impairments severe enough to be compensable, and that about 20 million were in jobs where noise levels were high enough to cause annoyance and to reduce efficiency.[12]

Apart from induced deafness, it is an open question at the moment whether or not noise plays a role in other medical difficulties such as heart trouble or gastric trouble, not to mention psychological problems. There is no doubt that loud noise is a major irritant. Aircraft—especially jets—have been particularly implicated, but loudly roaring car, bus, or truck engines, honking horns, and screeching tires are scarcely less of a problem in cities where there is already a high background noise level caused largely by the tires and engines of motor vehicles.

In order to quantify the discussion at all, one needs an appropriate measure of loudness. The unit universally used is decibels (db), which is defined as follows:

$$\text{Intensity level (db)} = 10 \log_{10} \frac{\text{measured intensity (in } 10^{-6} \text{ watts/cm}^2)}{10^{-16} \text{ watts/cm}^2}.$$

The logarithmic scale comes close to representing the way the human ear responds to stimuli: a 10 db difference is roughly equivalent to a doubling of apparent loudness, whereas a doubling of actual sound intensity results in an increase of only 3 db (i.e., $10 \log_{10} 2 \approx 3$). Since sound intensity decreases as the square of the distance from a "point source," and linearly as the distance

---

[12]G. Bugliarello et al., "Noise Pollution," an Internal Staff Memorandum to Resources for the Future, Inc., Carnegie-Mellon University, 1968.

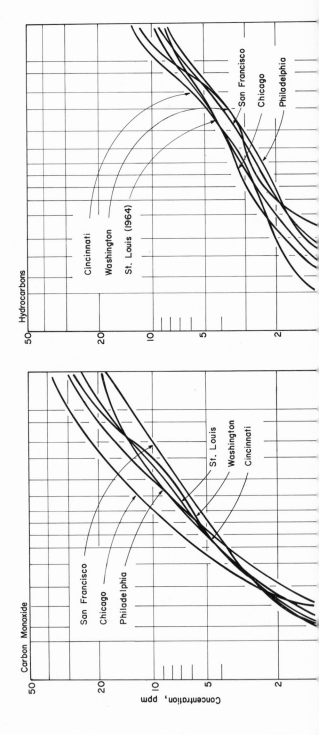

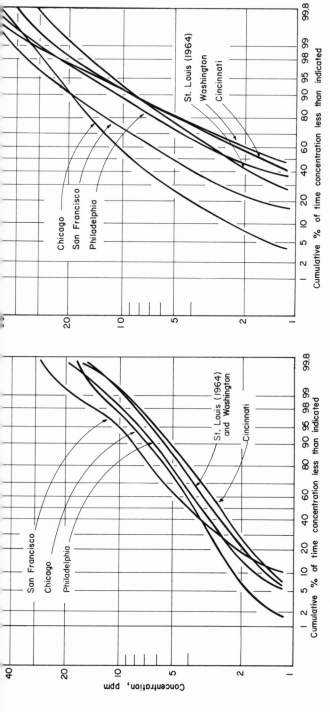

Figure 2–6. Frequency-time distribution of gaseous pollutants. (From U.S. Department of Commerce, "The Automobile and Air Pollution: A Program for Progress," Part II, December 1967.)

Table 2-7. Concentrations of Emissions in Los Angeles for 1947, 1955, and 1965, and Projections for 1980 and 2000

Year	Percent of days when 8-hr concentration exceeds:							Total emissions (million tons)
	1 ppm	2 ppm	5 ppm	10 ppm	15 ppm	20 ppm	30 ppm	
Carbon monoxide								
1947			67	33	0.07	<0.01	–	22.6
1955			97	33	3.6	0.3	–	34.4
1965			99.6	64	15.5	2.4	2.05	43
1980[a]			92.5	20	1.5	0.09	–	30
2000[a]			99.99	93	54	20	1.4	60
Hydrocarbons								
1947		55	2.8	0.03	–			3.5
1955		81.5	12.5	0.36	0.02			5
1965		93.5	31	2	0.17			6.6
1980[a]		75	9	0.2	<0.01			4.5
2000[a]		98.6	57	8.5	1.2			9
Nitrous oxides								
1947	23	3.3		0.045	–			1.2
1955	41	9		0.24	<0.01			1.65
1965	66	25		1.6	0.055			2.5
1980[a]	77	35		3.2	0.2			3.0
2000[a]	96.5	76		23	3.2			6.0

Source: Compiled by authors, based on frequency-time distribution data from C. E. Zimmer and R. I. Larson, "Calculating Air Quality and Its Control," *Journal of the Air Pollution Control Association*, Vol. 15 (1965), and APCO fuel consumption projections (unpublished).

– None.

[a]It is assumed that 60% of all vehicles will have control devices after 1968.

from a "line source"—such as a freeway—each doubling of the distance results in a 6 db or a 3 db drop in noise level, respectively.

Most of the noises typically encountered tend mainly to fall into the 50-100 db range. The benchmark level—below which hearing damage would be negligible—is usually taken as 80 db. The annoyance threshold is nearer 75 db. Regular exposure to 85 db or more is likely to result in serious hearing damage.

A single modern automobile cruising at moderate speed does not produce a particularly high level of noise. The primary noise source under these conditions is contact of the tires with the roadway. Tire noise has a continuous spectrum over the audible range, and rough pavement tends to increase the noise in the middle frequency ranges (100-1,000 cycles per second) by about 12 db. The overall noise level from an automobile under constant speed conditions and traveling on rough pavement may be about 80 db, which is enough to be irritating but not high enough to be by itself damaging to hearing. During acceleration, engine noise and vibration tend to predominate.

Diesel-powered road vehicles produce higher levels of noise. Ignition in a diesel engine occurs at a much higher pressure than in a gasoline engine. Engine

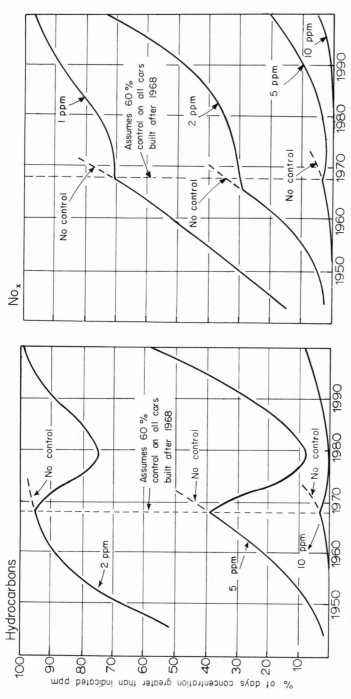

Figure 2–7. Projections of hydrocarbon and NO_x concentrations in Los Angeles. (From U.S. Department of Health, Education, and Welfare, NAPCA, *Summary of Emissions in the United States*, 1970 edition, May 1970.)

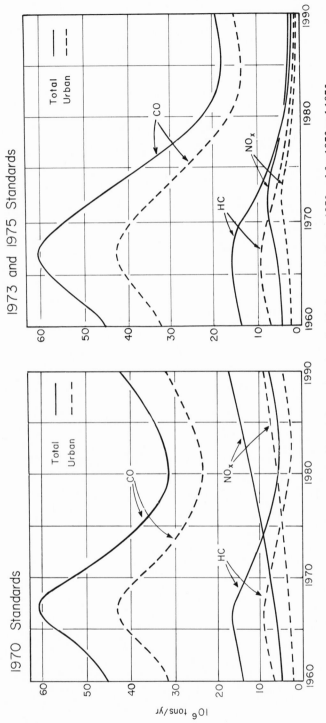

Figure 2–8. Emission estimates for the United States based on legislative standards for 1970 and for 1973 and 1975.

Table 2-8. Emission Requirements for United States and California

(grams per mile)

Date	HC	CO	NO$_x$	Evaporation control	Particulates
			Federal		
Prior to control	11.0	80.0	4.0		0.1–0.4
1968	3.4	34.0			
1970	2.2	23.0			
1971	2.2	23.0		Yes	
1973	2.2	23.0	3.0		
1974	1.5	23.0	3.0		
1975	0.5	11.0	0.90	Yes	0.10
Goals to 1980	0.25	4.7	0.40	Yes	0.03
			California[a]		
Prior to control	11.0	80.0	4.0		
1966	3.4	34.0			
1970	2.2	23.0		Yes	
1971	2.2	23.0	4.0	Yes	
1972	1.5	23.0	3.0	Yes	
1974	1.5	23.0	1.3	Yes	
Proposed for 1975	0.5	12.0	1.0	Yes	

Source: The Implications of Lead Removal from Automotive Fuel, An Interim Report of the Commerce Technical Advisory Board Panel on Automotive Fuels and Air Pollution, U.S. Department of Commerce, June 1970.

[a]As of June 15, 1970.

noise and vibration are more prominent especially under heavy load and/or acceleration conditions. When diesel-powered vehicles accelerate from a standstill, they produce a noise level of about 88 db, with the level rising to 95 db at about 80 seconds after start and then dropping quickly to a level of about 80 db.

CONCLUDING COMMENT

In this chapter we have presented information about emissions from internal combustion engines without sophisticated special control devices and about how these emissions are influenced by various factors—primarily driving modes and carburetion. In Part II we step back and take a look at energy requirements and energy storage for automotive vehicles and then review a number of technological alternatives from the point of view of performance and cost, and emissions of gaseous residuals and noise. Among the alternatives we discuss are improved or unusual versions of internal combustion engines; special control devices (some of which are already being used); and technologies that differ from conventional internal combustion engines in more fundamental ways.

PART II

Analysis of the Internal Combustion Engine and Its Alternatives

Chapter Three

Energy Conversion
Requirements for
Automotive Purposes

Before we can realistically assess and compare the capabilities of alternative power sources—or, for that matter, the external effects that may result—it is necessary to understand the "driving cycle," or "duty cycle" in the case of fleet vehicles, which governs the pattern and sequence of power and energy demands by automotive vehicles in urban conditions. The nature of this cycle is determined partly by the physical characteristics of the vehicle, partly by its relationship with the road and with other traffic, and partly by the behavior of the driver.

PATTERNS OF VEHICLE UTILIZATION

A "driving cycle," strictly speaking, is an artificial sequence of well-defined and reproducible actions (acceleration, cruising at various speeds, deceleration, idling, etc.) designed to *approximate* average driving conditions in a particular area. Obviously, such a cycle will vary from place to place and even with the season or time of day within a given city. Not much information is available as yet on driving cycles applicable to a wide range of urban areas or vehicle types. The first well-defined automotive cycle, called the "California Cycle," was described in 1962 and used to develop federal emissions standards.[1] More recently considerable reliance has been placed on what has come to be called the "LA-4 Cycle," which is supposedly typical of driving conditions for private automobiles in Los Angeles.[2] A dynamometer sequence (denoted XC-15) designed to simulate this cycle in the laboratory is reproduced as table 3-1.

---

[1] G. C. Hass and M. L. Brubacker, "A Test Procedure for Motor Vehicle Exhaust Emissions," *Journal of the Air Pollution Control Association*, Vol. 12 (November 1962), pp. 505-9.

[2] J. N. Pattison and M. P. Sweeney, "A Study of Los Angeles Driving as It Relates to Peak Photochemical Smog Formation," Paper No. 66-68, National Air Pollution Control Association Meeting, June 1966.

Table 3-1. LA-4 Synthetic Driving Cycle (XC-15 Dynamometer Sequence)

Mode	Speed range (mph)	Mode duration (sec)	Cumulative time (sec)
Accelerate 2.9 mph/sec	0-20	7	7
Accelerate 2 mph/sec	20-30	5	12
Decelerate - 1.1 mph/sec	30-0	25	37
Accelerate 2.9 mph/sec	0-20	7	44
Accelerate 1.3 mph/sec	20-30	8	52
Accelerate 1.0 mph/sec	23-33	10	62
Accelerate 2.0 mph/sec	35-55	10	72
Decelerate - 2 mph/sec	52-30	11	83
Decelerate - 0.86 mph/sec	30-0	26	109

Source: G. C. Hass, M. P. Sweeney, and J. N. Pattison, "Laboratory Simulation of Driving Conditions in Los Angeles Area," SAE Paper No. 660546, Society of Automotive Engineers, West Coast Meeting, Los Angeles, August 1966.

To analyze power and energy demands we must have a generalized prescription for the cycle that can be parameterized to accommodate local conditions. It is mathematically convenient to represent a "model" driving cycle in terms of a trapezoidal speed-time profile, as shown in figure 3-1. It can be shown that the average speed corresponding to this profile is given by:

$$\bar{v} = \frac{v_c}{N_s/3{,}600 \, v_c[(v_c/a) + T_s] + 1} \, , \tag{3.1}$$

where $\bar{v}$ is the average speed in mph, v_c is the cruise speed of the cycle in mph, N_s is the number of stops per mile, a is the average acceleration or deceleration in mph/second, and T_s is the average idling period in seconds.[3]

Average driving speed is shown in figure 3-1 as a function of the number of stops per mile at various cruise speeds, using an average acceleration of 2.2 mph/second (3.2 feet per second per second, or 3.2 ft/sec^2), and an average idle time of 30 seconds per stop. It can be seen that for more than about six stops per mile, the curves tend to approach each other closely, indicating the importance of the idling period in determining the average speed.

When the number of stops per mile is sufficiently large, the trapezoidal speed profile becomes a triangle. In this limiting case, v_c is no longer an independent parameter and the average speed becomes:

$$\bar{v} = \frac{60 \sqrt{a/N_s}}{2 + (N_s/60) \sqrt{a/N_s T}} \, , \tag{3.2}$$

---

[3] M. L. Walker, Jr., *A Methodology for Estimating Fuel Consumption for Various Engine Types in Stop-Go Driving*, IR&T-R-20, International Research and Technology Corp., Washington, D.C., March 1970.

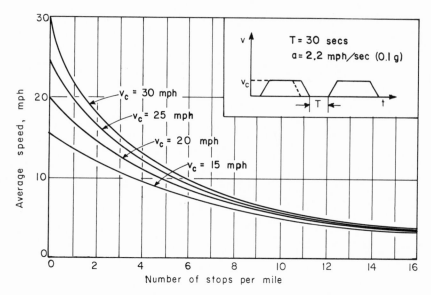

Figure 3-1. Influence of stop-and-go driving on average driving speed. (From M. L. Walker, Jr., *A Methodology for Estimating Fuel Consumption for various Engine Types in Stop-Go Driving*, International Research and Technology Corp., IRT-R-20, March 1970.)

in which the factor $60\sqrt{a/N_s}$ is the maximum speed achievable during any acceleration interval for a given number of stops per mile. This factor is plotted in figure 3-2, along with the maximum possible average speed (corresponding to zero idling time and equal to one-half of the peak speed between stops).

For a very large number of stops (i.e., the factor $N \gg 1$) the average speed approaches a limit that is strictly dependent on the number of stops per mile and on the idling time, as follows:

$$\bar{v} \to \frac{3,600}{N_s T} \, . \tag{3.3}$$

The use of equation (3.1) yields results that agree reasonably well with measured values. For example, data obtained by an instrumented test car traveling in and around Pittsburgh with approximately one stop per mile over 47 miles and with an assumed average cruise speed of 35 mph yielded a calculated average speed of 24 mph compared with a measured average of 21.7 mph.[4]

Driving cycles may be approximated by a sequence of discrete segments, as in table 3-1. However, it is equally reasonable and mathematically more con-

---

[4]F. Beckley Smith, Jr., W. A. P. Meyer, and R. U. Ayres, "A Statistical Approach to Describing Vehicular Driving Cycles," SAE Paper No. 690212, International Automotive Engineering Congress, Detroit, January 13-17, 1969.

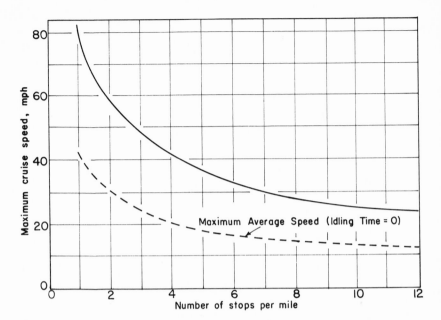

Figure 3-2.   Limiting cruise and average speeds in stop-and-go driving. (From M. L. Walker, Jr., *A Methodology for Estimating Fuel Consumption for Various Engine Types in Stop-Go Driving*, International Research and Technology Corp., IRT-R-20, March 1970.)

venient to introduce continuous frequency distributions, as illustrated by the plots in figure 3-3.

The top graph in figure 3-3 shows the Pittsburgh cycle acceleration data, plotted on probability paper. The solid line is a straight line fit, which seems to come reasonably close to reproducing the data, and which corresponds to a normal probability distribution with standard deviation $\sigma_a \cong 1.98$ (mph/sec.). (This can be read directly off the chart as the acceleration corresponding to the 84th "percentile.")

Figure 3-3 also gives the distribution of gradients (i.e. slopes) measured at 431 successive 15-second time intervals along the Pittsburgh route (excluding idling times). The solid line is a statistical fit which corresponds to a normal probability distribution having the standard deviation $\sigma_a \cong 3.75\%$.[5] In other parts of the country the standard deviation could be larger or smaller, the latter being more likely.

The distribution of values of the logarithm of velocity (log v) deduced from the Pittsburgh data is also shown in figure 3-3. The curve in log v exhibits a characteristic convexity, which suggests that speeds probably cannot be ade-

---

[5]Unfortunately the data are badly skewed toward positive slopes. This discrepancy has not been fully explained as yet, but it appears to be due to an unfortunate bias in the instrumentation.

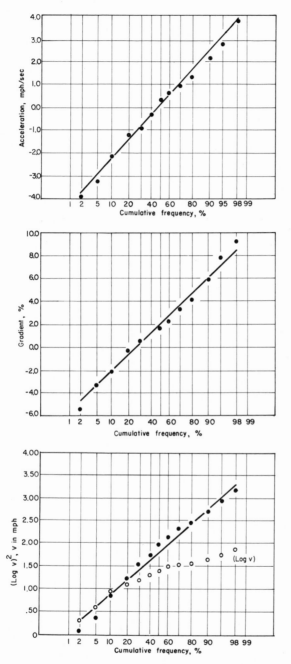

Figure 3-3. Marginal distributions for acceleration, gradient, and logarithm of speed, Pittsburgh cycle. (From F. Smith, W. Meyer, and R. U. Ayres, "A Statistical Approach to Describing Vehicular Driving Cycles," SAE Paper No. 690212, International Automotive Engineering Congress, Detroit, January 13-17, 1969.)

quately represented by simple log-normal distributions. However, it has been found that when values of $(\ln v)^2$ are plotted, as in the figure, the results do fall reasonably close to a straight line that would correspond to a normal distribution of the square of the logarithm of velocity.

Up to the present time no statistical data are available for special classes of vehicles such as taxis, city buses, delivery vans, and heavy trucks, which may be extremely important in urban traffic. It is certainly very desirable that the statistical characteristics of the driving patterns of these types of vehicles should be studied in depth.

AUTOMOTIVE VEHICLE PERFORMANCE

Obviously, there is a strong link between the purposes for which trips are made and the way vehicles are actually used. Data were presented in the last section on the basic elements of individual (automobile) driving patterns, although the latter are clearly determined by what the automobile is capable of doing as much as by its mode of interaction with the larger system or by the needs or habits of its driver. It seems worth discussing the physical capabilities of the vehicle in somewhat more detail here, inasmuch as the performance characteristics of the standard internal-combustion-engine–powered automobile of today will be the bench mark against which any "automobile-of-tomorrow" must inevitably be measured, and because these characteristics are intimately related to the production of residuals and their associated external effects.

Briefly, a certain amount of work (E) must be expended to accelerate a mass (W) to a given velocity v, regardless of the method of propulsion. For example, if a 3,200-lb automobile is to be accelerated to 60 mph (88 feet per second), the final kinetic energy $(E = \frac{1}{2} m_v^2)$ is numerically equal to 387,200 ft-lb whether this speed is reached in one second or in 100 seconds. The rate at which the instantaneous kinetic energy is increased is a measure of the power required. Thus, the minimum power necessary to accelerate our 3,200-lb car to 60 mph in 10 seconds—a classic measure of automobile "getaway" capability—is 38,720 ft-lb/sec, which is equivalent to an average power of 70 hp or 52.5 kw (disregarding frictional losses and air resistance for the moment).

To propel a wheeled vehicle, this power must be delivered as a torque (twisting force) at the axle. The relationship between axle torque T (ft-lb) and power delivered at the wheels P_w (hp) is:

$$P_w = 0.684 \, N_r T, \tag{3.4}$$

where N_r is axle rotational speed in revolutions per minute. A useful approximate expression for torque T is:

$$T = \frac{D}{2Re} \left[(C_1 a + \sin \tan^{-1} \theta + k_1) W + k_2 A v^2 \right], \tag{3.5}$$

where D is the effective wheel diameter in feet, R is the gear ratio (between drive shaft and wheels), e is the mechanical efficiency of the transmission, W is the total mass of the vehicle (lb), θ is the slope measured in percent, v is the velocity (ft/sec), a is the acceleration (ft/sec^2), A is the frontal cross-sectional area of the vehicle (ft^2), and C_1, k_1, k_2 are constants (with appropriate units). Actually $C_1 = g^{-1}$ where g is the acceleration of gravity (32 ft/sec^2).

As noted above, a real automobile operating on actual roads must use power not only to accelerate, but also to climb hills, and to overcome internal frictional and parasitic losses, plus rolling (tire) resistance and air drag. Internal parasitic losses in an internal-combustion-powered car arise from mechanical inefficiencies (especially in the transmission), power used to operate the fan, generator, oil pump and distributor, plus losses in the carburetor and muffler, etc., but *not* including auxiliary power for heating, air conditioning, radio and so forth, which would account for an additional 5%–10% of the power produced by the engine. Internal losses average 40% of rated engine brake horsepower (bhp).[6]

The ground friction or rolling resistance for modern passenger cars, which arises mainly from the constant flexing of the tires, increases linearly with the vehicle's speed in the lower speed ranges. It has been found that the coefficient C_1 (equation 5) is extremely sensitive to tire design,[7] load and degree of inflation, as well as speed, as shown in figures 3–4 and 3–5. This suggests that there may be considerable room for reducing vehicle power requirements, especially at low speeds, without sacrificing performance. Drag due to air resistance increases with the square of the speed; it is thus the principal component of power loss, particularly at speeds above 25 or 30 mph. Air resistance is proportional to vehicle frontal area and the air resistance coefficient is sensitive to variables in auto body design. For instance, the so-called "fastback" torpedo-shaped body styles, which have recently become popular, allow reduction of air drag by about 15% as compared to older styles.

Since the torque available is a function of the rotational speed N_r (rpm) of the axle, which can easily be related to an equivalent linear velocity of the vehicle, there is a maximum possible cruising speed for any given motor/gear ratio and vehicle beyond which no further torque is available because all the power output is being used to overcome drag. The power required to cruise at a given constant vehicle speed, taking into account rolling and air resistance, is plotted in figure 3–6 for two vehicle weight classes (2,000 lb and 4,200 lb) that are typical of personal vehicles now in service. In computing the values for these graphs, the value of the frontal area A was taken to be 20 ft^2 for case I (2,000 lb) and 25 ft^2 for case II (4,200 lb).

---

[6] George A. Hoffman, *Automobiles, Today and Tomorrow*, Rand Corporation Memorandum RM–2922, November 1962.

[7] Thus, so-called "radial ply" tires, first introduced a few years ago in Europe, yield much less friction loss than conventional tires.

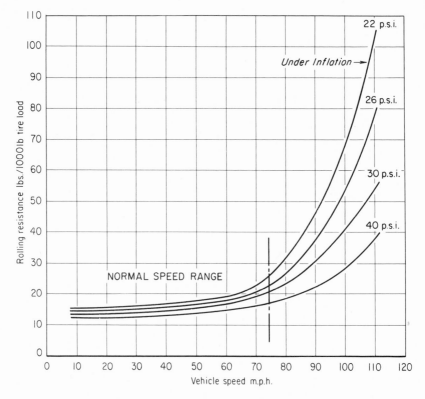

Figure 3-4. Rolling resistance versus speed as a function of tire pressure. (From George A. Hoffman, *Automobiles Today and Tomorrow*, Rand Corporation, RM-2922, November 1962.)

Figure 3-6 refers only to the power necessary simply to overcome drag at a constant speed; it does not include the power required for hill climbing or for acceleration (discussed earlier). The power *available* to the driver for accelerating on an upgrade is equal to the maximum power available at the drive wheels minus that required simply to maintain the speed of the car. This difference as a function of speed is shown in figure 3-7 for a typical 1960 compact car.

The torque available at a given speed naturally determines maximum acceleration capability at that speed. Despite the large drag forces at high speeds, it is obvious that the greatest torque is needed when the vehicle is standing still, to overcome inertia and start it moving. A conventional (Otto-cycle) internal combustion engine unfortunately generates no torque when its shaft is not rotating; hence the need for an electric motor to get it started, a clutch to permit the engine to continue rotating (idling) at a few hundred rpm even when the car is not moving, and a mechanical transmission with a variable gear ratio to permit the engine to operate at high speeds even though the drive shaft is running

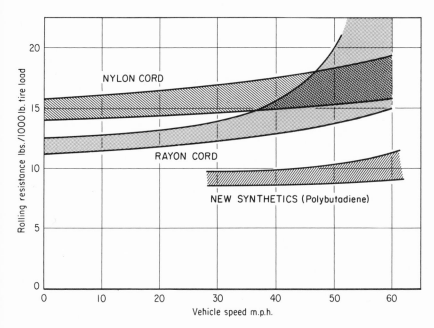

Figure 3-5. Rolling resistance versus speed for different tire materials. (From George Hoffman, *Automobiles Today and Tomorrow*, Rand Corporation, RM-2922, November 1962.)

slowly. Diesel engines and simple gas turbines have similar characteristics. This is one of the major weaknesses of existing propulsion systems. On the other hand, Rankine cycle (steam) engines and electric motors can deliver a maximum torque at zero velocity—an inherent advantage since this is precisely when the greatest torque is needed.

Figure 3-8 is a graph showing the maximum acceleration as a function of speed for an average 1960 compact car. It can be seen that there is a brief acceleration "spike" in low gear to 0.4 g, but the average time for such cars to accelerate from 0 to 60 mph is from 15 to 20 seconds, corresponding to an average acceleration (0-60 mph) of 0.14 to 0.18 g. Accordingly, 0.2 g can be set as a modest average accelerating capability, corresponding to 0-60 mph in about 14 seconds, while 0.1 g and 0.3 g define "low" and "high" average accelerations corresponding to 27 seconds (Volkswagen) and 9 seconds (sports car) respectively.

Equations 3.4 and 3.5 can be combined to give an expression for instantaneous vehicle power requirements as a function of velocity, acceleration, weight, and other parameters as follows:

$$P_w = \frac{0.0057}{e}\left[(k_1 + \sin \tan^{-1} \theta + a/g) Wv = k_2 Av^3\right], \qquad (3.6)$$

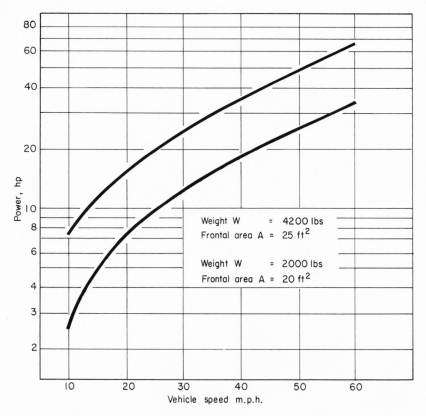

Figure 3-6. Power required to cruise, for typical vehicle weights.

where terms are measured in units previously given, and speed is related to drive shaft revolutions per minute (rpm) as

$$v = \frac{ND}{R} \cdot \qquad (3.7)$$

For the small gradients encountered in normal driving one can approximate

$$\sin \tan^{-1} \theta \sim \theta$$

where slope is measured in percent, this simplification will be used hereafter. It is usually convenient to express power in hp rather than ft-lb/sec, velocities in mph rather than ft/sec, and accelerations in mph/sec rather than ft/sec². Thus, 10 fps = 6.8 mph and 3.2 ft/sec² (= 0.1 g) is equivalent to very nearly 2.2 mph/sec. In these units, the power required at the rear wheels becomes:

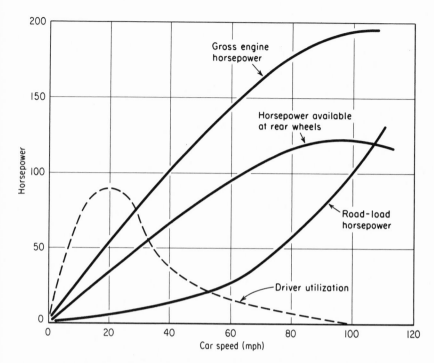

Figure 3-7. Power available to the driver at various speeds (1960 compact car). (From George A. Hoffman, *Automobiles Today and Tomorrow*, Rand Corporation, RM-2922, November 1962.)

$$P_w = \left\{ [\alpha + \beta\theta + \gamma]\, Wv + \delta\, Av^3 \right\}, \tag{3.8}$$

which is the same form as equation (5). As before, θ is the slope in percent, W is the mass, A is the cross section of the vehicle, and v is the velocity. Numerical values of the coefficients for the two classes of automobiles used to plot figure 3-8 are shown in table 3-2.

It is clear from figure 3-8 that the power delivered at the wheels required for acceleration at any given speed v is determined by adding the road-load power requirements at that speed (equation 8) to the rate of change of the kinetic energy of the vehicle. Because road-load demand increases sharply (in proportion to the cube of the absolute speed) it takes much more delivered power, for example, to produce a 2 mph/sec acceleration at 60 mph than it does to achieve the same acceleration from a standing start. The power required at the rear wheels to cruise and accelerate a 2,000-lb car at various rates (2, 4, 6 mph/sec) is shown in figure 3-9. Engine power must be sufficiently larger, of course, to compensate for internal losses (40%) and auxiliary equipment (5-10%).

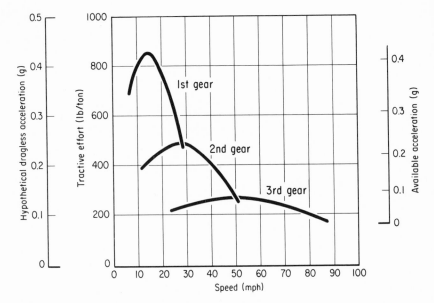

Figure 3–8. Typical acceleration characteristics (1960 compact car). (From George A. Hoffman, *Automobiles Today and Tomorrow*, Rand Corporation, RM-2922, November 1962.)

The average power used and the energy consumed per mile by a vehicle over an actual urban driving-cycle can be computed by integrating equation (8) over the appropriate distribution functions corresponding to the actual frequency of occurrences of various speeds, accelerations, and slopes, on the assumption that these are independent variables. In practice, the assumption of statistical independence is not strictly valid.[8] However, we can obtain simple formulas by

---

[8] A more general form of the trivariate density function would be:

$$f(x_1 s_2 s_3) = \frac{\sqrt{B}}{(2\pi)^{3/2}} \exp\left[-\frac{1}{2} \sum_{i,\,j\,=\,1}^{3} C_{ij}\,(x_i - \mu_i)\,(x_j - \mu_j)\right]$$

where
$$B = \sigma_1^2 \sigma_2^2 \sigma_3^2 \left[1 + 2\rho_{12}\rho_{13}\rho_{32} - \rho_{12}^2 - \rho_{13}^2 - \rho_{32}^2\right]$$

$$C_{ii} = B\sigma_j^2 \sigma_k^2 \,(1 - \rho_{jk}^2) \qquad i \neq j \neq k$$

$$C_{ij} = B\sigma_i \sigma_j \sigma_k^2 \,[\rho_{ik}\rho_{jk} - \rho_{ij}] \qquad i \neq j \neq k$$

μ, σ, and ρ are conventional mean, standard deviation, and correlation coefficients, respectively.

Table 3-2. Numerical Values of Power Coefficients

Coefficient	Volkswagen (2 passengers) $W = 2,000$ lb $A = 20$ ft^2	Standard car (2 passengers) $W = 4,214$ lb $A = 25$ ft^2
a	12.15×10^{-5}	12.15×10^{-5}
β	2.67×10^{-5}	2.67×10^{-5}
γ	5.76×10^{-5}	5.024×10^{-5}
δ	0.292×10^{-5}	0.432×10^{-5}

Source: Calculated by the authors.

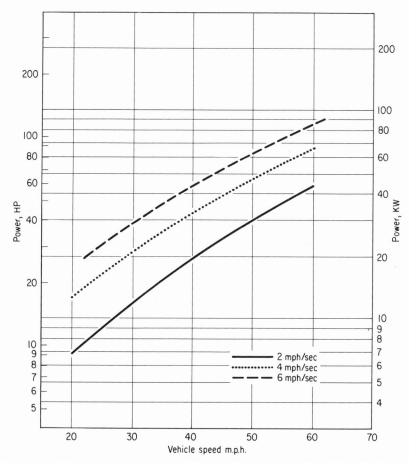

Figure 3-9. Power required at rear wheels to accelerate a 2,000-lb vehicle at various rates.

approximating the curves exhibited in figure 3-3, by normal or log normal functions as follows:

$$f_1(a) = \frac{2}{\sqrt{2\pi}\sigma_a} \exp\left\{-\frac{a^2}{2\sigma_a{}^2}\right\} (a > 0),\qquad (3.9)$$

$$f_2(\theta) = \frac{2}{\sqrt{2\pi}\sigma_\theta} \exp\left\{-\frac{\theta}{2\sigma_\theta{}^2}\right\} (\theta > 0),\qquad (3.10)$$

$$f_3(v) = \frac{1}{\sqrt{2\pi}\sigma_v v} \exp\left\{-\frac{(\ln v/v^o)^2}{2\sigma_v{}^2}\right\}.\qquad (3.11)$$

Then the average power required for vehicular operation will be given by the expression

$$\bar{P}_w = F \underbrace{\int_0^\infty dv f_3(v) \int_{-\infty}^\infty d\theta f_2(\theta) P_w(v,\theta,a=0)}_{\text{cruise}}$$

$$+ \underbrace{\frac{1}{2}(1-F)\int_0^\infty dv f_3(v) \int_{-\infty}^\infty d\theta f_2(\theta) \int_{-\infty}^0 da f_1(a) P_w(v,\theta,a)}_{\text{deceleration}}\quad (3.12)$$

$$+ \underbrace{\frac{1}{2}(1-F)\int_0^\infty dv f_3(v) \int_{-\infty}^\infty d\theta f_2(\theta) \int_0^\infty da f_1(a) P_w(v,\theta,a)}_{\text{acceleration}},$$

where F is the fraction of driving time spent cruising and $1 - F$ is the fraction spent (equally) decelerating and accelerating. During idling, of course, no power is used at the wheels. In the case of an internal combustion engine we can take $a = 0$ in the second (deceleration) term, thus treating cruise and deceleration conditions in the same manner as regards power requirements.[9] The indicated integrations are straightforward, yielding

---

[9]The equation is retained in the more general form to permit its application later to other power sources—notably electric and steam. In an electric car, energy may actually be recaptured during deceleration (by regenerative brakes) thus providing a partial battery recharge capability. Both electric and steam propulsion systems permit "dynamic braking"—the use of engine power to produce reverse torque.

$$\bar{P}_w = \left[\left\{ \gamma + \frac{\beta}{\sqrt{2\pi}} \sigma_\theta + \frac{1}{2}(1 - F)\frac{\alpha}{\sqrt{2\pi}} \sigma_a \right\} Wv_0 \exp\left(\frac{1}{2} \sigma_v{}^2\right) \right.$$

$$\left. + \delta S v_0{}^3 \exp\left(\frac{9}{2} \sigma_v{}^2\right) \right]$$

(3.13)

Energy consumption per mile traveled by a vehicle weighing W lb is a useful measure of efficiency. It is the expectation-value of the instantaneous power divided by the speed, which (using the above approximations) takes the form

$$\frac{\bar{P}_w}{v} = \left\{ \gamma + \frac{\beta}{\sqrt{2\pi}} \sigma_\theta + \frac{1}{2}(1 - F)\frac{\alpha}{\sqrt{2\pi}} \sigma_a \right\} W$$

(3.14)

$$+ \sigma A v_0{}^2 \exp(2\sigma_v{}^2).$$

The energy consumption in kilowatt-hours (kwh) per ton mile can be derived by multiplying equation (3.14) by the quantity 0.746 (2,000/W).

Applications of these approximate formulas are reserved for later chapters, particularly chapter 10. To use the more accurate law implied by figure 3–5 requires integrations that cannot be carried out in simple closed form for all values of the various parameters, but which obviously should be carried out numerically in any calculation done with serious intent to represent the real world.

An alternative approach is to compute the empirical distribution of P_w values from the observed distributions for a, θ, and v (figures 3–3 to 3–5). This cannot be done analytically from the data given, but it is a simple matter for a computer.

Figure 3–10 shows the cumulative percentage of time during which the vehicle power was less than the amount indicated in the graph during the Pittsburgh cycle. Thus, for 90% of the time the power was less than 50 hp (from a peak of approximately 185), and for 40% of the time it was negative—i.e., it could in principle have been used for regenerative braking. Conversely, however, for approximately 2% of the time, the power demands were above 100 hp. Any new type of propulsion system must be able to provide power for these brief peaks, unless it is to suffer by comparison with conventional automobiles.

Energy consumption per unit distance can also be expressed in terms of engine performance by averaging engine fuel consumption over the driving cycle. This may be expressed by:

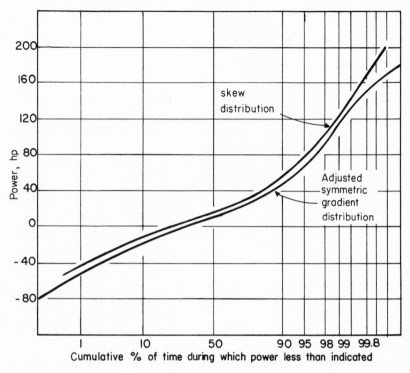

Figure 3–10.   Distribution of power during the Pittsburgh cycle.

$$E_c = \frac{H \int_0^t C_f P_b dt}{WS}.$$

(3.15)

where E_c is the average energy consumption per ton-mile, H is the heating value of the fuel in Btu/lb, C_f is the specific fuel consumption in lbs/bhp-hr, P_b is the brake hp of the engine, W is the vehicle weight in tons, S is the distance in miles over which C_f is averaged, and t is the time required to travel over S.[10]

It is clear that this parameter represents a virtually ideal standard of comparison for vehicular energy conversion techniques. It will be discussed again at greater length in chapter 14.

In any case, whatever the propulsive technique, the basic energy and power requirements must somehow be met. It remains now to consider the details of those techniques.

[10] M. L. Walker, Jr., *A Methodology for Estimating Fuel Consumption.* . . .

Chapter Four

Energy Conversion

Although energy conversion, broadly speaking, covers all possible combinations of initial and final states, the discussion in this chapter is confined to cases pertinent to self-contained automotive power plants. In short, we consider only energy conversion systems in which the initial state is chemical energy stored in a "fuel" and the final state is mechanical or electrical energy.

In principle, the fuel may be coal, wood, ammonia, hydrogen gas, aluminum or magnesium powder (or ribbon), as well as the more familiar hydrocarbons such as propane, gasoline, kerosene, or diesel oil. The more exotic possibilities should not be dismissed out of hand, especially in conjunction with so-called "fuel cells" or refuelable primary electrochemical cells.

Energy conversion devices can be divided into combustion engines (either internal or external) and direct conversion schemes. Internal combustion engines differ from the external combustion variety in that the fuel is burned in the "expander." (By this definition a gas turbine is on the borderline.) "Direct conversion" is simply a catch-all name for energy conversion processes not involving combustion in the usual sense.

A somewhat more fundamental, but slightly different, organizing principle can be derived from thermodynamic analysis. If the mode of operation of the device inherently requires that the energy of the fuel be converted to heat and transferred to a working fluid which goes through a cycle, it is denoted a "Q-engine" or heat engine.[1] If the internal energy need not be converted first to heat, then we are dealing with an "E-engine." Examples of the Q-engine are Rankine-cycle engines with fossil fuel or nuclear boilers, Stirling-cycle engines, and thermoelectric and thermionic devices. Examples of the E-engine include spark-ignition or compression-ignition internal combustion engines, gas turbines, electrochemical cells, and living tissue.

---

[1] F. Lauck, O. A. Uyehara, and P. S. Myers, "An Engineering Evaluation of Energy Conversion Devices," *SAE Transactions* (1963), pp. 41–50.

In spite of the lesser degree of sophistication implied thereby, we shall use the "combustion" and "direct conversion" classification in the remainder of this chapter.

INTERNAL COMBUSTION E-ENGINES

Internal combustion engines do not require that the internal energy of the chemical (or nuclear) fuel be converted first to heat. (Momentum transfer, for example, or electrical potential are other possible conversion mechanisms.) E-engines thus do not require "cycles" but (ideally) depend on continuous processes.[2] Nevertheless, so-called "air-standard cycles"—the Otto cycle, the Diesel cycle, and the Brayton (or Joule) cycle on which various E-engines operate—have been customarily defined as introductory concepts to help describe the various species of internal combustion engines.

The most familiar of all engines—the reciprocating (piston), spark ignition ICE—was first built in 1876 by Karl Otto, and its operational scheme has become known as the Otto cycle. In idealized form, illustrated in figure 4-1, it consists of the following four processes:

Process 1-2. Isentropic compression of the working medium (fuel-air mixture) by the piston.

Process 2-3. Constant volume heating of the medium (i.e., combustion).

Process 3-4. Isentropic expansion of the medium, doing work on the piston.

Process 4-1. Constant volume rejection of heat from the medium.

In practice, the "rejection" phase means the combustion products are physically exhausted. The expansion and compression are carried out in a cylindrical cavity in which a piston moves back and forth, or up and down. Each backward or forward motion of any reciprocating piston engine is called a stroke. The basic four-stroke Otto-engine cycle operates as follows:

1. Intake of air and fuel on the piston's outward stroke.
2. Compression of the mixture on the inward stroke.
3. Ignition at the maximum compression (dead center) followed by expansion on the outward stroke.
4. Exhausting of the combustion products from the cylinder on the next inward stroke.

This sequence is illustrated schematically in figure 4-2. It can be seen that the cycle affords one power stroke (no. 3) for every four strokes of the piston. Since

---

[2]In a continuous combustion (constant pressure) gas turbine, this is evident; actually, one of the fundamental drawbacks of reciprocating (piston-type) internal combustion engines is that the process is constantly interrupted and restarted.

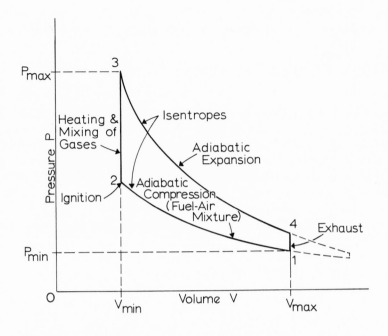

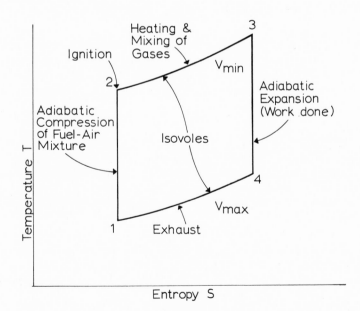

Figure 4-1. The air standard Otto cycle.

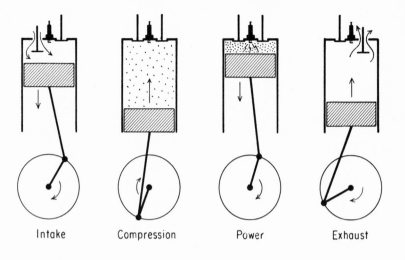

| Intake | Compression | Power | Exhaust |

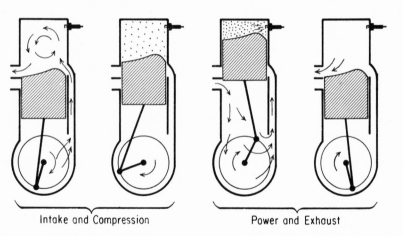

| Intake and Compression | Power and Exhaust |

Figure 4-2. Sequence of events in 4-stroke Otto cycle (above) and 2-stroke Otto cycle (below).

each piston is connected to an eccentric crankshaft to produce rotation, this implies one power stroke for every two complete crankshaft rotations. Hence, to operate reasonably smoothly, the engine should have at least four cylinders, and preferably six or eight.

The Otto-cycle can also be adapted to a two-stroke version, which provides one power stroke for each crankshaft rotation, or one for every two piston strokes (see figure 4-2). In this case, the fresh charge of fuel and air is compressed (typically by the piston itself when it is at the bottom of its stroke) and forced into the cylinder; this mixture is then compressed and ignited as in the four-stroke case. Near the end of the power stroke, a new fuel-air charge sweeps the exhaust out and the process starts anew. Superficially, it would seem that the two-stroke engine would surely be a more desirable device than the four-stroke, since it wastes less motion and requires fewer cylinders. Also, no elaborate valves are needed and only a small flywheel suffices to sustain the compression part of the cycle.

On the other hand, the fresh fuel-air charge does not sweep out the exhaust completely, and the power stroke is shorter, since exhaust gases are being expelled during the last part of the downstroke. Hence, the two-stroke version is actually inherently less efficient, consumes more fuel and produces more unwanted exhaust emissions per unit of work done (hp-hr). Furthermore, as the two-stroke engine inherently cannot be lubricated as easily as the four-stroke version, lubricating oil is generally mixed directly with the fuel. This eventually fouls the engine with a residue of partially burned oil; it also causes smoky exhaust fumes. This combination of conditions, in fact, has relegated the two-stroke engine mainly to applications where operating life need not be particularly long, such as motorboats, power lawn mowers, motorbikes, and model airplanes. The conventional prime mover for automobiles is the four-stroke Otto-cycle engine.[3]

What has been described above is the operation of a single piston. A complete engine will consist of a number of pistons—normally from four to eight—all connected to the same crankshaft in a "V," "Y," or "in-line" configuration and operating in a predetermined sequence so that an essentially constant torque is provided and inertial forces are as nearly balanced as possible. The rest of the mechanically complex hardware of the power train is concerned with matching the torque-speed output of the engine to the requirements of the vehicle (transmission, clutch) ensuring the proper air-fuel ratio, timing the cylinder-firing sequence, and cooling the engine, plus driving assorted electrical auxiliaries (starter, voltage regulator, generator, etc.).

Typical modern Otto-cycle engines operating under full load at optimum speed, without auxiliaries, achieve thermal efficiencies ranging from 20% for small engines to about 30% for larger ones. Fuel consumption ranges from 0.4 lb/hp-hr to 0.5 lb/hp-hr under the same conditions.

The behavior of an idealized Otto-cycle engine is shown in figures 4-3 through 4-5. "Theoretical" thermal efficiencies fall considerably beneath the "air standard" curves due to heat losses, dissociation of molecules, and variable specific heats. Mechanical losses are not taken into account in these curves. It is

---

[3]The Swedish Saab was an exception prior to 1968; concern for air pollution forced a change.

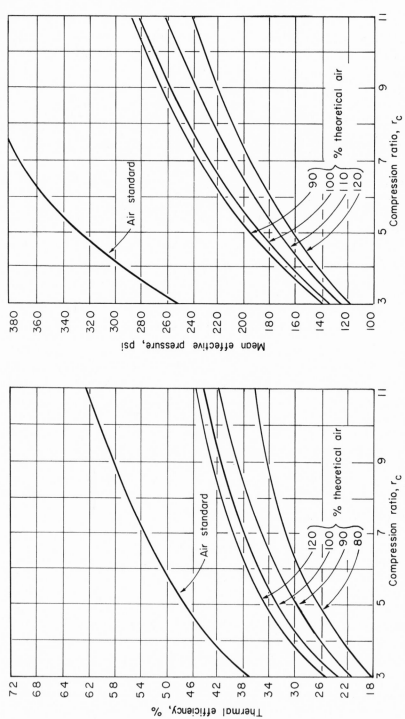

Figure 4–3. Thermal efficiency (left) and mean effective pressure (right) for the Otto-cycle internal combustion engine. (From T. Baumeister and L. S. Marks, *Standard Handbook for Mechanical Engineers*, 7th ed. Copyright © 1967 McGraw-Hill, New York. Used with permission of

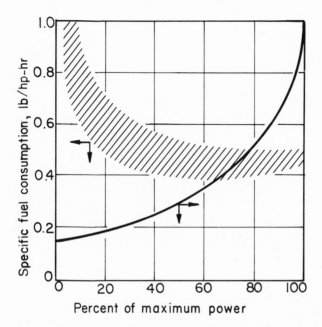

Figure 4-4. Variation of specific fuel consumption with power for Otto-cycle engine. (From M. L. Walker, Jr., *A Methodology for Estimating Fuel Consumption for Various Engine Types in Stop-Go Driving*, International Research and Technology Corp., IRT–R–20, March 1970.)

important to note that maximum power and maximum economy (thermal efficiency) do not coincide. The choice of a 15 : 1 air-fuel ratio is, in effect, a compromise between the two.

Indicated horsepower (ihp) may be computed directly from the mean effective pressure (*MEP*) in the cylinder, viz.,

$$\text{Power (ihp)} = (MEP)\frac{LAN}{33,000}$$

where

L = stroke (in),
A = area (in^2),
N = cycles/min,
MEP = pounds per square inch (psi).

Horsepower delivered at the shaft, known as brake horsepower (bhp) is smaller, due to mechanical losses due to friction, leakage, etc. in the engine. In fact, the mechanical efficiency coefficient is defined as the ratio of brake horsepower to indicated horsepower. "Air standard" curves are computed directly

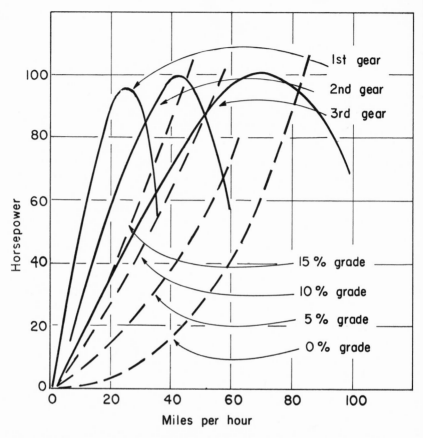

Figure 4-5.   Internal combustion engine horsepower characteristics. (From F. L. Schwartz, "Vehicular Gas Turbines," *Gas Turbine Engineering Handbook*, Gas Turbine Publications, Inc., Stamford, Conn., 1966.)

from the Otto-cycle points, assuming air as a working fluid.[4] For an automobile there are additional losses to consider. The power actually available at the rear wheels, however, is roughly 40% less than the advertised or "Detroit" horsepower of the engine; the rest goes into operating auxiliaries such as the fan, generator, water pump, and oil pump, and overcoming transmission losses, overcoming back-pressure in the muffler, etc. Efficiency also drops sharply as the engine speed differs significantly from optimum, due to changing load. (See figure 4-6.)

---

[4] The formula for the air standard Otto-cycle is given in many textbooks as:

$$\eta_a = 1 - (1/r_c)^{0.4}$$

where η_a is the indicated thermal efficiency, and r_c is the volumetric compression ratio.

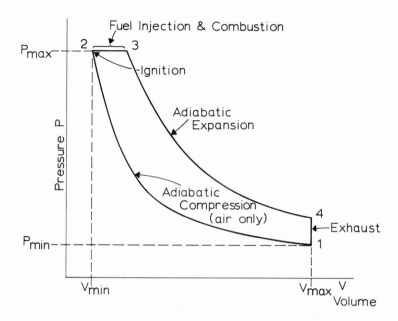

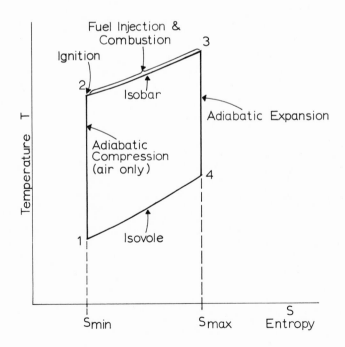

Figure 4-6. The air standard diesel cycle.

￼The second type of internal combustion engine in wide use today relies on ignition by compression alone rather than by means of a spark; this engine was originally devised by Rudolph Diesel in 1892. Basically, it is a piston engine much like the spark-ignition engine described above. Its essential feature, however, is the carefully timed injection of fuel into previously compressed air; the heat resulting from very high compression ratios (15 : 1 up to 20 : 1) ignites the mixture without a spark.

The diesel air-standard cycle (figure 4-6) is similar in most respects to the Otto cycle, differing primarily in the circumstances in which combustion occurs. The following four processes take place (see also figure 4-7).

Process 1-2. The piston sucks in air and compresses it adiabatically (isentropically), the temperature rising high enough in the process to ignite the fuel.

Process 2-3. Fuel is injected into the hot compressed air, and is ignited as it enters. Injection and combustion occur continuously throughout part of the stroke, heat being supplied essentially under constant pressure (not at constant volume, as in the Otto cycle). The motion of the piston compensates for the buildup of pressure that would otherwise be produced by the heat of combustion.

Process 3-4. After fuel injection ceases and combustion is complete, the gases expand and cool adiabatically (isentropically), doing further work on the piston.

Process 4-1. The exhaust valve opens and most of the gases blow off while the combustion chamber remains at roughly constant volume and the pressure drops rapidly to atmospheric. The returning piston itself then pushes the remaining exhaust gases out (in the four-stroke version).

Air alone is compressed in the diesel engine.[5] The heat developed is sufficient to ignite the fuel when injection begins at the point of maximum compression. Thus, both the sparkplug and the carburetor are eliminated, but a controlled fuel injection system is essential. Moreover, compression ignition obviously requires higher pressures (about twice as high) than occur in Otto spark-ignition engines and thus diesels tend to have much heavier construction.

Like the Otto engine, the basic four-stroke diesel described above is also adaptable to a two-stroke cycle. In this case, one power stroke per revolution is provided, rather than one every two revolutions as in the four-stroke case. In general, two-stroke diesels offer greater power per unit of weight at a given operating speed. The shorter time available for each operation, however, causes

---

[5] A minor advantage is that blow-by past the piston rings is mostly air.

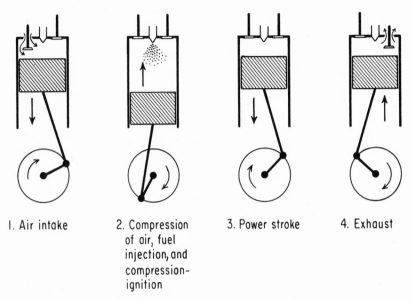

| 1. Air intake | 2. Compression of air, fuel injection, and compression-ignition | 3. Power stroke | 4. Exhaust |

Figure 4-7. Sequence of events in 4-stroke diesel engine.

poor "scavenging" of air. Thus, two-stroke diesels are often equipped with scavenging blowers to provide air at 2–5 psi above atmospheric pressure at the inlet, which hastens the removal of burned gases and provides a fresh air charge for the next cycle. Both types of diesel engine frequently use superchargers (precompression of the fuel-air mixture) to increase power output (up to 50%); the use of supercharging on the two-stroke diesel also improves its air scavenging characteristics.

Theoretical performance curves are shown in figures 4-8 through 4-10. The "air standard" curves for thermal efficiency are calculated directly from the cycle, using thermodynamic data for air.[6] So-called "theoretical" curves take into account heat losses, dissociation, etc., but exclude mechanical losses. Because of the higher compressions (required to initiate combustion without a spark) thermal efficiencies tend to be higher than in an Otto-cycle engine. Torque speed curves are qualitatively similar, however.

Diesel engines normally operate on a refined paraffin hydrocarbon fuel similar to tetradecane ($C_{14}H_{30}$). Diesel fuel is now cheaper than gasoline, but it

---

[6]The formula in this case is:

$$a = \frac{(r_d^{1.4} - 1)}{1.4 r_c^{0.4} (r_d - 1)},$$

where r_c is the volumetric compression ratio and r_d is the expansion ratio during constant pressure combustion.

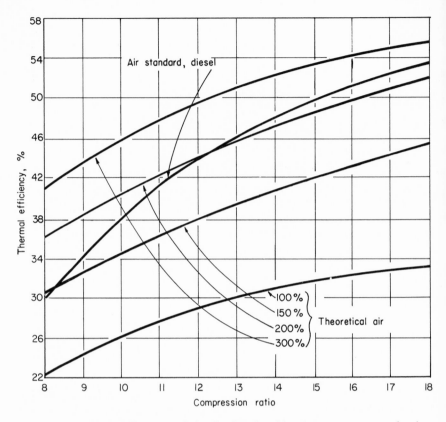

Figure 4-8. Thermal efficiencies of the diesel cycle with constant pressure combustion. (From T. Baumeister and L. S. Marks, *Standard Handbook for Mechanical Engineers*, 7th ed. Copyright © 1967 McGraw-Hill, New York. Used with permission of McGraw-Hill Book Company.)

might not continue to be cheaper if it were to capture any sizable fragment of the U.S. automotive market, since the amount that can be produced from a barrel of crude petroleum is strictly limited. At the moment, the supply exceeds the demand. Diesel engines are currently used primarily in applications where fuel economy and long life are of prime importance. Disadvantages are extra cost, bulk, weight, noise, and hard starting. Also, diesel engines tend to get out of adjustment rather easily and, when not perfectly tuned, give rise to a smoky and unpleasant-smelling exhaust; although unburned hydrocarbons and carbon monoxide emissions are actually much less than from spark-ignition engines.

The third basic type of internal combustion engine is the gas turbine that operates on an air-standard cycle which was first described by James Joule in England and George Brayton in the United States (with reference to recip-rocating engines). Initially, the air-fuel mixture is compressed adiabatically (isentropically) (1-2) as in the Otto and diesel cycles; combustion then occurs

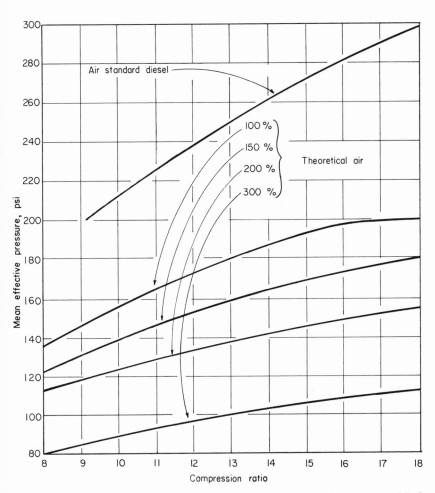

Figure 4-9. Mean effective pressure for the diesel cycle. (From T. Baumeister and L. S. Marks, *Standard Handbook for Mechanical Engineers*, 7th ed. Copyright © 1967 McGraw-Hill, New York. Used with permission of McGraw-Hill Book Company.)

and further heating takes place at constant pressure but increasing volume (2-3) as in the diesel; the "power stroke" (3-4) is an adiabatic (isentropic) expansion of the exhaust gases against the turbine blades; the final exhaust stroke (4-1) also takes place at constant (atmospheric) pressure. Pressure-volume (P-V) and temperature-entropy (T-S) diagrams are shown in figure 4-11. The last step is the only difference between the Brayton and diesel cycles, although the method of implementation varies greatly between the two.[7]

---

[7]The Brayton (or Joule) cycle is in principle applicable to external combustion (Q) closed-cycle engines, in which the working fluid is recycled, as well as to the more familiar open-cycle internal combustion engines.

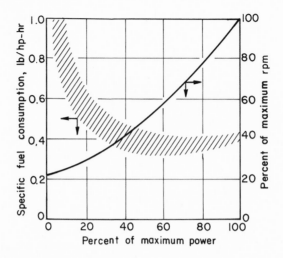

Figure 4-10. Variation of specific fuel consumption with power for the diesel cycle engine. (From M. L. Walker, Jr., *A Methodology for Estimating Fuel Consumption for Various Engine Types in Stop-Go Driving*, International Research and Technology Corp., IRT-R-20, March 1970.)

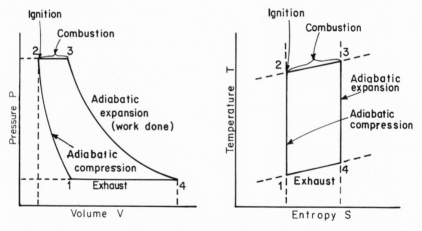

Figure 4-11.   The air standard Brayton cycle.

The basic gas turbine configuration is generally familiar: it consists of a turbocompressor, a combustor—where heat of combustion is added to compressed air—and a turboexpander, as illustrated by figure 4-12. The turbocompressor may (in principle) be on the same shaft as the expander, which is the power unit. However, this simple arrangement is almost useless for variable speed operation. At low shaft speeds, in particular, the compressor will be inefficient, and the output torque will therefore be very low. Thus, a more practical arrange-

Simple

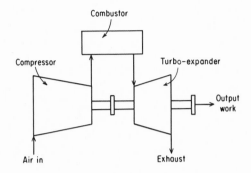

Compound

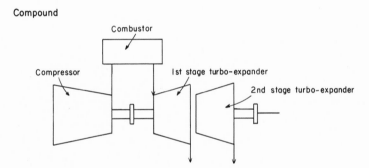

Figure 4-12. Schematic diagram of simple (above) and compound (below) gas turbines.

ment for an automotive application, where variable loads and variable speeds are required, is a compound (split shaft) or "free" turbine with a two-stage expander. The first stage is simply the power supply for the compressor, which operates at roughly constant speed regardless of load. The second stage is the output power, which can be regulated by varying the fuel supply in the combustor.

It can be shown that work done per "cycle," i.e., the net area enclosed by the P-V or T-S curves, ideally increases with increasing pressure.

For relatively low compression ratios (of the order of 5 : 1) the temperature at state 4 (after expansion) is higher than that of state 2 (after compression but before combustion). This circumstance allows the net efficiency of the cycle to be improved by the use of a heat exchanger or regenerator, reversibly absorbing the waste heat until the exhaust temperature has dropped to that of state 2. Then, after the isentropic compression (phase 1-2) in the next cycle, the heat stored in the heat exchanger or regenerator can be transferred to the fuel-air mixture prior to combustion. Thus, regeneration raises the average temperature of the gas during the heat input phase and lowers it during the heat rejection phase. However, as the pressure ratio rises, the compressor discharge temperature approaches the turbine discharge (exhaust) temperature, and eventually exceeds it. Under these circumstances the regenerator would transfer heat in the opposite direction (from combustor to the exhaust), which would drop the efficiency below that of the simple cycle. However, the better the heat exchanger, the higher the efficiency and the lower the optimum compression ratio. Thus, for an 80% efficient heat exchanger (assuming a compressor efficiency of 80%, an expander efficiency of 85%, an overall mechanical efficiency of 95%, and a combustor efficiency of 100%), the optimum compression ratio is r_c = 7.4 and the achievable overall thermal efficiency will be roughly 34%. Improving the heat exchanger from 80% to 90% increases overall thermal efficiency from 34% to 40% with r_c = 5. These interrelationships are shown graphically in figure 4-13. Typical variation of specific fuel consumption versus percentage of maximum power is shown in figure 4-14.

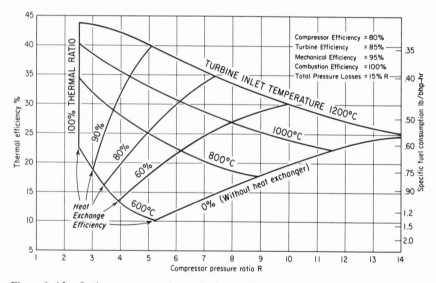

Figure 4-13. Optimum compression ratio for peak efficiency for various inlet temperatures and heat exchanger efficiencies. (From Noel Penny, "Gas Turbines for Land Transport," *Science Journal*, April 1970.)

Additional cycle variations are also possible, such as intercooling and reheating, as well as semi-closed and completely closed cycles. In the first, the turbocompressor is divided into two parts, linked by an "intercooler" that cools the partially pressurized air back to atmospheric temperature before maximum

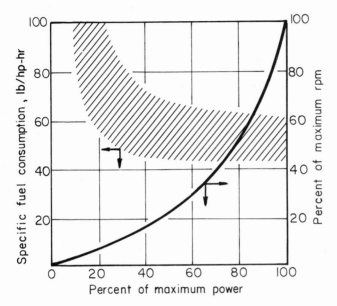

Figure 4-14. Variation of specific fuel consumption with power for the gas turbine. (From M. L. Walker, Jr., *A Methodology for Estimating Fuel Consumption for Various Engine Types in Stop-Go Driving*, International Research and Technology Corp., IRT-R-20, March 1970.)

pressure is reached at the outlet of the second compressor. In a "reheater" system, a similar arrangement is made in the turbine unit. The turboexpander is divided into two parts separated by a second combustor that heats the partially expanded gases from the first turbine back to their maximum temperature and allows complete expansion through the second turbine. The purpose of these additional complexities is, of course, to increase the overall efficiency, although they do so at considerable added cost and weight of equipment.

For any gas turbine, the efficiency achieved is a function of both the pressure ratio, as outlined earlier, and the "machine efficiency" of the components; for example, if the compressor and turbine efficiencies are each 90% and the combustor efficiency is 100%, the overall system machine efficiency is 80%. Actual machine efficiencies range between 80% and 90%.

In practice, since machine efficiencies are difficult to improve, the turbine inlet temperatures and pressure ratios control the overall achievable efficiencies in both simple-cycle and regenerative-turbine systems.

In the closed cycle, the working fluid (not air) is continuously recycled. The heat, from an external source, is transferred through the walls of a closed heater.

This cycle has been proposed for a nuclear power plant in conjunction with a gas-cooled reactor. The advantages of the closed cycle are: clean, noncorrosive, working fluid, control of the pressure, composition and heat transfer characteristics of the working fluid, high absolute pressure and density of the working fluid, and constant efficiency over a wide load range.

A precooler is required to reduce the temperature of the working fluid before recompression. The higher fluid densities make it possible to reduce the size of compressor and turbine per horsepower and at the same time increase the maximum capacity of the power plant. By varying the absolute pressure at the compressor inlet, the weight of working fluid circulated may be varied at will without changing the compression ratio or the temperatures, so that a wide range of load can be carried at practically constant speed and efficiency. The major disadvantage of this cycle is the size and cost of the required high-temperature heater.

In the semi-closed cycle, approximately two-thirds of the working fluid (air) is recirculated. This cycle, like a closed cycle, requires a precooler for the re-circulated gas; it also requires a "charging" compressor to provide the necessary air for combustion. The compressor may be on a common or split shaft. The semi-closed cycle is similar to the closed cycle in that it can be operated at high densities and approximately constant efficiency over a wide load range. The major disadvantage of this cycle is the corrosion and fouling problems that occur with the recirculation of the products of combustion, particularly when sulfur or ash-containing fuels are used.

HEAT (Q) ENGINES

Heat is the most degraded (or least recoverable) of energy forms. The production of heat is a measure, often, of the inefficiency of energy conversion. However, a fraction of heat energy can be recovered by a suitable type of engine in which a working fluid is successively compressed, heated to a high temperature, and expanded (doing work in the process), then further cooled and finally recompressed.

In a practical heat engine, the working fluid will always be a gas due to the requirement for high temperatures and large volume and pressure changes. Work is extracted from heated compressed gas by letting it expand against a piston (or turbine blade). The gas is then cooled, compressed, and finally heated again, and the cycle begins anew. An open-cycle system, in which "exhausted" working fluid is rejected into the environment, has equivalent limitations, although the description is more complex. The maximum percentage of the total heat energy that can be converted into work by means of such a cycle—i.e., the maximum theoretical efficiency η—is a function of the temperature change of the gas before and after expansion, viz.,

$$\eta = \frac{T_m - T_a}{T_m}, \tag{4.1}$$

where T_m is the maximum temperature reached in the cycle and T_a is the ambient temperature of the environment, both measured in degrees Rankine (R) or Kelvin (K) above absolute zero. This is the well-known Carnot-cycle efficiency.[8] To maximize η it can be seen that a heat engine should operate with T_m as high as possible and T_a at ambient temperature (since lower temperatures can only be reached by refrigerators which require power to operate). It is to reduce the temperature of the working fluid to T_a after the expansion phase that heat engines—such as fossil fuel electric power plants—paradoxically require cooling.

The maximum temperature T_m that actually can be reached in practice depends on the energy release mechanism and the energy content of the fuel, and also on the strength of the materials from which the engine is constructed, especially at high temperatures. The engine materials requirements are one of the limiting factors for practical Q-engines.

While it is not intended to enter into an extensive thermodynamic discussion here, a quick review of the nature of expansion-compression cycles will be useful. Generally speaking, in the course of a working cycle, pressure (P), volume (V), enthalpy (h), entropy (S), and temperature (T), are all changing simultaneously.

Any one of the five basic variables can be held constant while allowing changes in the other four. Any two of the five can be chosen as independent. One idealized sequence of thermal events, by means of which heat energy is transformed into useful work as efficiently as possible, is known as the Carnot cycle. An ideal heat engine based on the Carnot cycle would consist of: (1) a perfect gas in a cylinder containing a piston, (2) a reservoir of high temperature T_m (a heat source), and (3) a reservoir of lower temperature T_a (a heat sink).

In such an engine, the gas would be initially at temperature T_a. In the first step of the cycle, it would be compressed from its initial volume V_1 to a smaller volume V_2 *isothermally* (i.e., at constant temperature), heat being rejected into the "heat sink" in the process. This is shown schematically in figure 4-15. Approximately isothermal compression would occur, for instance, if a hot gas were to pass through a radiator immersed in running water. Next the gas is further compressed *adiabatically* (and *isentropically*, i.e., with heat being neither added nor removed) to V_3, the temperature rising to its maximum T_m in this phase. Adiabatic processes generally occur very rapidly, before there is time for heat loss to occur. In the third step, the gas expands isothermally to V_4 (equal to V_2), drawing heat from a source at constant temperature, such as a burner, and doing work. Finally, the gas is expanded adiabatically (i.e., with no loss of heat) to its original condition. When the pressure-volume and temperature-

---

[8]The Carnot limit does not apply to E-engines (discussed previously), of which internal combustion engines and gas turbines are examples. Fuel cells are also classed as E-engines.

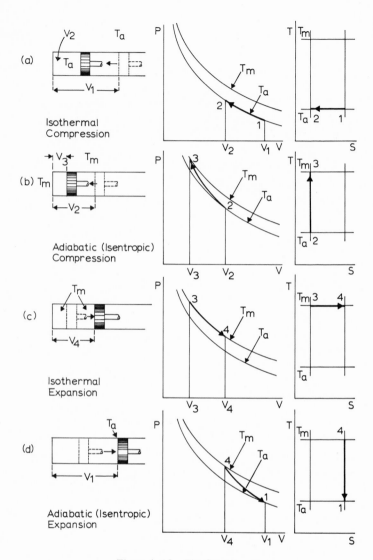

Figure 4-15. The Carnot cycle.

entropy change curves of figure 4-15 are combined, the results are the so-called Carnot-cycle P-V and T-S diagrams, familiar to students of physics and engineering. This particular cycle is called ideal because it is reversible, which means no net entropy change occurs during the cycle and the only heat that is wasted—the irreducible minimum—is that which is rejected into the "sink" (i.e., the environment).

The so-called Stirling cycle, which will be discussed again later in connection with the Stirling cycle engine, is also theoretically reversible and therefore equivalent to the Carnot cycle in efficiency. As illustrated in figure 4-16, the cycle consists of an isothermal compression (1-2) via a coolant bath or the equivalent, as in the previous case. During the next phase (2-3) heat is added but the working fluid is held at constant volume. This is followed by isothermal expansion (3-4) during which time the gas expands against a piston but does not drop in temperature because it is in thermal contact with a heat reservoir (e.g., a burner). During the fourth phase (4-1), the gas remains at constant volume while it is physically transferred to a "cold" space where it loses heat and drops in pressure to the original condition.

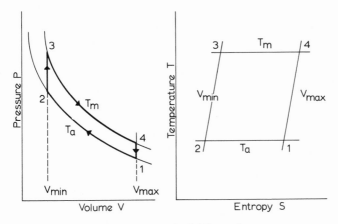

Figure 4-16. The Stirling cycle.

The Ericsson cycle, shown in figure 4-17, is also theoretically reversible and therefore "ideal." It begins, like the Carnot and Stirling cycles, with an isothermal compression (1-2), where thermal contact with a low-temperature reservoir is maintained. The second phase is a constant pressure (isobaric) heating (2-3), during which contact with a high-pressure reservoir is maintained. This is followed by an isothermal expansion (3-4) during which thermal contact with a heat source is again maintained. The cycle is completed by a phase in which heat is lost at constant pressure.

There is no practical example of a Carnot-cycle engine, but both the Stirling and Ericsson cycles were incorporated (approximately) into experimental engines in the nineteenth century using air as the working medium. However, air engines did not prove to be competitive in practice with steam (Rankine cycle) engines.

The Rankine cycle is a thermodynamic cycle adapted to condensing fluids (such as steam). The six basic steps, which vary slightly depending on the fluid,

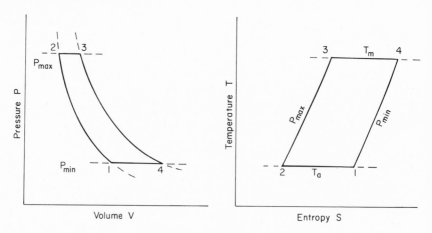

Figure 4-17. The Ericsson cycle.

are: (1) adiabatic compression (in the liquid phase), (2) heating, (3) further heating (and boiling) at constant temperature and pressure, (4) superheating of the vapor at constant volume, (5) adiabatic expansion during which work is done by the vapor on a piston (or turbine), and (6) condensation and heat rejection at constant pressure and temperature.

The sequence described above applies to steam and certain other "wetting" fluids for which isentropic expansion intersects the saturated vapor line, viz., condensation occurs spontaneously if the superheated vapor expands adiabatically (see figure 4-18, top). If "drying" fluids are used, the opposite is true, and adiabatic expansion carries the vapor away from the saturation line. Strangely enough, the vapor actually becomes more superheated (by definition) as it expands and cools. A Rankine cycle utilizing a drying fluid requires no superheating stage; instead, it requires a desuperheater to get rid of unwanted (waste) heat in the gas phase, so that condensation can take place (see figure 4-18, bottom). Heat rejected in the desuperheater may, of course, be recovered (in part) by a heat exchanger and utilized after the condensation and compression stage.

The thermal efficiency of a Rankine power cycle is defined as

$$\eta = \frac{W_{out}}{h_{in}}, \tag{4.2}$$

where, for a superheat cycle (wetting fluid) corresponding to the top drawing in figure 4-18,

$$W_{out} = \Delta h_{56} - \Delta h_{12}, \tag{4.3}$$

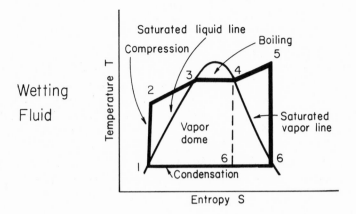

Wetting
Fluid

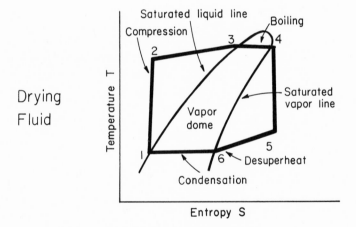

Drying
Fluid

Figure 4-18. The Rankine cycle.

and

$$h_{in} = \Delta h_{24} + \Delta h_{45} . \tag{4.4}$$

For a drying fluid (also shown in figure 4-18) the expressions are:

$$W_{out} = \Delta h_{45} - \Delta h_{12} , \tag{4.5}$$

$$h_{in} = \Delta h_{24} - \alpha \Delta h_{56} , \qquad (4.6)$$

where α is the regeneration efficiency.

It is convenient to use a terminology applicable to both cases. Thus, the term Δh_e is the enthalpy drop (work done) during isentropic expansion through a reciprocator or turbine, while Δh_{12} is the pump work required to move the fluid from the condenser to the boiler against boiler pressure. Δh_d is the amount of desuperheat that must be removed from the expanded vapor, α being the regenerator efficiency. Two alternative values of α are used in our subsequent calculations: $\alpha = 0.50$ and $\alpha = 0.80$. All the above quantities except pump work, Δh_{12}, can be calculated from published thermodynamic data for various fluids. Results are plotted on figure 4-19.

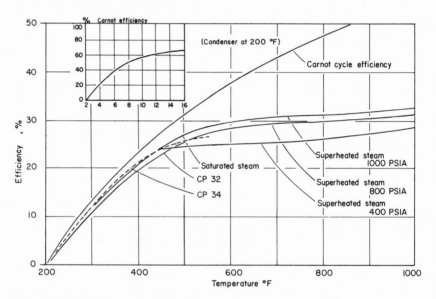

Figure 4-19. Rankine cycle engine efficiencies for various conditions and working fluids.

The theoretical pump work is the sum of two terms:

$$\text{pump work} = \int P dV + \int V dP , \qquad (4.7)$$

where P and V are pressure and specific volume, respectively. The first term is the nonflow contribution due to compression, and the second term is the work necessary to move the fluid into the boiler against the existing pressure. For a relatively incompressible liquid the $P dV$ term is negligible in relation to the $\int V dP$ term.

$$\int V dP \cong V_o \int_{P_{min}}^{P_{max}} dP. \qquad (4.8)$$

While past Rankine cycle applications have mainly been concerned with superheated steam as the working fluid, it is now becoming apparent that there are a variety of working fluids that may ultimately be superior to steam in a Rankine cycle depending on the application. In order to develop a rationale for comparison of working fluids, it is useful to consider potential candidates as "wetting" or "drying" depending on the behavior of the vapor as it expands adiabatically.

If the maximum and minimum temperatures are *fixed*, it can be shown by a simple geometrical argument that a fluid intermediate between the "wetting" and "drying" cases offers maximum efficiency with the simplest arrangement, since the drying fluid cycle requires a regenerator or a desuperheater, while the wetting fluid must be superheated before expansion to avoid premature condensation.[9]

It can be shown, also, that cycle efficiency increases with the slope of the saturated *liquid* line on a T–S diagram. As this line approaches vertical, the Rankine cycle more and more closely resembles a Carnot cycle. A vertical saturated liquid line would correspond to a liquid specific heat of zero—that is, an infinitesimal increase in heat content (enthalpy) results in an infinite temperature rise. Hence, other things being equal, the lower the liquid specific heat, the better the cycle efficiency. Characteristics of a number of actual fluids are discussed in chapter 7.

Although a reciprocating steam engine resembles a reciprocating Otto-cycle or diesel engine—indeed, "conversions" are not uncommon—the physical configuration of the expander is not functionally related to the nature of the energy conversion mechanism in a Rankine-cycle engine, as it is in an internal combustion engine. In fact, the "expander" in a steam engine is analogous to the torque-converter (transmission) of a conventional automotive power plant, and the transfer of energy from the fuel to the working fluid actually takes place in the vapor generator. As there is no difference between a steam turbine and a steam reciprocator in terms of thermodynamics, discussion of detailed configurations is deferred to chapter 8.

ELECTROCHEMICAL CONVERSION

As noted previously, in any chemical reaction, the energy content of reaction products is different from that of the original reactants. If this difference (Δh) is negative, the reaction is exothermic, and energy is liberated; if it is positive, the

[9] J. Bjerklie and S. Luchter, "Rankine Cycle Working Fluid Selection and Specification Rationale," Paper No. 690063, SAE Annual Meeting, January 1969.

reaction is endothermic and energy is absorbed. The expression Δh thus represents the heat content of the products minus that of the reactants.

In an exothermic reaction, the liberated energy normally appears as heat. However, it is not necessary that this be so. With proper conditions and hardware, many reactions will channel their energy as electricity. For example, if a zinc plate and a copper plate are placed in a copper sulfate solution and connected externally by a wire, the zinc will still go into solution replacing the copper (which will be deposited on the copper plate); the energy released will generate an electric current in the wire, which may be used to perform work or dissipated as heat anywhere in the electric circuit. The reaction is

$$Zn + CuSO_4 \rightarrow ZnSO_4 + Cu.$$

Basically, the generation of electrical energy from chemical reactions requires the separation of the reactant molecules into oppositely charged ions (in a suitable medium) and the utilization of the resulting potential to drive an electric current through an external circuit. The charge transfer medium is an ionic conductor called an electrolyte; usually it is an ionic liquid, although it may be a plastic membrane or even a rigid solid such as ceramic. The charges accumulate on suitable electrodes, which form the interface terminals between the internal electrochemical action of the cell and the external electric circuit. An external voltage applied at the electrodes will bring about a chemical reaction in a suitable electrolyte and the physical aggregation of reactants within the cell, usually at the electrodes.[10] If the voltage and current are reversed in polarity, the chemical reactions work in the opposite direction. Thus, the application of external voltages to some cells that normally produce electricity will—apart from losses due to internal resistance and other irreversible processes—restore the original conditions of the reactants, thus recharging the cell. An obvious case in point is the standard lead-acid automobile battery. This battery, which supplies the current for starting, lighting, and ignition, is partially charged and discharged hundreds of times during its lifetime.

The complete cell consists of two half-cells; in the previous example, a zinc half-cell and a copper half-cell. The cell reaction can also be divided into two parts—oxidation and reduction (electron loss and gain, respectively). At the surface between the solution and each electrode (whether the electrode material participates in the reaction or not), there exists a potential difference, which is called the electrode potential. (Table 4-1 lists standard electrode potentials at $25°C$.) The full potential of any cell is the sum of its individual electrode potentials.

---

[10]This phenomenon is also responsible for the various uses of electrolysis, electroplating, and electrowinning, which form the basis for a sizable industry.

Table 4-1. Standard Electrode Potentials, 25°C

Electrode	Reaction	E (Volts)
Li; Li^+	$Li \longrightarrow Li^+ + e$	+2.958
Rb; Rb^+	$Rb \longrightarrow Rb^+ + e$	+2.924
K; K^+	$K \longrightarrow K^+ + e$	+2.922
Na; Na^+	$Na \longrightarrow Na^+ + 2e$	+2.714
Zn; Zn^{++}	$Zn \longrightarrow Zn^{++} + 2e$	+0.761
Fe; Fe^{++}	$Fe \longrightarrow Fe^{++} + 2e$	+0.441
Cd; Cd^{++}	$Cd \longrightarrow Cd^{++} + 2e$	+0.402
Co; Co^{++}	$Co \longrightarrow Co^{++} + 2e$	+0.283
Ni; Ni^{++}	$Ni \longrightarrow Ni^{++} + 2e$	+0.236
Sn; Sn^{++}	$Sn \longrightarrow Sn^{++} + 2e$	+0.140
Pb; Pb^{++}	$Pb \longrightarrow Pb^{++} + 2e$	+0.126
Pt; H_2; H^+	$\frac{1}{2}H_2 \longrightarrow H^+ + e$	±0.000
Pt; Ti^{+3}, Ti^{+4}	$Ti^{+3} \longrightarrow Ti^{+4} + e$	−0.040
Ag; AgBr(s); Br^-	$Ag + Br^- \longrightarrow AgBr + e$	−0.073
Pt; Sn^{++}, Sn^{+4}	$Sn^{++} \longrightarrow Sn^{+4} + 2e$	−0.150
Pt; Cu^+, Cu^{++}	$Cu^+ \longrightarrow Cu^{++} + e$	−0.167
Ag; AgCl(s), Cl^-	$Ag + Cl^- \longrightarrow AgCl + e$	−0.222
Saturated Calomel	$Hg + Cl^- \longrightarrow \frac{1}{2}Hg_2Cl_2 + e$	−0.242
Normal Calomel	$Hg + Cl^- \longrightarrow \frac{1}{2}Hg_2Cl_2 + e$	−0.280
0.1-N Calomel	$Hg + Cl^- \longrightarrow \frac{1}{2}Hg_2Cl_2 + e$	−0.334
Cu; Cu^{++}	$Cu \longrightarrow Cu^{++} + 2e$	−0.340
Pt; $Fe(CN)_6^{-4}$, $Fe(CN)_6^{-3}$	$Fe(CN)_6^{-4} \longrightarrow Fe(CN)_6^{-3} + e$	−0.356
Pt; $I_{2(3)}$; I^-	$I^- \longrightarrow \frac{1}{2}I_2 + e$	−0.536
Pt; Fe^{++}, Fe^{+3}	$Fe^{++} \longrightarrow Fe^{+3} + e$	−0.771
Ag; Ag^+	$Ag \longrightarrow Ag^+ + e$	−0.799
Hg; Hg_2^{++}	$Hg \longrightarrow \frac{1}{2}Hg^{++} + e$	−0.799
Pt; Hg_2^{++}, Hg^{++}	$Hg^{++} \longrightarrow 2Hg^{++} + 2e$	−0.906
Pt; Br_2(l); Br^-	$Br^- \longrightarrow \frac{1}{2}Br_2 + e$	−1.066
Pt; Cl_2(g); Cl^-	$Cl^- \longrightarrow \frac{1}{2}Cl_2 + e$	−1.358
Pt; Ce^{+3}, Ce^{+4}	$Ce^{+3} \longrightarrow Ce^{+4} + e$	−1.610

Source: Scott L. Kittsley, *Physical Chemistry* (New York: Barnes & Noble, 1955).
Note: (s) = solid; (l) = liquid; (g) = gas.

If a cell operates at constant temperature and pressure, the electrical work that is done by the system, per mole of reactant consumed, is $W_{out} = nFE$ where n is the valence change or number of electrons involved in the reaction, F is the Faraday unit,[11] and E is the electromotive force of the cell in volts.[12]

The heat of reaction (energy content) per unit mass is frequently tabulated in terms of "fuel" only, although energy densities for electrochemical cells must clearly be amortized over the weights of both the fuel and the oxidizer if the latter is carried with the battery (i.e., if air is not used for this purpose). Thus, depending on the molecular weight of the oxidizer, the real theoretical energy density for any reaction may be considerably less than the "fuel only" value. Table 4-2 lists a number of voltage-producing reactions for both fuel only and fuel plus oxidizer. The difference in the two values is significant.

Table 4-2. Molar Free Energy Values of Various Reactions

Reaction	W_{out} (cal/g mole of fuel)	W_{out} (whr/lb, fuel only)	W_{out} (whr/lb, fuel and oxidant)
$H_2 + \frac{1}{2}O_2 \longrightarrow H_2O$	+56,690	14,900	1,658
$CH_4 + 2O_2 \longrightarrow CO_2 + 2H_2O$	+196,500	6,475	1,295
$C_3H_8 + 5O_2 \longrightarrow 3CO_2 + 4H_2O$	+503,926	6,034	1,302
$2Li + \frac{1}{2}O_2 \longrightarrow Li_2O$	+133,684	5,169	2,349
$NH_3 + \frac{3}{4}O_2 \longrightarrow \frac{1}{2}N_2 + 3H_2O$	+81,090	2,515	1,045

Note: Technically W_{out} is defined as the change in the Gibbs Free Energy for the reaction.

In addition, when any electrochemical reaction is actually reduced to practice, the available energy (W_{out}) per unit mass must be adjusted to take account of the physical structure of the reactor unit, which typically weighs four or five times as much as the reactants alone.

Batteries and fuel cells are classed as E-engines, whose operating efficiencies are not functions of temperature and therefore not subject to the "Carnot limit." The efficiencies of electrochemical systems, however, depend on temperature. In some reactions, efficiency increases with increasing temperature, and in other cases the reverse is true. The $O_2 + 2H_2 \rightarrow 2H_0$ reaction is among the latter, as shown in figure 4-20.

In electrochemical systems, maximum efficiency also requires that the processes occur extremely slowly, implying a correspondingly small current

---

[11] The Faraday unit (F), a fundamental unit in electrochemistry, is the charge of a monovalent positive ion—96,500 coulombs per gram equivalent. Bivalent ions are designated $2F$, and n-valent ions are nF.

[12] *Handbook of Chemistry and Physics*, 45th ed. (Cleveland: Chemical Rubber Publishing Co., 1964).

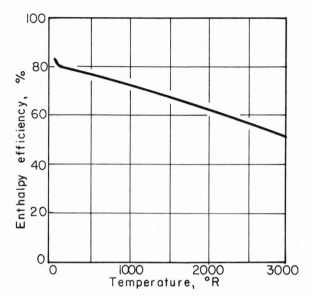

Figure 4-20. Enthalpy efficiency variation with temperature of O_2-H_2 reaction. (From F. Lauck, O. A. Uyehara, and P. S. Myers, "An Engineering Evaluation of Energy Conversion Devices," *SAE Transactions*, 1963.)

flow. Departure from this condition results in irreversibilities and internal losses and, incidentally, in many cases in physical changes in the reactants. It is these losses and changes which give the battery a finite lifetime.

In practice, of course, the rate at which the energy is extracted (i.e., the available power) is of paramount importance in automotive applications. In this regard, high potential differences between electrodes often tend to be correlated with high current densities and thus high power and energy densities. Other things being equal, if the halves of an electrochemical couple are taken from opposite ends of the standard potential table and thus have a relatively high potential difference, the couple will offer better energy and power densities than one with roughly the same energy content but a lower potential difference. In part, this is because the effects of polarization tend to reduce actual cell voltages by a certain amount, especially at high discharge rates, relatively independently of the couple.

Polarization is the difference between the potential of an electrode when electrons are flowing and the reversible open-circuit potential of the electrode and is a measure of the inefficiency of the charging or discharging process at a particular current density. (Charging and discharging overvoltages will have opposite absolute values.) This potential difference is caused by three distinct polarization processes. The first is due to the electrical resistance of a thin layer of electrolyte at the electrode surface. Because this resistance obeys Ohm's law in polar electrolytes, it is usually referred to as the "IR drop."

The second process, concentration polarization, occurs because the concentration of the reactants becomes depleted at the surface of the electrode during the flow of electrons. The reaction rate at the electrode surface depends on the surface concentration and not on the bulk concentration of reactants.

The third process, activation polarization, results from the energy needed to initiate the rate-determining chemical reaction. Even the simplest chemical "macroreaction" involves a sequence of "microreactions." These generally proceed at different rates, and the slowest obviously governs the overall rate. Since all three types of polarization phenomenon are proportional to electron flow or current density, it is not surprising that the inefficiency of the electrode process also increases with the current density or the charge and discharge rates.

Basically, there are two possible methods of restoring the "fully charged" condition in an electrochemical cell: (1) replace the reactant (fuel) and/or oxidizer, either continuously (fuel cells) or in batches (replaceable anodes, or primary cells), or (2) reverse the cell voltage and recharge the system electrically (secondary cells).

A typical primary cell configuration (see figure 4-21) consists of two electrodes separated by an electrolyte, with compartments for introduction of the reactant or "fuel" and oxidizer separately, as shown. The reactant is fed to the anode, the oxidizer to the cathode; current flows when ions from the fuel and oxidizer diffuse through the electrodes and the electrolyte. In principle, only the

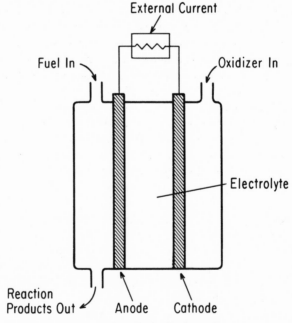

Figure 4-21. Fuel cell schematic.

fuel and oxidizer participate in the reaction, while the electrodes and the electrolyte remain inert; but in many cells one or both electrodes actually participate in the reaction. For instance, zinc, sodium, magnesium, lithium, and calcium, which have been used as anodes in conjunction with air cathodes, are also the fuels in the particular cases. The electrolyte generally serves as the ionic conductor. The reactions occur at the electrode-electrolyte interface. When the external circuit is open, the anode accumulates a layer of negative charges on its electrolyte interface, which attracts positive ions in the electrolyte; at the same time, the cathode accumulates a similar layer of positive charges, which in turn attracts negative ions in the solution. These layers act to prevent any further reaction. On the other hand, when the circuit is closed, the electrons flow through the external circuit, the charge balance is altered, and the reactions proceed until the fuel is exhausted or the circuit is once more opened.

From these processes and the general remarks above, certain technical requirements become clear:

1. The electrolyte, in addition to being chemically compatible with the electrodes, must be a good electrical conductor.
2. The products of the reaction in question should be safe and nontoxic. The ideal product is water (although carbon dioxide or gaseous nitrogen is normally acceptable).
3. Unless the electrodes themselves participate in the reaction, they must be of a material that is inert to both the reactants and the electrolyte under all conditions likely to occur. Yet they must facilitate the migration of ions from the fuel and oxidizer, respectively, through the electrode material to the electrolyte. In practice, this usually requires a catalyst. Since it is at the electrode-electrolyte interface that the reactions occur, the electrodes must offer as large an "effective" surface as possible.
4. During the ionic migrations, the electrolyte should remain inert and unchanged.

These technical requirements have been summed up very well by Liebhafsky and Cairns under three general headings: conductivity, reactivity, and invariance.[13]

In addition, there should be no costly or exotic catalysts and the cell should utilize cheap reactants if it is to be used in automotive applications. In the final analysis, cost (capital and operational) will be all important in determining the commercial future of the fuel cell.

There are, of course, a large number of possible fuel and oxidizer combinations and reactions, but the ones of major interest are hydrogen-oxygen,

---

[13]H. A. Liebhafsky and E. J. Cairns, "Hydrocarbon Fuel Cells–A Survey," AIEE Pacific Energy Conversion Conference, 1962.

hydrocarbon-oxygen, hydrocarbon-air, hydrazine-air, ammonia-air, and various alkali metals (particularly zinc) with air. The characteristics of the most studied fuel cell couples are summarized in table 4-3. Successful fuel cell operation requires that the fuel and oxidizer be "activated" at the respective electrodes and that the ions, created at the electrode-electrolyte interfaces, migrate through the electrolyte to the opposite electrode. To facilitate this, fuel cell

Table 4-3. Energetics of Selected Primary Cell Systems

Fuel (g) = gas (l) = liquid	Oxidant	Cell voltage	Energy density of fuel + oxidant, or fuel alone if oxidant is air	Estimated weight of reactant storage containers	Weight of fuel + oxidant	Total weight
					280 watts, 35 days	
			whr/lb	lb	lb	lb
$NH_3(g)$	O_2	1.124	1,000	130	235	365
$NH_3(g)$	Air	1.124	2,410	9.5	97.5	107
$N_2H_4(g)$	O_2	1.56	1,185	96.2	198	294.2
$N_2H_4(g)$	Air	1.56	2,370	8.7	99.1	107.8
$H_2(g)$	O_2	1.23	1,661	223.5	141.5	365
$H_2(g)$	Air	1.23	14,950	102.3	15.7	118
$CH_3OH(g)$	O_2	1.19	1,086	123	216	339
$CH_3OH(g)$	Air	1.19	2,710	9	86.7	95.7
$CH_3OH(l)$	O_2	1.185	1,080	123	218	341
$CH_3OH(l)$	Air	1.185	2,700	9	87	96
$C_4H_{10}(g)$	O_2	1.12	1,335	129.6	176	305.6
$C_4H_{10}(g)$	Air	1.12	6,120	8.1	38.4	46.5

Metal-air couple	Voltage		Energy density of active material		
	Open circuit	Discharge	Theoretical	Ultimate design	Actual design
			whr/lb	whr/lb	whr/lb
Zinc–air	∿1.5	1.2–1.3	560	120	60
Magnesium–air	3.0	1.2	1,800	350	50
Aluminum–air	2.7	1.2	2,400	400	200
Cadmium–air	∿1.2	1.0	260	70	50
Sodium–air	2.6	2.3	930	300	200
Iron–air	∿1.25	0.8–0.9	410	125	40

electrodes are usually constructed from some porous but nonreactive material with a suitable catalyst embedded in it. (Up to the present time the catalyst has generally been one of the noble metals, such as platinum or palladium.) The fuel and oxidizer molecules diffuse through the electrode structure, and, dissociated or ionized by the catalyst, are adsorbed on the interface surface.

Electrodes may be constructed of porous carbon, porous metal (e.g., nickel) or compacted metal powder. Porous carbon, in particular, is often used in low temperature hydrogen-oxygen cells; carbon itself is catalytic enough so that little or no additional catalyst is usually needed on the oxygen cathode. Porous metal, usually nickel or stainless steel, has the advantage of being easily formed into thin flexible sheets of high conductivity. Powdered metal electrodes usually consist of finely divided catalytic particles, generally pressed or bonded to a screen of the same metal by some moldable inert material. All porous electrodes have the problem that the electrolyte or water produced in the reaction may be drawn into the pores by capillary action and "drown" or deactivate the electrode. This is one of the problems that limits the life and the overload capability of fuel cells.

The electrolyte, which is the ionic charge carrier in the system, is perhaps more sensitive to chemical and temperature conditions of the reaction, since it must be kept invariant. It may be an alkaline such as potassium hydroxide, an acid, a molten salt, or even a solid. Some high-temperature fuel cells use fused salts of alkalis or carbonates, which are often impregnated into a suitable matrix or paste—the so-called "fixed" electrolytes. At low temperatures, alkaline electrolytes have been the most successful, at least for hydrogen cells, although water formed in the reaction tends to dilute the electrolyte. If CO_2 is produced, however, a potassium carbonate precipitate is formed that can destroy the activity of the electrodes. Acids, on the other hand, have the advantage of being unaffected by the presence of carbon dioxide. Thus all attempts to utilize carbon-based fuels (hydrocarbons or alcohols) have been forced to rely on acid electrolytes. A variant approach is the "vehicle held" electrolyte system, where the electrolyte is held between the electrodes in a porous sheet of asbestos. This system permits the use of very thin electrodes, resulting in compact and lightweight cells.

In most cases, as we shall see in chapter 10, it has proven very difficult to achieve a practical level of performance in fuel cells at ambient pressures or temperatures, but high pressures or temperatures bring substantial problems of their own. The controls and auxiliaries required to overcome these difficulties and achieve stable operation add greatly to the complexity of the hardware.

The need for cheaper, more effective, and less easily "poisoned" catalysts is one of the major problems confronting fuel cell research. The list of suitable catalytic materials is restricted so far largely to platinum and other noble metals and their compounds, which command a royal price. Many experimental cells (notably hydrocarbon cells) have required up to several hundred grams of platinum per generated kilowatt.[14] The best results achieved to date are of the

---

[14] H. A. Liebhafsky, "The Electrocatalyst Problem in the Direct Hydrocarbon System," and Galen R. Frysinger, "Low Cost Fuel Cell Electrodes," *Proceedings of the 20th Annual Power Sources Conference*, 1966.

order of 50 grams per kilowatt.[15] At a price of approximately $5.50 per gram, this implies a cost of $275 to $1,500 per kilowatt for the catalyst alone. Costs of this magnitude may be tolerable for certain special applications, such as military and space hardware development, but not for the everyday transportation market. Apart from being costly, the noble metals are extremely scarce. For example, if fuel cells producing 100–200 amps/ft^2 and using only a monatomic layer of platinum on the surface of the electrodes were installed in all the cars and trucks produced in the United States in 1960, it has been calculated that the platinum required would exceed the world's annual production by a factor of 2,000.

An alternative, of course, is to use electrodes that participate in the cell reaction and are thus physically consumed. This technique eliminates the necessity for a catalyst on the consumable electrode and has, in fact, been successful. However, in such a case, the electrodes must be easily replaceable and they must react only when the external load is imposed, which may be only a small fraction of the time in the case of an automotive vehicle. (Brief descriptions of the major types of fuel cells are given in the appendix.)

OTHER DIRECT CONVERSION SCHEMES

Among the forms of direct conversion that have been studied there are several electrothermal devices such as the *thermionic* generator, in which electrons are boiled off to a heated metal surface and collected for use as current, and the *thermoelectric* generator, which utilizes the effect of temperature difference on two different metals to produce a voltage and a current.[16] Both of these are formally heat (Q) engines (and therefore Carnot limited) except that the working fluid is an electron gas. They are also "external" combustion systems, meaning that heat can be supplied by means of a simple burner, which would be either "on" or "off" and thus cause very little pollution. One could theoretically envision a heat source and a thermionic or thermoelectric generator, in parallel with a storage battery to handle peak loads, linked to an electric drive.

Unfortunately, efficiencies are still quite low, despite years of research and development. The thermoelectric cell should theoretically be capable of about 12% efficiency, but 5% is a limit seldom exceeded in practice. With one exception, all experimental prototypes have been in the subhorsepower range to date. A 40-hp(e) thermoelectric generator with 6.5% efficiency and weighing 25 lb/hp(e) seems to be feasible in the near future, but these operational charac-

---

[15] E. L. Simmons, E. J. Cairns, and D. J. Surd, "The Performance of Direct Ammonia Cells," *Journal of the Electrochemical Society*, Vol. 116, No. 5 (1969), p. 556.

[16] There are two important phenomena: (1) the Seebeck effect, which is a voltage differential across a junction between two dissimilar metals or semiconductors where the opposite ends of each are held at a temperature different from that of the junction; and (2) the Peltier effect, which is a current flow across a junction proportional to a temperature differential.

teristics would not be competitive for automotive applications. Battelle Institute has projected a cascaded system with approximately 7.7% efficiency and a specific weight of 13 lb/hp(e) by 1980.[17]

Thermionic conversion is based on the "evaporation" of electrons from a hot surface (Edison effect). When these electrons are recaptured, a current flows. Unfortunately thermionic diodes tend to be low-voltage, high-current devices, which mean a number of cells must be "cascaded" in series. The state of the art at present is roughly 5% overall efficiency and a specific weight of 20 lb/hp(e) in reasonable sizes. The Battelle Institute has projected improvements by a factor of 2 in both figures of merit by 1980.[18] One major difficulty with both thermoelectric and thermionic conversion, is that very high temperatures, from about 2,400°F to 3,600°F, are required. Flame-heated cathodes invariably oxidize very rapidly at these temperatures, and a flame-heated "container" for the cathode would obviously have the same problem. Thus there is a strong inherent inverse relationship between performance and lifetime.[19] A long-lived thermionic convertor therefore requires an alternative heat source, presumably nuclear, or a radical improvement in the technology of refractory materials.

Another electrothermodynamic scheme that seems worthy of mention is a heat engine approximating the ideal Carnot cycle in thermodynamic efficiency and yielding an electric rather than a mechanical output. The working fluid in the system is a charged aerosol in a moving gas stream, and there are no moving parts except for the gas and liquid components which are recycled. It is claimed by the developer that power-to-weight ratios as high as 500–50,000 watts/lb can be achieved in principle. Such ratios, if actually achievable, combined with the essentially static mode of operation (except for the gas stream), would make this a potentially rewarding system for further development.[20]

---

[17] J. A. Hoess et al., "Study of Unconventional Thermal, Mechanical, and Nuclear Low-Pollution-Potential Power Sources for Urban Vehicles," Summary Report to National Air Pollution Control Administration, March 15, 1968.

[18] Ibid.

[19] N. Sanders et al., "Electric Power Generation," NASA Conference on New Technology, SP-5015, June 1964.

[20] A. T. Marks, "Heat Electrical Power Transducer," U.S. Patent No. 3,297,887 (1967).

Chapter Five

Energy Storage

In this chapter we are concerned with the on-board storage of usable energy in some form other than the chemical energy of fuel. The distinction is admittedly difficult to draw. In general, however, it can be said that the fuels can simply be physically replaced, while the other storage mechanisms require that the system be returned to its initial energetic state in some energy-consuming manner. In brief, they must be "recharged" somehow. Three general classes of such systems suggest themselves: electrochemical, thermal, and mechanical.

ELECTROCHEMICAL ENERGY STORAGE

In chapter 4 a distinction was made between two different methods of restoring the potential energy of an electrochemical system: (1) replacement (e.g., of reactive anodes and/or cathodes) or continuous refueling from an out-side reservoir and (2) electrical recharging. In practical terms the latter—considered in the present chapter—corresponds roughly to the usual notion of secondary cells (batteries), while the former obviously describes primary cells or "fuel cells," which were discussed previously.

There is no need to discuss the general principles of battery operation or the historical background. This review is confined to those battery concepts which are commercially important for heavy-duty vehicular applications or which seem to offer a promise of playing such a role in the future.

Commercially Available Storage Cells

The Lead-Acid Cell

The so-called lead-acid battery is standard equipment in almost every one of the millions of automobiles manufactured annually.[1] It is highly engineered,

---

[1] Basic sources for this section are G. W. Vinal, *Storage Batteries*, 4th ed. (New York: John Wiley and Sons, 1955); and Vinal, "Electrochemical Sources of Electric Power,"

reliable, and relatively inexpensive. It offers more energy storage capacity per unit cost than any other existing commercial electrochemical system. However, from the standpoint of vehicle propulsion its performance capability leaves much to be desired. Maximum energy density available in commercial versions is on the order of 10-13 whr/lb, if a long-life design is important. Thus, about 1,000 pounds of lead-acid cells would be needed to provide 10 kw to a small electric runabout for one hour of operation (in other words, a maximum range of 40-50 miles at 30 mph). Still, because of its low cost and its well-developed technology, the lead-acid cell is used to power some vehicles, including mail trucks, delivery vans, golf carts, forklift trucks, and a few experimental electric automobiles.

The lead-acid cell consists of a metallic lead anode and a lead dioxide (PbO_2) cathode immersed in a solution of sulfuric acid (H_2SO_4) and water. When the cell is discharging, the reactant material on both electrodes is transformed into lead sulfate. This can be seen from the following equation, which should be read from left to right for the discharge reaction and from right to left for the charge.

$$PbO_2 + Pb + 2H_2SO_4 \rightleftharpoons 2PbSO_4 + 2H_2O.$$

During discharge, the passage of one faraday of electricity (96,500 coulombs) consumes one gram-equivalent each of lead and lead dioxide and two equivalents of sulfuric acid, while producing two equivalents each of lead sulfate and water. The opposite reaction occurs on recharge. However, as the process is not 100% efficient, more energy must be put in than can be recovered on discharge. The actual energy conversion efficiency depends on the rates of charge and discharge, of course, but a fair average for the lead battery under normal operating conditions is about 75%. This means that 4 kwh must be stored, in general, if 3 kwh are to be taken out of the battery as work. (Actually, there are charging losses as well as discharge losses; during a slow charge, efficiency may be 80%, which implies that 5 kwh must be supplied to store 4 kwh in the battery.)

Calculation of the free energy available in the above reaction reveals that the discharge liberates approximately 80 watt-hours of electrical energy per pound of reactants. Thus, the currently available figure of 10-13 whr/lb for commercial long-life lead-acid batteries will serve as some indication of the amount of deadweight involved in a practical configuration. However, higher proportions of "live" to "dead" weight are possible, and it is certainly conceivable that the lead-acid cell can be improved to offer perhaps 25-30 whr/lb at reasonable discharge rates with reasonable lifetimes.[2]

Conference on Energy Sources (1947-49), published as *Sources of Electric Energy*, AIEE, January 1951.

[2] Raymond Jasinski, Tyco Laboratories, Inc., Testimony at the Hearings of the Senate Subcommittee on AntiTrust and Monopoly, "Economic Concentration," Part 6, 1968.

Lead-acid cells, though electrochemically similar, are manufactured in a great variety of sizes and shapes, depending on the intended application. The three major types are: starting-lighting-ignition (SLI), traction, and stationary.

The SLI lead-acid cell is used primarily in conventional, ICE-powered automobiles. Its pattern of use requires high currents (up to 300 amperes over a wide temperature range) for a few seconds at a time, which has led to a specific type of thin plate construction with lead dioxide on the negative plate and sponge lead on the positive. Costs in mass production (with a very highly developed technology) average $0.25 per pound.

Traction-type lead-acid cells, used for golf carts, forklift trucks, and the like, are designed to withstand repeated deep discharges and to supply power continuously for long periods (three to ten hours). The principal point of difference from the SLI type in construction is that the positive plates are explicitly designed to prevent the loss of reactant material, which is what would otherwise occur in frequent deep charge-discharge cycles. One method devised to prevent losses is the so-called "ironclad" construction, which encases the active materials in a perforated plastic armor. With this type of construction, the traction cells have longer lives (three to six years), but they are more expensive than the SLI type. Costs of $0.55 per pound are about average.

Stationary cells, unlike the other two types, are designed for low-output, long-life applications, and thus stress stability and high efficiency. A comparison of energy density versus discharge capacity is plotted in figure 5–1 for the three types of lead-acid cells.[3] More recently, permanently enclosed maintenance-free systems have been developed. Because of the sealed construction, however, energy densities do not exceed 10 whr/lb, or 1 whr/cu/in.[4]

Another new and perhaps more significant development is a high-capacity, deep-discharge, tripolar lead-cobalt battery being produced by Electric Fuel Propulsion, Inc., in Ferndale, Michigan. The tripolar construction facilitates high charging and discharging currents. Cobaltous sulfate added to the electrolyte forms a protective coating and reduces polarization at the positive grid. The manufacturer claims this battery has an energy density of nearly 30 whr/lb and a specific power of 38 watts/lb and that the battery can be completely discharged and recharged 800 times. Furthermore, recharge to 80% of capacity is said to be effected in 46 minutes and to a full 100% in 90 minutes.[5] Recent trends in lead-acid battery energy density are shown in figure 5–2.

The Nickel-Iron Cell

The second major type of storage battery of long-standing is the nickel-iron alkaline cell, or Edison cell, which was developed by Thomas A. Edison around

---

[3] C. K. Morehouse, R. Glicksman, and G. S. Lozier, "Batteries," *Proceedings, IRE*, Vol. 46, No. 8 (August 1958).

[4] J. R. Smyth, "The Status of Battery Power," Presentation by the Electric Storage Battery Company to Federal Power Commission, May 21, 1965.

[5] R. Aronson, "The MARS II Electric Car," SAE Meeting, Detroit, May 1968.

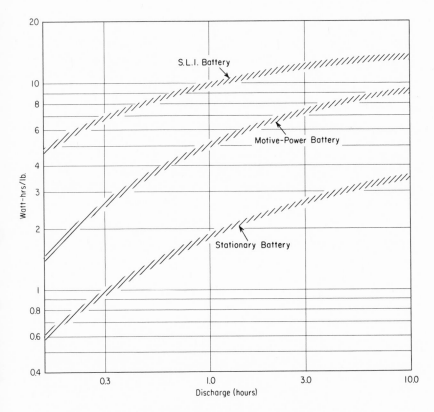

Figure 5-1. Energy storage capabilities of various lead-acid cells.

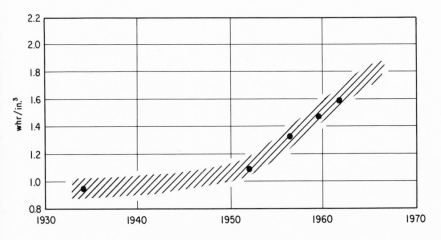

Figure 5-2. Lead-acid cell trends.

the turn of the century and first marketed in 1908. The present version is virtually identical. It consists of an iron anode and an assemblage of nickel tubes filled with nickel hydroxide and nickel, forming a cathode, immersed in an alkaline (KOH) electrolyte.[6]

On the standard five-hour discharge rate, the nickel-iron cell exhibits a slightly greater capacity for a given weight than does the lead-acid cell, although the energy densities of the two batteries are of the same order of magnitude. At high discharge rates, however, the total capacity of the nickel-iron cell remains close to its rated value whereas its voltage drops sharply; on the other hand, the lead-acid cell does not lose voltage even though the capacity drops at high discharge rates.

Yardney[7] has under research and development a heavy industrial-type battery which will deliver 16–20 whr/lb at discharge rates of one to five hours. Similar results have been obtained with a sintered nickel-iron battery developed by General Telephone and Electronics.[8] The high-rate characteristics of this battery depend on the large active iron electrode area provided by the sintering of a low-carbon, high-purity carbonyl iron onto a nickel mash substrate.

Nickel-iron cells have several advantages over lead-acid cells. They operate at higher temperatures and have longer lives. Also, they can be overcharged, or completely discharged for long periods of time, without damage. There are also disadvantages, however. Nickel-iron cells perform poorly at low temperatures. Their charge retention is poor. They use more water and are more expensive to manufacture than the lead-acid batteries.

The widest use of nickel-iron batteries has been in heavy-duty industrial and railway applications, where discharges of one to eight hours at constant voltages are required. Because of their low energy density and sharp voltage drops at high discharge rates, these batteries are not well suited for automotive purposes.

The Nickel-Cadmium Cell

The nickel-cadmium cell was invented about 1900 by Junger and Berg, but was not put to use until after World War II, when development was spurred by heavily funded military R&D programs. Sales grew from $5 million in 1954 to $20 million in 1961, mainly for cells for starting military aircraft engines. Probably only the high cost ($3.80/lb) and relative scarcity of cadmium have prevented the nickel-cadmium cell from being applied to large-scale commercial uses and encroaching on the territory of the conventional lead-acid cell. The nickel-cadmium battery is the most versatile commercial battery available at the present time. It offers high power in fast discharges when needed. Some cells can be recycled many thousands of times and have the same mechanical ruggedness

---

[6] For further details, see Vinal, *Storage Batteries*; and Morehouse et al., "Batteries."

[7] Yardney Electric Corporation literature.

[8] E. R. Bowerman, "Sintered Iron Electrode," *Proceedings, 22nd Annual Power Sources Conference*, May 1968.

and indifference to electrochemical abuse as the nickel-iron cell, and can be recharged in a short time if necessary. Energy density of 15–20 whr/lb can be achieved, which is somewhat higher than that typically available from the previous two systems. The major drawback is the high cost of cadmium and a frustrating lack of reproducibility; some commercial Ni-Cd cells simply cease to function properly after a few cycles.

The basic cell consists of a nickel cathode and a cadmium anode in an alkaline (KOH) electrolyte. There are two types of construction: the so-called pocket type and the sintered-plate type, which differ as to the method by which active material is supported to form the electrodes. Of the two, the sintered-plate cells, because of thinner plate construction and large electrode surface, offer better performance at higher current drains and at lower temperatures.

A recently developed "bipolar" configuration[9] offers the possibility of achieving extremely high discharge rates (competitive with capacitors). In a bipolar version, the positive electrode material of one cell and the negative material of another cell are mounted like a sandwich around a thin conducting substrate. Nickel-cadmium batteries of this type can be designed to be discharged in one-second bursts, yielding a power density as high as 180 watts per pound. Trends in Ni-Cd cell performance are shown in figure 5–3.

Ni-Cd cells can also be designed to withstand very high charging rates. In most sealed systems, charging at high rates causes a buildup of gas pressure that will eventually cause a rupture. This problem can be eliminated by the addition of a third or auxiliary electrode. While there are other factors that prevent the charging rate from being increased (i.e., the charging time from being shortened) indefinitely, the use of a third electrode permits the nickel-cadmium battery to be fully charged in as little as 10 to 15 minutes, which is adequate for automotive applications.

The Silver-Zinc Cell

Of all the secondary batteries in commercial production, the silver-zinc system offers the highest energy density; indeed, it is the only one whose energy density, in practice, begins to approach that required for electric vehicle propulsion. The theoretical possibilities of the silver-zinc couple have been known for decades, but not until after World War II did the cell actually become available commercially—again in response to military needs.[10]

The silver-zinc system consists of a zinc anode, a silver oxide cathode, and an alkaline (KOH) electrolyte. This electrochemical pair offers a theoretical energy density of 222 whr/lb; 40–55 whr/lb can be achieved in a practical configura-

---

[9]H. Seiger, A. Lyall, and S. Charlip, "Bipolar Nickel Cadmium Cells for High Energy Pulses," Intersociety Energy Conversion Engineering Conference (IECEC), September 1966.

[10]For further information see G. A. Dalin and M. Sulkes, "Design of Sealed Ag-Zn Cells," *Proceedings, 20th Annual Power Sources Conference*, May 1966. See also *Energy Conversion Systems Reference Handbook*, Vol. 6, WADD Technical Report 60–699, Commerce Clearinghouse, September 1960.

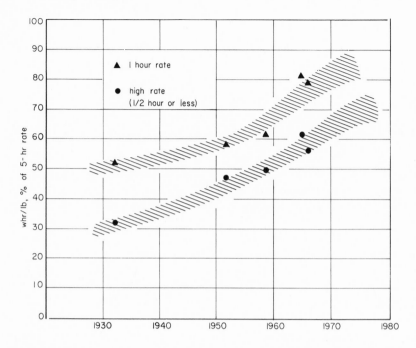

Figure 5-3. Improvements in specific energy as percentage of maximum for nickel-cadmium batteries.

tion; and values approaching 100 whr/lb have been attained at very low discharge rates. In addition, the silver-zinc couple can supply high currents at high discharge rates of 30 minutes or less on demand; it offers power densities up to 150 watts per pound.

However, there are serious shortcomings. The cell has a rather short cycle life: the plates need replacing after perhaps 100 charge-discharge cycles, or after one or two years at most in an automotive application. Various developments could conceivably alter this picture somewhat. (The Astropower Laboratory of McDonnell Douglas Corporation claims to have achieved 1,100 charge-recharge cycles, with a possibility of reaching 3,000 to 4,000 cycles ultimately.)[11] The initial cost is so high, however—about $30 per pound, attributable largely to the value of the silver—that the silver-zinc cell is not likely to be used for vehicular propulsion in spite of its attractive energy and power characteristics. Although silver-zinc cells have been used to power experimental electric automobiles, including the Yardney Electric and the General Motors "Electrovair," present applications are largely in the missile-space field where light weight is far more important than a low price.

[11]*Industrial Research*, April 1967, pp. 44-45.

The Silver-Cadmium Cell

This is another recent secondary cell that offers better performance than the standard lead and nickel cells. It has a longer life than the silver-zinc cell, but it utilizes even more costly material and has a lower energy density. Hence, it need not be considered further as a candidate for application to electric vehicles.

The Nickel-Zinc (Drumm) Cell

The nickel-zinc system deserves comment because it has been investigated recently by at least five companies (General Telephone and Electronics, General Electric, Texas Instruments, Eagle-Pitcher, and Yardney Electric). The system is similar to a nickel-cadmium cell. Replacing the cadmium with zinc doubles the energy density (in principle) and preserves the good power density characteristic, but introduces dendrite formation and other classic cycle-life problems associated with zinc anodes. General Telephone and Electronics has reported lifetimes of 100–200 cycles, but only for 50% discharges—which negates the theoretical advantage over the nickel-cadmium cell. $Ca(OH)_2$ added to the electrolyte limits the concentration of soluble zincate, but also reduces the energy density. According to Yardney Electric literature, the Yardney "Nyzin" cell is at the development stage and has shown an energy density of up to 25 whr/lb. It can be charged to full capacity and left in that condition for extended periods without deterioration. It has a cycle life of up to 250 cycles and a charge rate of 2–3 hours. Production costs are hard to estimate until the design is stabilized; however, the basic materials are inexpensive.

High-Energy Cells under Development

The cells discussed above are commercially available and their performance characteristics are well-known, but, except perhaps for the nickel-cadmium and silver-zinc cells, they are grossly deficient from the standpoint of automotive vehicular requirements. Furthermore, the silver-zinc and nickel-cadmium couples are prohibitively expensive. While special purpose or experimental vehicles can (and have) been designed to use lead-acid or nickel-cadmium batteries, the development of an electric vehicle more nearly comparable to the ICE-powered car (or bus) in performance depends on achieving a breakthrough in terms of energy density, power, and cost.

A large number of electrochemical pairs exist which do offer the possibility of higher energy densities. A table of various theoretical couples is exhibited in table 5-1. Some of these systems, in fact, are currently under development at the present time. They divide naturally into three groups: metal-air cells (semi-fuel cells), lithium-organic electrolyte cells, and molten electrolyte cells.

The Metal-Air Cells

In theory, an air electrode can be used in conjunction with many metals: in fact, systems using lithium, magnesium, aluminum, zinc, sodium, calcium, and

Table 5-1. Theoretical Energy Densities of Possible Battery Couples

Electrochemical couple	Watt-hours per pound of reactant
Li-F	2,850
$Be-O_2$	2,820
$Li-O_2$	2,480
Mg-F	2,120
$Mg-O_2$	1,710
$Al-O_2$	1,430
$Ca-O_2$	1,350
Mg-Cl	1,250
Li-Cl	1,140
$Na-O_2$	1,010
Na-Cl	825
Mg-S	780
Na-S	345-605 (depending on temperature)

iron are all under active development. The system that has received by far the most attention is the zinc-air cell, which uses cheap and plentiful materials, is technologically quite simple, and offers a high energy density. (Organizations in the United States that are, or have been, engaged in research on zinc-air batteries include General Electric, General Motors, Yardney Electric, Leesona-Moos Laboratories, Electric Storage Battery Co., and the General Atomic Division of Gulf Oil Corporation; in England, Energy Conversion Ltd. is also active in the field.) The zinc-air system consists of a zinc anode and an inert air cathode, at which oxygen is catalyzed from the environmental air, in an electrolyte of potassium hydroxide (KOH). The reaction that takes place at the anode during discharge is:

$$2Zn^{++} + 4KOH \rightarrow 2ZnO + 2H_2O + 4K^+$$

while at the cathode:

$$O_2^{--} + 2H_2O + 4K^+ \rightarrow 4KOH.$$

The theoretical energy density of the zinc-air couple is 595 whr/lb (based on the weight of the zinc only, since oxygen is extracted from the surrounding air). Of course, packaging, structural members, electrolyte, faradaic inefficiency, and so on, reduce this value by a considerable fraction for practical cells.

Until recently, metal-air cells were not designed to be electrically recharged. During recharge, the zinc in secondary cells utilizing zinc electrodes tends to form dendrites, and the long, spiky crystals puncture the separators and cause internal short circuits. This tendency can be alleviated to a considerable degree by using separators or by continuous circulation of the electrolyte. In the secondary zinc-air cell being developed by Gulf General Atomic, for instance, the electrolyte containing the dissolved zinc oxide (ZnO), which gradually precipitates out, is pumped out of the cell stack, and the zinc oxide separated by a filter and stored as sludge.[12] During recharge, the zinc oxide is reduced by electrolysis, and the zinc anode is replated and restored to its original state.

The prototype design utilizes solid zinc plated on an inert conducting substrate, as a "front surface" electrode, so that all of it can be used during discharge. On recharge the zinc is replated on the substrate, and oxygen is evolved at the surface of the porous nickel cathode. The cathode is a porous nickel structure with an air plenum. During discharge, pressurized air seeps through the porous nickel into the electrolyte, where its oxygen combines with the dissolving zinc to form zinc oxide. The electolyte is circulated during both charge and discharge.

A schematic layout of this system is shown in figure 5-4. It can be seen that several parasitic auxiliaries are required in this circuit besides the basic cell stack—for example, an electrolyte circulation pump, an air separator that removes spent air from the electrolyte leaving the cell, an air compressor, and a zinc oxide separator and storage device.

It is obvious that there will be some reduction in energy density, relative to what can be achieved in a primary cell, because of the extra components required.

In an early prototype of the Gulf General Atomic system, the overall weight for a 14-kwh unit was 300 pounds, including all hardware, electrolyte, and so on. This implies an energy density of 48.8 whr/lb, considerably less than the best that has been achieved with the inherently simpler primary zinc-air cell (or 150 whr/lb), but still comparable to the best result currently available from commercial secondary batteries such as the silver-zinc system. Furthermore, calculations suggest a marked improvement in energy density as the overall system size is scaled up, due largely to the smaller relative weight of peripheral equipment such as pumps and blowers. Conceivably, energy densities of 70-80 whr/lb can be expected for rechargeable zinc-air batteries in larger units. However, with present technology the peak power output is low in comparison to the total energy available. Since high power densities are fully as important as high energy densities for automotive applications, work is currently being concentrated on methods of improving the ratio of peak power to energy capacity, and on increasing the cycle life, which is currently low.

---

[12]D. V. Ragone, "An Electrically Rechargeable Zinc-Air Battery for Motive Power," in *Power Systems for Electric Vehicles*, PHS Publ. No. 999-AP-37, U.S. Public Health Service, National Center for Air Pollution Control, Cincinnati, Ohio, 1967; and P. R. Shipps, "A Zinc-Air Secondary Battery System for Vehicular Propulsion," *Proceedings, 20th Annual Power Sources Conference*, May 1966.

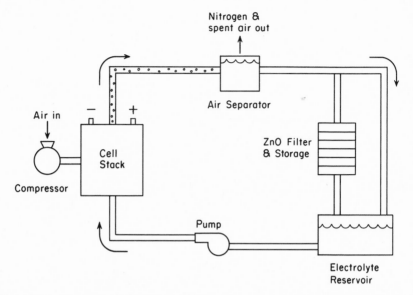

Figure 5-4. Schematic of General Atomic's zinc-air battery. (Adapted from P. R. Shipps, "A Zinc-Air Secondary Battery System for Vehicle Propulsion, Report No. GA-7188, General Atomic, Division of General Dynamics, now Gulf General Atomic Co., a division of Gulf Energy and Environmental Systems.)

The possible trends in zinc-air power density projected by General Atomic for various battery sizes are shown in figure 5-5. It seems possible that zinc-air rechargeable systems having energy densities of 70-80 whr/lb and power densities of 40-50 watts per pound may be available within a decade or so for vehicle propulsion purposes. Cost estimates vary between $1.50 and $3.30 per pound, depending on the energy density and power density achieved.

Leesona-Moos Laboratories have been experimenting with a rechargeable version of their successful primary zinc-air cell, which consists of a porous zinc anode impregnated with KOH electrolyte, sandwiched between two hydrophobic air cathodes that need be wetted only at their surfaces. An inorganic separator (developed by the McDonnell Douglas Astropower Laboratory) inhibits dendrite formation. The straightforward design of the Leesona-Moos cell is likely to cost $1.50-$2.00 per pound, depending on the amount of catalyst required on the air cathode.

Some work has been done at Yardney Electric on the use of calcium with an air electrode,[13] although this is not an active project. On a purely theoretical basis, calcium should be superior to zinc as an air cell anode; it offers a higher electromotive force for a lower equivalent weight. However, there have been some problems in practice. Calcium (like sodium and lithium, which are also

---

[13]A. Charkey and G. A. Dalin, "Metal-Air Battery Systems," *Proceedings, 20th Annual Power Sources Conference*, May 1966.

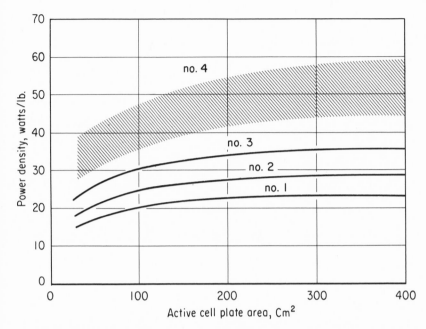

Figure 5–5. Performance characteristics of rechargeable zinc-air cell. Curve number 1 represents the 1965 state of the art. Curve number 2 represents systems using the same components but carrying smaller quantities of reactants and electrolyte to decrease the overall weight. Curve number 3 represents the same system electrochemically but optimized for power rather than energy. The shaded curve, number 4, represents predicted power densities based on methods of improvement that are possible but not yet demonstrated. (Adapted from P. R. Shipps, "A Zinc-Air Secondary Battery System for Vehicle Propulsion," Report No. GA-7188, General Atomic, Division of General Dynamics, now Gulf General Atomic Co., a division of Gulf Energy and Environmental Systems.)

more electronegative than hydrogen) tends to combine spontaneously with water to form the hydroxide, releasing hydrogen in the process. This makes it difficult to use calcium with an aqueous electrolyte, although an aqueous electrolyte is preferable for reduction of oxygen at the cathode. A considerable research effort has resulted in a partially aqueous electrolyte in which both electrodes are compatible, consisting of two parts of methanol to one of water with a solute of 10% $(NH_4)_2SO_4$ and 5% LiCl. While reasonably high energy densities were obtained (80 whr/lb of active cell pack), the performance of calcium-air cells has tended to be poor in comparison with their theoretical limits. A marked voltage degradation occurs at anything but very low drainage rates. Development efforts continue, however.

A lithium-moist air cell utilizing an organic electrolyte is currently undergoing exploratory development by Globe-Union Corporation, with support from the U.S. Army.[14] This system has a very high theoretical energy density (2,566

[14]J. E. A. Toni, G. D. McDonald, and W. E. Elliott, Globe-Union Corporation, "Lithium Moist Air Battery," Final Report under Contract DA-44-009-AMC-1522 (T), July 1967.

whr/lb), but a number of problems must be resolved before anything more positive can be said about its actual performance. The feasibility of a lithium anode in a nonaqueous electrolyte and the reduction of oxygen in a nonaqueous electrolyte, must be established, for example, although earlier results offer limited encouragement.

A rechargeable sodium-air battery has been built by the Atomics International Division of North American Aviation.[15] This system is, in effect, a double cell. A molten sodium salt electrolyte (at 130°C) is used with the sodium anode, and an aqueous NaOH electrolyte with the air cathode. The two electrolytes are separated by a sodium amalgam which acts as a cathode to the sodium–sodium-salt cell and as an anode to the NaOH-air cell. While this system is also in early stages of development, the possible performance may be as high as 200–300 whr/lb (including the weight of auxiliaries) depending on the discharge rate. Maximum attainable power densities appear to be in the range of 40 to 50 watts per pound. Costs have been estimated in the neighborhood of $2 per pound for large-scale production.

A proprietary iron-air cell is being developed by Westinghouse Research Laboratories. General Telephone and Electronics also has such a program. The theoretical energy density (based on iron alone) is about 650 whr/lb, assuming the oxidation reaction proceeds as far as possible. If 20 or 25 percent of this can be achieved in a practical configuration, the cell would have a maximum energy density of 120–150 whr/lb. An aqueous electrolyte can presumably be used, but no details of performance or construction have been published. Two of the great attractions of iron as an anode element are its extremely low cost and a well-developed technology derived from experience with nickel-iron cells.

Aluminum is another possible high-energy anode material for a metal-air cell. Aluminum has exhibited either excessive corrosion or excessive passivity, depending on the type of electrolyte, and as a result, little recent work has been done on it. However, the Zaromb Research Corporation of Passaic, New Jersey, has apparently succeeded in inhibiting the corrosion of aluminum in an alkaline electrolyte and anticipates the possibility of achieving energy densities as high as 240 whr/lb.[16] However, what little work has been done to date has focused on a "refuelable" primary aluminum-air cell, rather than on a secondary (rechargeable) cell.

Alkali-Metal Anodes with Organic Electrolytes

The increasing need for high energy densities in a portable electric power source points quite naturally to the light electronegative (alkali) metals, which have both very high energy contents and low molecular weights.

---

[15] L. A. Heredy, H. L. Recht, and D. E. McKenzie, "The Atomics International Sodium-Air Cell," in *Power Systems for Electric Vehicles*.

[16] S. Zaromb, "Aluminum Fuel Cell for Electric Vehicles," in *Power Systems for Electric Vehicles*.

The use of lithium anodes in combination with various cathode materials and nonaqueous electrolytes has been explored by Gulton Industries, Electric Storage Battery Company, and Globe-Union, with Whittaker Corporation, Electrochimica, Lockheed Missiles and Space Company, Livingston Electronics, General Motors, and the U.S. Army Electronics Laboratory, among others.

Lithium is the lightest of the alkali metals and the most active metal in the periodic table; it is also reasonably abundant and cheap. Since it reacts spontaneously with any aqueous solution, with oxygen, the halogens, and even nitrogen, it is necessary to find some suitable nonaqueous electrolyte. In practice, this implies an electrolyte that contains no hydrogen ions and yet offers good conductivity; which is to say, it must be compatible with a solute salt that will not introduce a prohibitive internal resistance in the cell.

A number of nonreactive electrolyte compounds have been tried, including propylene carbonate (PC), butyrolactone (BL), dimethylformamide (DMF), dimethyl sulfoxide (DMSO), acetonitrile (AN), nitromethane (NM) and methylformate (MF), containing various salts of lithium. The actual performances of several of these experimental batteries are tabulated in table 5-2.

Table 5-2. Various Lithium Organic Systems

System	Type	Energy density (whr/lb)
$Li/LiClO_4-PC/CuF_2$	Flat plate primary	223
$Li/LiClO_4-MF/CuF_2$	Flat plate primary	156
$Li/LiClO_4-BL/CuCl_2$	Flat plate primary	125
$Li/LiCl, AlCl_3-PC/AgCl$	Flat plate secondary	30

Source: K. H. M. Braeuer, "Organic Electrolyte, High Energy Density Batteries: A Status Report," *Proceedings, 20th Annual Power Sources Conference*, May 1966.

Note: PC = propylene carbonate; MF = methyl formate; and BL = butyrolactone.

It can be seen that relatively high actual energy densities can be achieved in lithium cells. However, all the above suffered from two drawbacks: a short shelf life and low efficiencies at medium and high discharge rates.

A rechargeable lithium cell consisting of lithium and silver difluoride (AgF_2) in an electrolyte of butyrolactone saturated with KPF_6 has been explored by the Whittaker Corporation.[17] Laboratory models of this system have undergone 50 recharging cycles at greater than 90% depth of discharge, with high charge-discharge efficiency. However, lithium utilization in early work has been relatively low. The theoretical limit for the lithium-silver difluoride couple is 678 whr/lb.

---

[17]M. Shaw, "Lithium Silver Difluoride Secondary Battery," *Proceedings, 20th Annual Power Sources Conference*, May 1966.

A lithium-copper fluoride cell tested by Lockheed Missiles and Space Company[18] has achieved energy densities ranging from 57 to 80 whr/lb. The goal in this program is the development of a 300-whr/lb battery for the National Aeronautics and Space Administration. However, this system is being developed specifically to operate at low discharge rates. A similar cell built by Electric Storage Battery Company has given 110 whr/lb in a laboratory model.

Successful lithium organic cells have also been developed by Gulton Industries. These cells use a nickel halide cathode—specifically nickel chloride ($NiCl_2$) and nickel fluoride (NiF_2)—in a nonaqueous electrolyte such as propylene carbonate (PC).[19] Dissolved salts, such as potassium hexafluorophosphate are added to raise the electrical conductivity of the solution. Both of these are rechargeable systems having high energy densities. The theoretical energy for the Li-NiF_2 couple, for example, is 620 whr/lb; 90 to 100 whr/lb has actually been attained at low currents. Present work on these systems is directed at improving their performance at high discharge rates. Electrochimica Corporation is actively developing nonaqueous lithium-silver chloride, lithium-copper chloride and lithium-copper fluoride cells. Specific energies up to 82 whr/lb have been obtained at a 10-hour discharge rate.[20] The P.R. Mallory Company has also examined a PC lithium-copper chloride system.[21] Work on developing a practical battery is not being continued because of the problems in finding appropriate separator materials,[22] but the company is sponsoring research on new ion-exchange membranes to be used as separators and is making an intensive proprietary effort to develop an advanced lithium-organic cell. No details are available on the latter. Cost estimates for lithium-organic cells in mass production range between $4 and $5 per pound.

Liquid Metal Cells/Molten Salt Electrolytes

As noted earlier, the battery anode materials that are most attractive from an energy standpoint are so reactive that they tend to be incompatible with aqueous electrolytes, while systems incorporating organic electrolytes have not yet been able to achieve high discharge rates and long lifetimes.

A different approach, which has already been moderately successful in the laboratory, is to utilize a molten salt of the anode materials as an electrical

[18]H. F. Bauman, "Development of Lithium Cupric Fluoride Batteries," in *Power Systems for Electric Vehicles.*

[19]R. C. Shair, A. E. Lyall, H. N. Seiger, "Lithium Nickel-Halide Batteries," in *Power Systems for Electric Vehicles.*

[20]M. Eisenberg, "High Energy Non-Aqueous Battery Systems for Electric Vehicles," in *Power Systems for Electric Vehicles.*

[21]M. L. B. Rao, "Formation and Discharge of Copper Chloride in an Organic Electrolyte," *Electrochemical Society Journal*, Vol. 114, No. 1 (1967), p. 13.

[22]A. N. Dey, "Ion Exchange Membrane Separators for Organic Electrolyte Batteries," *Electrochemical Society Journal*, Vol. 115, No. 2 (1968), p. 160.

conductor and plating bath. The anode itself may also be molten. Advantages are that very high charging and discharging rates become possible, since electrolyte polarization virtually ceases to be a consideration. Cycle life can be very long because the physical characteristics of the cell (at a given temperature and state of discharge) do not vary with time for liquids as they do for solids.

One concept that has received a good deal of attention is the General Motors lithium-chlorine cell, utilizing a molten lithium anode, a molten lithium-chloride electrolyte and a chlorine (gas) cathode.[23] In the current configuration, cylindrical electrodes of liquefied lithium and lithium chloride are stored separately.[24] During discharge, lithium is transported via capillary action down to a power production chamber through a metal fiber wick. Chlorine is fed in from outside the cell to a porous graphite center electrode, whereupon the lithium and the chlorine react to produce energy and lithium chloride. A schematic of the lithium-chlorine cell is shown as figure 5-6.

In the recharge cycle, the lithium chloride is reduced by electrolysis to chlorine gas and lithium, the former being ducted out to be reprocessed and stored and the latter rising again into its storage region.

So that both lithium and lithium chloride will be molten, the operating temperature is approximately 615°C, which is slightly above the melting point of lithium chloride. Theoretical energy density is 1,000 whr/lb at operating temperatures, and GM has projected 100-200 whr/lb for a future hypothetical vehicular application.[25]

In conceptual engineering studies by the Allison Division of GM under U.S. Army sponsorship, a total energy capacity of 200 kwh was estimated for a system weighing 1,250 pounds, equivalent to an energy density of 160 whr/lb. This system would have a rated power output of 50 kw, with peaks to 75 kw, and a rated power density of 40 watts per pound with a peak power density of 60 watts per pound.[26]

Work on a lithium-chlorine system is also being pursued at the laboratories of the Standard Oil Company of Ohio (SOHIO) under an Army contract.[27] The cathode in this system is formed of an active microcrystalline carbon plate, which is then further conditioned by a proprietary process converting the carbon to a high-surface-area polymer of carbon, chlorine, and alkali metals (either

---

[23] See J. Werth, J. Kennedy, and R. Weaver, "Lightweight Lithium Chlorine Battery," AIAA Winter Meeting, Philadelphia, September 1964. See also R. Eliason, J. Adams, and J. Kennedy, "Design Features and Performance of a Lithium-Chlorine Cell," American Chemical Society Meeting, Detroit, April 1965.

[24] J. P. Powers, "Regenerable Lithium Chloride System Development," Vols. I and II, Final Technical Report, September 1968, AD 842, 463.

[25] D. A. J. Swinkels, "Electrochemical Vehicle Powerplants," *IEEE Spectrum*, May 1968.

[26] J. P. Powers, "Lithium-Chlorine Electrochemical Energy Storage System," Final Technical Report, March 1967, AD 822, 442.

[27] R. A. Rightmire et al., "A Sealed Lithium Chloride Fused Salt Secondary Battery," Presented at the SAE Meeting, Detroit, January 1969.

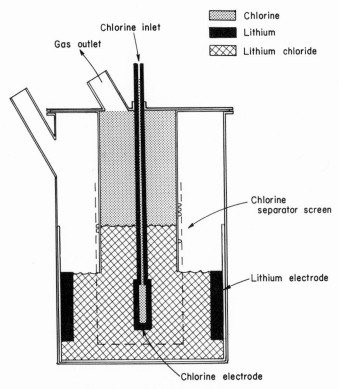

Figure 5-6. Schematic of the GM lithium-chlorine cell. (Adapted from H. Wilcox, "Electric Vehicle Research," in *Power Systems for Electric Vehicles*, PHS Publ. No. 999-AP-37, U.S. Public Health Service, National Center for Air Pollution Control, Cincinnati, Ohio, 1967.)

lithium or potassium). This electrode is designed to partially discharge the chloride ions in the fused salt and does not require an external gas feed. The anode is composed of a solid aluminum-lithium alloy that reduces anodic corrosion by the fused LiCl or KCl.

An energy density of 34 whr/lb has been achieved, and one of 60-70 whr/lb is extrapolated for an engineering prototype. Discharge rates of 150 watts per pound have already been achieved. As with all molten salt batteries, fabrication must be conducted in a pure inert atmosphere to avoid contamination by oxygen, water, or nitrogen.

Costs for production of this battery have been estimated at $1.50-$3.00 per pound. The basic raw materials are inexpensive: LiCl costs 85 cents a pound at present, and the price will probably decline as demand increases.

The sodium-air cell using a molten sodium salt electrolyte has already been described. Another concept utilizing molten sodium for an anode is the well-known Ford sodium-sulfur battery.[28] In this case, however, although both anode and cathode, as well as all reaction products, remain liquid (at 300°C) in separate compartments, the "electrolyte" is a solid ceramic of specially treated β-alumina, which is permeable to sodium ions but which will not pass sodium or sulfur atoms or sodium-sulfur molecules. The arrangement is shown schematically in figure 5-7. During discharge, sodium ions pass from the sodium reservoir through the ceramic separator into the liquid sodium reservoir, forming (liquid) sodium sulfide which dissolves in the liquid sulfur. During the charge cycle, the current is again carried by the sodium ions, which pass back through the ceramic separator to the sodium side. There is apparently no permanent physical change in either component, and the charge-discharge cycle can theoretically be repeated indefinitely (although only a few hundred cycles have so far been verified).

Furthermore, the system has attractive energy and power densities. Theoretical energy density for sodium-sulfur (Na_2S_5, Na_2S_4, and Na_2S_2) is 345

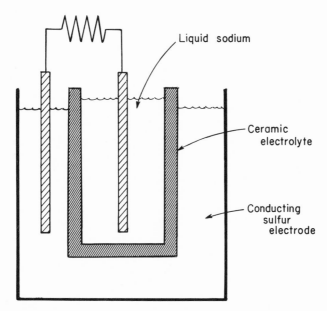

Figure 5-7. Schematic of the Ford sodium-sulfur cell. (Adapted from T. W. DeWitt, "A Sodium-Sulfur Secondary Battery," in *Power Systems for Electric Vehicles*, PHS Publ. No. 999-AP-37, U.S. Public Health Service, National Center for Air Pollution Control, Cincinnati, Ohio, 1967.)

---

[28]See J. T. Kummer and N. Weber, "A Sodium Sulfur Secondary Battery," *SAE Automotive Energy Conference*, Detroit, January 1967; also T. DeWitt, "A Sodium Sulfur Secondary Battery," in *Power Systems for Electric Vehicles*.

whr/lb; in fact, 150 whr/lb are projected for a practical system in large sizes, but only about 80–90 whr/lb have been achieved in the laboratory. Under rapid discharge (i.e., 15 minutes), the sodium-sulfur cell exhibits good power density of approximately 100 watts per pound without an unacceptable voltage drop.

The appealing features of this system are clear: it is a high-energy battery that discharges rapidly on demand, and (presumably) can be recharged an indefinite number of times; it is mechanically simple and hermetically sealed; and its component materials are very cheap and easy to purify. There are a few serious problems, however. From a cold start it takes several hours to reach the fairly high temperatures required for system operation (but once the system is started the temperature is self-sustaining due to internal resistive losses). Mechanical strength and physical integrity of the thin ceramic electrolyte shield must be maintained. There is always a finite risk that an external shock will crack or rupture the shield and ruin the battery. Although Ford scientists discount any possibility of a violent explosion, there is inevitably some hazard attached to a system utilizing any molten reactant, whether it be sulfur, lithium, or sodium. Nevertheless, the sodium-sulfur battery offers one of the best combinations of attractive features available among the cells whose characteristics are generally known.

A variety of bimetallic galvanic cells have been developed at the Argonne National Laboratory. The most intensive work has been done on lithium-bismuth, lithium-tellurium,[29] and lithium-selenium[30] couples, with limited research on a lithium-sulfur cell.[31] These cells consist of a lithium anode, a fused lithium halide electrolyte, and a lithium alloy cathode combining lithium with bismuth, tellurium, selenium, or sulfur. The operating temperatures vary from $360°C–470°C$, where anode metal, cathode alloy, and electrolyte are all in a molten state. In order to make a transportable cell, the electrolyte has been formed into a paste by 50% dilution with a ceramic powdered filler.[32] Excellent energy and power densities may be obtained with these paste cells as compared to other systems that have been studied. Thus a LiTe cell can be expected to exhibit a specific power in excess of 360 watts per pound with a specific energy of 80 whr/lb. The Li-Se system has a higher energy density, which is comparable to the LiCl system discussed previously. The relatively undeveloped Li-S system appears to be the most promising of all, potentially yielding energy densities up

[29]H. Shimotake and E. J. Cairns, "Bimetallic Galvanic Cells with Fused Salt Electrolytes," Intersociety Energy Conversion Engineering Conference, Miami, August 1967.

[30]H. Shimotake and E. J. Cairns, "A High Rate Lithium Selenium Secondary Cell with a Fused-Salt Electrolyte," Presented at the CITCE Meeting, Detroit, September 1968.

[31]H. Shimotake and E. J. Cairns, "A Lithium/Sulfur Secondary Cell with a Fused Salt Electrolyte," Presented at the Electrochemical Society Meeting, New York, May 1969.

[32]H. Shimotake, A. K. Fischer, and E. J. Cairns, "Secondary Cells with Lithium Anodes and Paste Electrolytes," Presented at the Intersociety Energy Conversion Engineering Conference, Washington, D.C., September 1969.

to 300 whr/lb, with power density above 100 watts per pound, and 200 whr/lb with power density of 200 watts per pound.[33]

Recharging of the bimetallic cells can be accomplished very quickly—in less than 15 minutes for cell capacities of 1 amp-hr/cm². The major problem (in addition to the inconvenience of all high-temperature molten salt batteries) is the extreme corrosiveness of the lithium and other cathode alloys, which makes it necessary to use expensive refractory metals and their alloys for construction of the cathode compartment and current collectors. Also, the metals used in the cathode are quite expensive. The LiS cell seems to be the most reasonable in price, though the high electronic resistivity and viscosity of the lithium-sulfur cathode greatly increases the complexity of the cell design. It must be emphasized that the present discussion is based on laboratory single-cell couples; developmental scaling up of these cells is only now under way.

General Comments on Batteries

An ideal secondary battery system would be one that is capable of high energy and high power densities at ambient temperatures, that has high electrochemical efficiency (low internal losses), and that can utilize cheap, safe materials in a simple configuration with a long life. Specific energy and specific power of seven systems are shown in figure 5-8.

In order to provide a comparison of the various possibilities in common terms, the principal battery systems of interest are listed in table 5-3, and ratings (excellent, good, fair, poor) are shown for certain salient characteristics. It can be seen that no one system meets all of the requirements, although several meet a number of them. Unfortunately, not enough is known about some cells (the Westinghouse iron-air cell, for example) to rate them, though many have intrinsic promise. The question is, which one of the present projected systems offers the best compromise?

This table is, of course, not quantitative. Its ratings are at best a matter of informed opinion, and one authority's "good" might be another one's "fair." Also, as the technology changes, some ratings might have to be changed. But the table does serve as a rough overview of the immediate prospects of these systems for vehicle propulsion.

Although the ultimate utility of an energy storage system clearly depends on economic factors, these are perhaps the least predictable aspects of many developmental systems. One relevant figure of merit for batteries in an electric vehicle would be the depreciation per discharge (i.e., the initial cost divided by the cycle life). Of course, higher energy densities also have a greater intrinsic value, not only because of the superior performance that would be allowed, but

---

[33]H. Shimotake and E. J. Cairns, "Lithium/Chalcogen Secondary Cells," Presented at the Fourth Advances in Battery Technology Symposium, Los Angeles, December 1968.

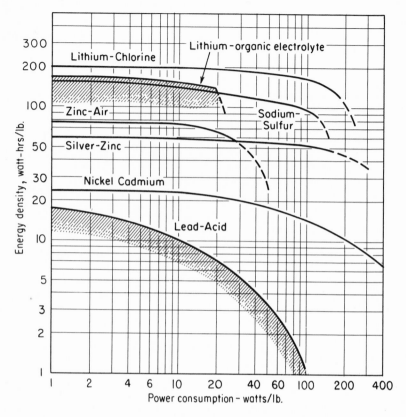

Figure 5-8. Energy storage capacity versus specific power for various battery types. (From U.S. Department of Commerce, *The Automobile and Air Pollution: A Program for Progress*, Part I, October 1967.)

also because of the reduced energy consumption (see chapter 10). Therefore an oversimplified, but still useful composite figure of merit might be something like the following:

$$\text{Utility} = \frac{\$/\text{lb}}{\text{cycle life} \times \dfrac{\text{watt-hour}}{\text{pound}} \times \text{discharge fraction}}.$$

Although numerical values for some of these variables have been given in the text (where available), there is really no solid basis at this time for relating cost estimates to cycle lifetimes. Also the latter are likely to change dramatically in the future, for at least some developmental systems. Thus, although a composite measure of "utility" could be arrived at on the basis of current data and crude manufacturing cost estimates based on materials costs and engineering com-

Table 5-3. Ratings for Various Batteries

E = Excellent; G = Good; F = Fair; P = Poor

System	Energy density	Power density	Ability to recharge or refuel	Service life	Material availability	Cost	Other problems
Pb-Acid	P	G	G	G	E	E	–
Ni-Fe	P	F	G	E	G	F	–
Ni-Cad	P	E	E	E	P	P	–
Ag-Zn	F	E	P	P	P	P	–
Zn-Air*	G	F	G	F(?)	E	G†	Complexity, maintenance(?)
Mg-Air*	E	F	F(?)	?	E	E†	Refueling is mechanical
Li-organic*	E	P	F–G	F(?)	G	G†	–
Na-S*	G–E	E	E	E	E	E†	Safety, high temperature, start-up
Li–Cl*	E	E	E	E	G	G	Safety, toxicity, complexity, start-up(?)
Li-Te*	E	E	E	E	P(?)	F(?)	Temperature, toxicity, start-up(?)

*Estimates.
†In production.

plexity, the results would have very little meaning. Since such figures are likely to be picked up and quoted out of context, regardless of how carefully they are originally qualified, it seems wiser to omit them. However table 5-4 lists costs for the relevant basic materials. There is an engineering rule of thumb to the effect that manufactured objects made of ordinary materials (e.g., steel) tend to cost "about" $1 per pound. If we assume this applies to the "package"—the

Table 5-4. Battery Materials Costs

Possible anode materials		Cathode materials	
	$/lb		*$/lb*
Aluminum	0.28	LiCl	0.85
Cadmium	4.00	Nickel	1.10
Calcium	1.00	Selenium	4.50
Iron	0.05	Silver	30–35
Lead	0.16	Sulfur	0.02
Lithium	7.75	Tellurium	6.00
Magnesium	0.35		
Sodium	0.24		
Zinc	0.16		

electrolyte, separators, and all structural elements, excluding special fabrication costs of electrodes, if any—then the total cost of a battery would be $1 per pound plus electrode fabrication plus the appropriate prorated cost of whatever exotic materials were used. For instance, if lithium metal accounted for 10 percent of the weight of the battery, a net cost of $1.70 per pound plus electrode fabrication costs might be expected.

THERMAL ENERGY STORAGE

In essence, thermal energy storage concerns the storage of heat energy via the physical condition of a material and its release by a change in that condition—for example, by a change of phase, a change from one crystalline form to another, or a change in temperature. For obvious reasons, it is necessary in any practical case that the change in the substance be reversible, that the composition of the material not be altered in the process, that the energy-releasing change occur within a suitable temperature range, and that a significant amount of energy be released.

Whatever the mechanism of heat energy storage, the energy would be exchanged with some suitable working fluid, which would then be expanded through a heat engine, giving up heat energy and doing work, and then recycling back through the heat storage mechanism.

The thermal substance comprising this mechanism can be "recharged" in a number of ways—for example, by passing a hot combustion exhaust through it, by an electric immersion heater, etc. In practice, however, it would seem necessary that the recharge source be available when needed, which essentially means that the most feasible vehicular application of the principle is in a hybrid system in conjunction with an external combustion engine (e.g., steam). In this case, the thermal storage system functions as the analog of a storage battery, providing a reserve capacity for peak load surges. It might be asked, why the thermal storage at all, then? Why not just a direct use of the working fluid? The answer is that a larger external combustion system would be needed to meet the peak power demands of a typical usage cycle (see discussion of hybrids in chapter 12).

Perhaps the most practical method of embodying the thermal storage principle is the use of either a molten salt or a eutectic. Generally speaking, the salt used should have a large latent heat-of-fusion, and a melting point somewhere near the temperature at which the fluid is to be delivered. If such a salt contained in a heat exchanger is imposed in a working fluid "ahead" of the system's expander, it gives up its heat of fusion when a sudden increase in output load occurs (i.e., if the working fluid temperature begins to drop).

After a careful analysis of possible salts for such parameters as suitable temperature ranges, corrosiveness, etc., a number of salts and various eutectics emerge as thermal-storage contenders. Table 5-5 lists the pertinent single (i.e., non-eutectic) salts.

Table 5-5. Single Molten Salts Suitable for Thermal Energy Storage

Salt	Melting point (MP)		Specific heat at MP (Btu/lb °F)	Heat of fusion (Btu/lb)
	°C	°F		
$LiNO_3$	254	490	0.39	160
$NaNO_3$	310	590	0.44	76
NaOH	318	604	0.50	69
KNO_3	337	640	0.30	51
KOH	360	680	0.36	73
LiOH	462	884	0.36	38

Source: Joel Jacknow, "Molten Salts for Heat Storage in a Rankine Cycle System," Paper No. IRT-N-86, International Research and Technology Corp., Washington, D.C., October 30, 1969.

MECHANICAL ENERGY STORAGE

Energy can also be stored mechanically by exploiting various inertial and deformation effects, notably in rotational and elastic systems, to wit: in deformable solid bodies (e.g., springs), in compressed gases, and in flywheels.

Generally speaking, springs are used only for the storage of very small amounts of energy. Their possible application to full-scale vehicular energy storage seems, at first glance, absurd. However, if the term is used to encompass all types of deformable solids, the aspect is perhaps less negative. The theoretical energy content for such deformable materials ranges from a small fraction of a whr/lb for steel springs to as much as 12-13 whr/lb for various polyurethanes, considering only the weight of the material itself. An installed system would probably not yield more than a few whr/lb.[34] Some of the deformable solids, at least, have potentially sufficient energy capacity for automotive use, although no such application has been tried as yet. Considerable effort would be required to design suitable conversion mechanisms and to solve bonding and other problems before such a vehicle could be built. In any case, actual implementation would doubtless make use of numerous small spring mechanisms rather than one large one.

Storage of energy in compressed gases presents a slightly more favorable case. Basically, the energy available is the work that can be done in expanding the gases to ambient pressure. The slower the compression and expansion, the more efficient the recovery. For example, in compressing 1,000 cubic feet of air to a pressure of 100 atmospheres (or 1,500 psi) and a volume of 10 cubic feet, we

[34]J. A. Hoess et al., "Study of Unconventional Thermal, Mechanical, and Nuclear Low-Pollution-Potential Power Sources for Urban Vehicles," Summary Report to the National Air Pollution Control Administration, Battelle Memorial Institute, Columbus Laboratories, March 15, 1968.

find that some 10 million ft-lb are expended, or about 3.76 kwh.[35] Since 1,000 cubic feet of air weigh about 85 pounds, this amounts to roughly 44 whr/lb. However, this value is the nominal energy density for the compressed air only; it does not allow for inefficiencies due to changes in entropy during compression and expansion, and does not allow for the weight of pumps, tubes, tankage, etc. By the time these are taken into account in any safe and practical container design, the energy density for the system is down to about 10 whr/lb or less. In fact, one analysis, based on 5,000 psi, revealed a capacity of only 1.5 whr/lb for the gas and container![36]

Of the mechanical storage techniques, the high-speed rotating flywheel offers perhaps the most convenient vehicular method. In fact, it has been used on occasion for such purposes, albeit rather specialized ones. In such systems, the kinetic energy built up in the wheel is transferred gradually as needed to an electric generator and thence to a motor; the system usually is regenerated by bringing the flywheel back up to maximum speeds at periodic stops. In addition, the gravitational potential energy recaptured during downhill runs can also be used to feed energy back to the flywheel. From a vehicular standpoint, an attractive feature of flywheel storage is its capacity for giving up all its energy very rapidly. The actual coupling mechanism by which the energy is transferred might be a traction drive, a torque converter, or a planetary gearset, for example; in any case, it should be as nearly steplessly variable as possible.

The constraints on flywheel storage are the stress limitations of the flywheel itself and the gyroscopic effects resulting from the conservation of angular momentum. A disc-rim flywheel, operating at an angular velocity resulting in a centrifugal force of 100,000 psi on the rim, yields about 10 whr/lb.[37] If the yield strength could be increased to 200,000 psi, the energy density would increase to about 40 whr/lb. However, this is just the density for the wheel itself. A practical flywheel system normally includes a bulky reduction gear system and other components, so that overall energy density—already marginally low by vehicular performance criteria—is considerably lower in practice for the system as a whole. Furthermore, because of inevitable frictional losses, a flywheel system is constantly dissipating energy, even when the vehicle is at a standstill. In addition, the incidental gyroscopic effects in driving up and down hills, for example, could be awkward. This difficulty could be eliminated by using opposed counter-rotating flywheels, but only at considerable cost in system weight and complexity.

In short, the flywheel can probably be ruled out in the near future as a primary energy storage mechanism for private—or general—use vehicles, although

[35]Bruce Chalmers, *Energy* (New York: Academic Press, 1963).

[36]Hoess et al., "Study of Unconventional . . . Power Sources."

[37]L. U. Kline, S. M. Marco, and W. L. Starkey, "Work Capacities of Energy Storage Systems on the Basis of Unit Weight and Unit Volume," *Transactions, ASME*, Vol. 80 (1958), p. 909.

not necessarily for specialized forms of mass transit vehicles. A flywheel system built by Maschinen Fabrik Oerlikon was at one time used to drive transit buses in some European cities. The flywheel, lying horizontally underneath the passenger deck, weighed three tons, nearly 25% of the total weight of the bus. The gyrobus system had a top speed of about 35 mph and an average traffic speed of 14 mph, but an effective range between "recharge" stops of only two or three miles.[38]

A possibility that might be worth exploring is the use of a flywheel as a reserve storage system where prime power is supplied by some other means, such as a small gas turbine operating at constant output. Here one could fully exploit the flywheel's most valuable attribute—the ability to release energy extremely rapidly without serious loss of capacity. The Oerlikon gyrobus, for instance, is said to have achieved a power density of about 150 watts per pound.

The gyrobus and several other applications of flywheels to relatively long-term energy storage are described briefly in table 5-6.

Table 5-6. Applications of Flywheels for Energy Storage

System	Energy capacity, whr/lb, wheel only	Drive system	Application
Oerlikon Electrogyro	2.7	Electric	Bus
General Dynamics	19.5	Electric	Aerospace power supply
North American Aviation	12.75	Hydrostatic	Remote power storage
Gyreacta and Hydreacta		Planetary	Vehicle transmission

Source: J. A. Hoess and R. C. Stahman, "Unconventional Thermal, Mechanical, and Nuclear Low-Pollution-Potential Power Sources for Urban Vehicles," SAE Paper No. 690231, International Automotive Engineering Congress, Detroit, January 13-17, 1969.

Recently further interest in flywheel possibilities has been expressed in various quarters. The Ground Vehicle Systems Group of Lockheed Missiles and Space Company, for example, has proposed to the San Francisco Municipal Railway a "kinetic wheel" generator consisting of a high-speed flywheel of tapered radial cross section driving a built-in electric generator. This wheel, made of maraging steel, is reputedly capable of storing at least ten times as much energy as the Oerlikon flywheel.[39]

In addition, the work of David W. Rabenhorst at the Johns Hopkins Applied Physics Laboratory in Silver Spring, Maryland, on the so-called super flywheel

---

[38]"The Oerlikon Electrogyro," *Automobile Engineer*, Vol. 45 (December 1955), p. 559.

[39]"High Speed Energy Wheel Offers Trolleys Portable Electricity," *Product Engineering*, July 20, 1970.

has attracted considerable attention. Using bar-shaped flywheels constructed of high-strength filaments, Rabenhorst's group claims that energy densities of 40 whr/lb are possible with present materials, and he projects future capabilities more than five times as high.[40] Should such high energy densities eventually become practical, the flywheel could indeed become a highly competitive energy storage technique.

[40]"Super Flywheel Configurations Form the Heart of Mechanical-Powered Drives," *Product Engineering*, April 12, 1971.

Chapter Six

New Developments in
Internal Combustion Engines

The major objectives of research and development in ICE technology in recent years have been to find ways of: (1) reducing engine size, weight, and cost in relation to power, and (2) increasing efficiency and reducing unwanted emissions.

The first objective is being pursued primarily by exploring various means of converting the energy released by combustion directly into rotational motion, thus dispensing with the elaborate and heavy crankshaft mechanism. The gas turbine and the Wankel rotary engine are only two of a number of possible approaches. The relevance of these rotary engines to eliminating the major weaknesses of existing engines—pollution and noise—is at best indirect. They suffer somewhat from the same basic weaknesses—inefficient and incomplete combustion under part-load conditions, too much noise, and lack of torque at zero speed resulting in a need for mechanical or other transmissions and torque converters—although recent research efforts have improved them considerably. Internal combustion engines—in small, simplified versions—might conceivably play a useful role in a "hybrid" ICE-electric system where the ICE would be used to drive a generator at constant speed, while batteries (or flywheels) and electric motors would provide the actual propulsion. The Wankel rotary engine, or a specialized nonregenerative gas turbine, might be applicable in this context. The rationale of a hybrid system is discussed in chapter 12.

The second major objective of current research on ICE technology is aimed at solving the problem of automotive air pollution by means of "fine tuning," or add-on devices. Even if this program is entirely successful in the terms in which it has been formulated, it will not eliminate most of the deficiencies mentioned above (inefficiency at part load, lack of torque at zero speed, etc.). If mild palliatives can pacify public outcries against the current situation and postpone the day when more fundamental changes are needed, such research will perhaps have been worthwhile from the standpoint of the industries that have most to lose by any alteration in the status quo. In any case, a brief summary of research on emissions control is included in this chapter.

ROTARY INTERNAL COMBUSTION ENGINES

Many efforts have been expended in the past century to develop internal combustion engines that will produce torque at a reasonably low speed without a crankshaft and its attendant complication. Within the last decade or so, these efforts have produced a wide variety of prototype rotary piston engines, as well as miniaturized turbines. These designs, however, are largely experimental.

The best-known and most advanced rotary engine—and the only one commercially available—is the Wankel engine. This was invented by Felix Wankel in 1953. The patent is owned by NSU Motor Werke of Germany, which has begun to manufacture the engine in small numbers, and which also licenses its manufacture by a number of other major companies, including Curtiss-Wright—and, more recently, General Motors—in the United States.

Basically, the Wankel engine consists of a single rotor with a rounded triangular shape, rotating eccentrically in a single casing with a double-lobed cross section—which is the path followed by the apexes of the convex triangle in its eccentric rotation. Thus, the apexes always brush the walls of the casing. The result is that there are three compartments which rotate around the chamber, changing size and shape as they move by virtue of the eccentricity of the rotation. Thermodynamically, the engine is based on the Otto cycle in its "two-stroke" form. On one side of the chamber casing is a spark plug. Each compartment passes through all the phases of the cycle once in each revolution. Since there are three compartments, three power strokes are produced with each shaft revolution; each compartment is always in some phase or another of the Otto cycle—and each in a different one, since they are 120° apart. This sequence is illustrated in figure 6–1.

It can be seen that precise construction of the central chamber is most important in the Wankel engine. The rapidly rotating eccentric rotor must fit closely, since this is what maintains the separation between the three compartments: with too loose a fit, leakage between successive chambers causes losses and inefficiency; with too tight a fit, the engine jams and perhaps destroys itself by overheating. Furthermore, this good fit must be maintained under widely varying conditions of pressure and temperature. Leakage, cooling, and lubrication are all serious problems. In the early version of the Wankel engine, oil was simply mixed with the fuel, much in the manner of a two-stroke reciprocating engine. However, this results in an intolerably dirty exhaust and cannot be seriously considered. In newer versions, NSU claims that a forced lubrication system to bring oil to the friction points keeps the exhaust clean. Fuel consumption and thermal efficiency are now said to be comparable with conventional Otto-cycle engines, although the problems of inefficiency under part-load conditions remain. Power-to-weight is still good. A two-cylinder Wankel engine weighing 239 pounds and built by Curtiss-Wright has produced 185 bhp at 5,000 rpm.

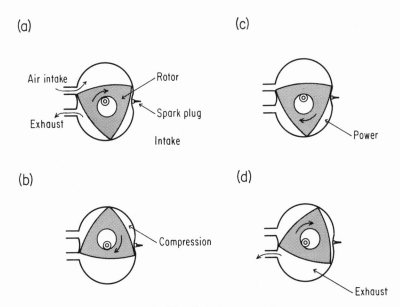

Figure 6-1. The Wankel rotary engine.

The Wankel is quite light in weight, cheap and simple to produce, and easy to maintain; and it can be free from vibration if the eccentric rotor is counter-balanced by a shaft. If an emission-free exhaust can be produced, the Wankel engine or one of its rivals may conceivably capture large segments of the auto-motive engine market. It was felt at first that the difficulties of cleaning up the exhaust would be even greater for this engine than for the reciprocating-type Otto-cycle engine, owing to the large surface-to-volume ratios in the combustion chamber and the basic two-stroke cycle. However, this appears not to have been the case. The quenching effect of the unfavorable surface-volume ratios results in cooler running conditions, which inhibit the formation of nitrogen oxides. Furthermore, it was found that the Wankel geometry allowed the use of a thermal reactor immediately adjacent to the combustion chamber ports and thus further combustion of the unburned hydrocarbons swept out of the chamber. Curtiss-Wright claims that such an exhaust reactor system is capable of reducing unburned hydrocarbons in the exhaust to 45–50 ppm, although typical HC test levels have averaged around 100 ppm (compared to a federal requirement of not more than 275 ppm of HC for 1968 cars, and 180 ppm for 1970). Their con-clusion is that the Wankel is indeed able to meet 1970 HC and CO_2 require-ments, although the NO_x level is still somewhat above the expected requirements.[1]

---

[1] David Cole and Charles Jones, *Reduction of Emissions from the Curtiss Rotating Combustion Engine with an Exhaust Reactor*, SAE Paper No. 700074, January 1970.

Wankel-powered cars are now beginning to appear in the marketplace. Toyo Kogyo in Japan and NSU in Germany are producing such vehicles for public sale; Mercedes-Benz has announced plans to develop experimental cars using Wankel engines; Rolls-Royce is developing a diesel version of the Wankel principle; and GM in the United States has recently concluded an ambitious licensing arrangement involving $50 million over a five-year period.

There are many other valveless rotary designs, generally more complicated than the Wankel, of which we need only mention a few.[2] Isuzu in Japan is developing a trochoidal design with a two-lobed rotor rotating eccentrically in a three-lobed housing. Renault in France has a four-lobed rotor engine which is licensed to American Motors Corporation in the United States (shown in figure 6–2).

An engine invented by Eugene Kauertz of Germany has two pairs of wedge-shaped rotating pistons rigidly attached to a pair of concentric central shafts. The leading and following pistons move like a cat and mouse, alternately approaching and moving away from one another, thus permitting intake, compression, expansion, and expulsion of gases in the space between them. This motion is controlled by an arrangement of gears. A prototype bench version, operating at 4,000 rpm, has reportedly produced 213 bhp. Because there are no unbalanced inertial forces, the engine runs very smoothly. A similar engine, known as the Virmel and invented by Melvin Rolfsmeyer, is reported to have achieved 337 bhp at 3,800 rpm from a 148-cubic-inch displacement prototype weighing 130 pounds; this engine incorporates a very ingenious built-in two-speed transmission.[3] The Tschudi Engine Corporation of Flushing, New York, has devised a related type of engine based on four toroidal segmented pistons running in a toroidal track. The pistons are affixed in pairs to concentric central shafts with phasing controlled by a set of rollers and cams rather than gears.[4] Because of the circular cross section of the toroidal track, compression can be maintained by means of conventional piston rings, unlike the Kauertz and Virmel engines. Simplified drawings of several "cat and mouse" schemes are shown in figure 6–3.

Another basic type of rotary engine is the so-called "revolving block" design. One such, invented by Austin Mercer (Bradford, U.K.), has a single pair of opposed pistons in a cylinder that rotates around a central axis. The pistons are fixed to rollers that follow a complicated path (two intersecting circles, symmetrically offset from the center), and as the block rotates the pistons are forced to move in and out. Porsche in Germany has patented a design with a cruciform revolving block containing four cylinders that move in and out radially, controlled by a large two-lobed cam attached to the center shaft.

---

[2] Wallace Chinitz, "Rotary Engines," *Scientific American*, February 1969.

[3] *Science and Mechanics*, October 1966.

[4] T. Tschudi, U.S. Patent No. 2,734,489 (1956).

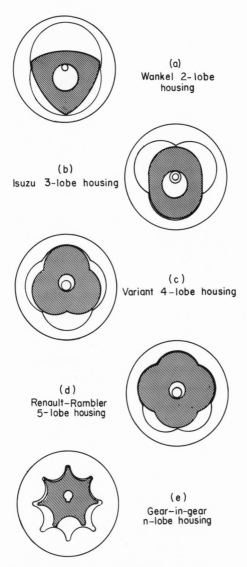

(a)
Wankel 2-lobe
housing

(b)
Isuzu 3-lobe housing

(c)
Variant 4-lobe housing

(d)
Renault-Rambler
5-lobe housing

(e)
Gear-in-gear
n-lobe housing

Figure 6-2.   Trochoidal rotor configurations.

An engine developed by Kal-Pac Engineering Ltd., of Vancouver, British Columbia, utilizes a pair of crossed pistons—a charging piston and a power piston—as shown in figure 6-4. The two pistons, intersecting at right angles, are connected at their midpoints to crankpins of the same crankshaft. As it moves down the cylinder, the charging piston draws in fuel from the carburetor. The left spark plug fires at mid-stroke, and the power piston moves to the right,

(a) Kauertz engine

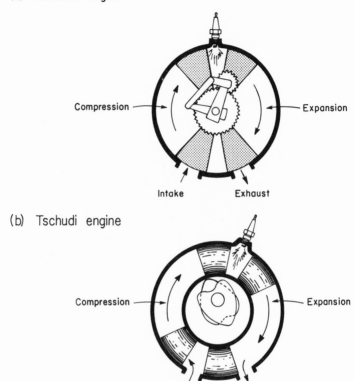

(b) Tschudi engine

Figure 6-3. The Kauertz (above) and Tschudi (below) rotary engine configurations.

closing the ports as it does so, until compression and ignition take place at the right end. The charging piston then moves up, beginning the cycle again at the top. Although considerable interest has been expressed in this engine, there are as yet no plans for marketing it in the United States.[5]

G. A. Dotto and Leon Linn of P. R. Mallory and Co. have developed a rotary engine based on the old "vane pump" concept, which consists of a set of 16 wedge-shaped combustion chambers, rotating on a shaft offset from the center of a cylindrical cavity.[6] The neighboring chambers are separated from each other by vane-like separators, which, controlled by an eccentric cam, slide in and out of slots in the central rotor to maintain a constant 0.015-inch clearance with

---

[5]"Novel Engine Concepts Raise a New Challenge," *Product Engineering*, February 24, 1969.

[6]G. A. Dotto and W. L. Linn, U.S. Patent No. 3,301, 233. Also personal communication from Leon Linn of P. R. Mallory Co.; and "Low-Weight Rotary Powerplant Developed," *Aviation Week and Space Technology*, March 13, 1967.

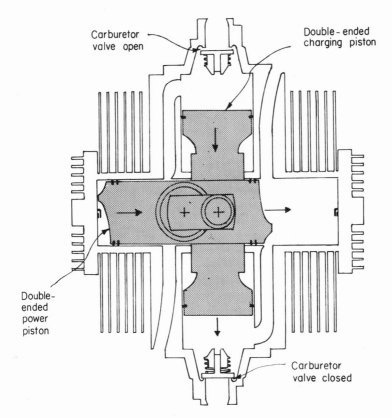

Carburetor valve open

Double-ended charging piston

Double-ended power piston

Carburetor valve closed

Figure 6–4. The Kal-Pac engine. (Adapted from *Product Engineering*, February 24, 1969, p. 82, copyright Morgan-Grampion, Inc., 1969).

respect to the outer wall. Since the rotor is offset from the center of the cavity, the volume in each wedge-shaped space varies with its rotational angle (or position on a hypothetical clock face) as shown in figure 6–5. Thus combustion chamber volume is maximum at six o'clock; air and fuel are injected through a port at seven o'clock, maximum compression (8 : 1) occurs at twelve o'clock and ignition occurs at two o'clock and continues to about four o'clock, followed by exhaust venting at five o'clock. Grooves cut in the outer wall housing connect three successive chambers in the combustion region (from two o'clock to four o'clock) so combustion is continuous and no spark is required after startup. Oil for lubrication is supplied to the crankshaft, and vane slots are lubricated by centrifugal force. A scavenger pump collects the oil at the outside and returns it to the crankshaft. In tests, oil was lost at the rate of 1 quart per hour. Fuel consumption at full load (400 bhp at 5,000 rpm) was 0.4 lb/hp-hr, corresponding to about 35% thermal efficiency. The torque-speed characteristics of the Mallory engine are similar to those of the gas turbine, but with a flatter curve

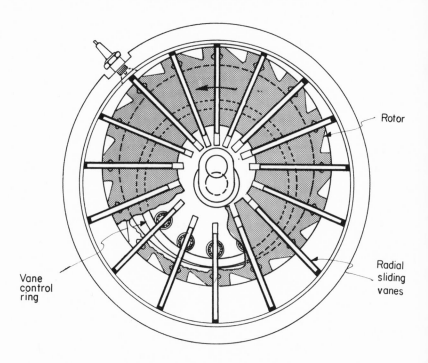

Figure 6-5. The Mallory rotary engine.

between 2,000 and 5,000 rpm. A unit weighing 158 pounds, cast of a copper-chromium alloy, produced 400 bhp at 5,000 rpm (with an effective displacement of about 450 cubic inches). With an aluminum casting, the engine weight would be about 80 pounds, resulting in a figure of merit of 0.2 lb/bhp, which is competitive with much larger aircraft gas turbines and probably considerably superior to anything else in its size range.

At low speeds, power output from the simple version decreases drastically because the compressor, attached to the same shaft, supplies inadequate air. A possible solution might be to pair two engines on a "split shaft," with the first unit operating simply as a compressor or supercharger for the second—a method similar to the one used in the turbocompound engine, which is described in chapter 7. Another approach would be to use the principle of the differential turbine, also discussed in chapter 7. With either arrangement maximum torque would be obtained at low speeds. The need for a transmission for automotive application would thus be eliminated, and total weight would still be extremely low.

The power-to-weight ratios of some rotaries are impressive. A conventional automotive reciprocating engine typically requires 3–5 lb/hp not including the transmission. The Wankel engine requires only 1.5 lb/hp or so, while the "bare"

Virmel engine achieved better than 0.4 lb/hp, as did an early working prototype of the Mallory engine.

Assuming that no insurmountable technical problems arise (e.g., lubrication or excessive wear), one or another of these new engines might be suitable for applications where ICE-electric hybrid systems could be considered. This combination is discussed in more detail in chapter 12.

THE CONSTANT PRESSURE RECIPROCATING INTERNAL COMBUSTION ENGINE

A promising system which adapts the Brayton cycle to a piston engine has been developed by Mechanical Technology, Inc., Latham, N.Y., using extensive computer design and analysis.[7] This device is something of a hybrid: combustion is external to the cylinder, but combustion products still constitute the "working fluid." The configuration is similar to a standard reciprocating internal combustion engine. The cylinders may be in either an in-line or V configuration, although the latter is perhaps more convenient for this application, and a constant pressure burner is situated in a separate chamber between the banks of opposed cylinders, as shown in figure 6-6. Expansion of the hot gases through

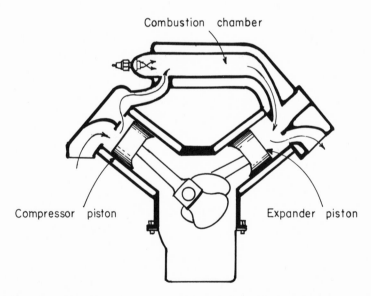

Combustion   chamber

Compressor   piston   Expander   piston

Figure 6-6. Reciprocating internal combustion engine with separate constant pressure combustion chamber. (Adapted from *Product Engineering*, February 24, 1969, p. 81, copyright Morgan-Grampion, Inc., 1969.)

---

[7]G. B. Warren and J. W. Bjerklie, "Proposed Reciprocating Internal Combustion Engine with Constant Pressure Combustion," SAE Paper No. 690045, Automotive Engineering Congress, January 1969.

one of each pair of cylinders moves the piston and provides power; the opposing piston in each pair serves to draw in air and compress it, then injects it into the combustion chamber.

The developers claim significant improvements over the conventional ICE, including a smoother power flow, a probable 20–30% reduction in specific fuel consumption, and greatly reduced exhaust emissions, with almost complete elimination of carbon monoxide, nitrogen oxides, and unburned hydrocarbons. The gases remain in the combustion chamber for a long time—three to five times that in gas turbines and twenty times that of the conventional ICE gases. This, coupled with leaner mixtures and lower temperatures, permits essentially complete combustion, so that there is little or no "quenching." In addition, the burner conditions will probably not permit the formation of nitrogen oxides. The lack of a carburetor prevents any emission of "hot-soak" gasoline fumes. The possibility also exists of cutting off fuel entirely during deceleration, which would serve both to reduce fuel consumption still further and to eliminate exhaust products completely under circumstances when they are normally high in pollutants.

A far from negligible advantage, as far as the industry is concerned, is the fact that the engine can be built with essentially the same tooling used in making present internal combustion engines.

CONTROL OF EMISSIONS DUE TO INCOMPLETE COMBUSTION IN CONVENTIONAL ICEs

There are two fundamental approaches to the problem of eliminating unwanted by-products of combustion in the engine: (a) preventing their formation in the first place, or (b) removing them from the gas stream before they leave the vehicle. Devices implementing both approaches have been tried.

It will be recalled that one reason for the appearance of hydrocarbons in the exhaust stream is the short time available for combustion and the quenching effect of the "cold" combustion chamber walls. An optimum combustion chamber design for minimizing unburned hydrocarbons would be one with a low surface-to-volume ratio. This suggests a spherical or at least hemispherical combustion chamber. (The Wankel rotary engine discussed at the beginning of this chapter is intrinsically bad from this standpoint, as figure 6–1 illustrates.) Surface-to-volume ratio is difficult to change in an existing design, however, and it cannot be reduced to zero for any internal combustion system as there must always be some wall surface in contact with the flame. Presumably, there will always be some unburned hydrocarbon in the exhaust from the chamber.

The reciprocating engine developed by Mechanical Technology, Inc., discussed above, in which a standard ICE is converted to the Brayton cycle by substituting a separate combustion chamber and using the piston cylinder only for expansion, rather effectively solves the quenching problem by avoiding it.

The long residence time of the gases, lower temperatures, and excess air effect complete combustion before the gases come into contact with the cooler walls.

Another simple design innovation is to move one of the piston rings from its traditional location to a point as near the piston head as possible in order to eliminate the cold space between piston and cylinder wall where unburned fuel and air can be trapped. This innovation has been proposed by Sealed Power Corporation,[8] but not yet by any automobile manufacturer.

Fuel Injection

The production of carbon monoxide can be minimized by maintaining the optimum air-fuel setting of the carburetor for all conditions of use (idling, cruising, acceleration, deceleration). Electronic sensing and feedback circuits have been devised to perform this function, but the cost is not negligible. Fuel injection is, in principle, more efficient than carburetion (premixing), and any serious attempt to reduce unwanted exhaust products would probably require a fuel-injection system or even a stratified charge system (discussed below), combined with a more sensitive means of matching the fuel input to the engine's demands. Unfortunately fuel injection is also expensive. Peugeot installs fuel injection systems on about 40% of its cars, but even in quantity production this adds appreciably to the cost of the car. Starting with the 1968 model-year, Volkswagen's Model 1600 (medium-priced) cars for export to the United States have been equipped with fuel injection systems, controlled by a shoebox-sized electronic computer which is said to cost $70. The movement may spread because fuel injection also offers the advantage of a 10% improvement in fuel economy.

Stratified Charge Systems

To vary the load of a conventional internal combustion engine, the quantity of fuel-air mixture is varied, but the ratio of air to fuel remains approximately constant. The efficiency of fuel utilization, particularly under partial load conditions, could be improved considerably if this ratio could be varied and a full charge of air used with a variable amount of fuel. Unfortunately, air-fuel mixtures more diluted than about 15 : 1 either do not ignite at all or ignite unreliably. It is the purpose of the stratified charge injection system to produce a "rich" mixture in the immediate vicinity of the spark plug; the "lean" mixture elsewhere in the combustion chamber is then consumed by the resulting flame. Overall air-fuel ratios achieved in this manner in laboratory tests range from 16 : 1 to 50 : 1.[9]

---

[8] W. J. Kalb, *Iron Age*, September 12, 1968.
[9] *Motor Vehicles, Air Pollution and Health*, A Report to Congress by the Surgeon General, Washington, D.C., 1962.

There are many possible design techniques by which the desired stratification effect can be achieved, but premixing (e.g., in a carburetor) is clearly ruled out. Perhaps the most work in this area has been done by Texaco on the TCP (Texaco Combustion Process) engine, although development has also been pursued by Pure Oil Corporation, Citroen, Renault, and Institut Français de Pétrole, as well as several universities. In a typical stratified charge engine, as shown in figure 6-7, a high velocity swirl is imparted to intake air through proper design of the air-intake port. The fuel is injected during compression at an angle back toward the swirl of the air, where it is entrained and carried toward the spark plug in the center so that the fuel is concentrated at the ignition point. When ignition occurs, the flame spreads outward in a concentric manner toward the cylinder walls. Under conditions of partial load, combustion will not spread to the outer areas of the cylinder because only air will be present there.[10]

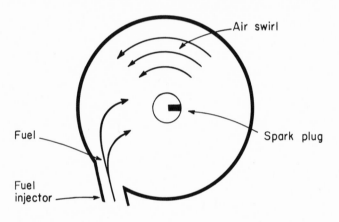

End view of cylinder

Figure 6-7. Schematic of the stratified charge principle. (Adapted from J. E. Witzky and J. M. Clark, "The Third Cycle—Stratified Charge," *Mechanical Engineering*, March 1969, pp. 29 ff.)

As an additional bonus, the stratification technique permits control of the preflame reaction time of the fuel by timing the fuel injection before the spark discharge. Thus, not only is autoignition[11] prevented by allowing insufficient time to establish the necessary conditions, but the sensitivity of the combustion to octane number is reduced.

---

[10]J. E. Witzky and J. M. Clark, Jr., "The Third Cycle—Stratified Charge," *Mechanical Engineering*, March 1969.

[11]Detonation—characterized by a "ping" or knocking sound—occurs when the unburned charge in a combustion chamber ignites by itself because of increased pressure and temperature propagated by the normal flame front. It produces severe pressure fluctuations in an engine cylinder, and can cause serious damage, e.g., to the valves and cylinder rings.

The advantages of the stratified charge approach can be summarized briefly as follows:

(a) The compression ratio can be increased indefinitely without regard to the octane number of the fuel.[12] Since knocking is not a problem, nonleaded fuel could be used, and fuel requirements would be less stringent. The Texaco stratified charge engine will operate on diesel fuel, for example.

(b) The lean overall fuel-air mixture ratios available in this engine result in more favorable part-load fuel economy and reduced emissions.

(c) The ability to control the engine load by air-fuel mixture ratio alone makes the stratified charge technique especially adaptable to the two-stroke engine cycle.

(d) Supercharging (whereby air is compressed before being mixed with the fuel, thus permitting the consumption of a greater mass of charge by a given piston stroke, with resulting improved performance) can be effectively used with stratified charge engines. Octane number, and the increased possibility of detonation or preignition which usually accompanies supercharging in a spark-ignition engine, can be disregarded.

(e) Insensitivity to octane number permits the use of lower-grade fuels without lead additives, which will reduce chamber deposits and spark-plug fouling. The ability to use lead-free fuel also opens up the possibility of using a catalytic converter to eliminate unburned hydrocarbons, since it would not be inactivated by deposits of lead.

Generally speaking, the stratified charge principle allows more efficient combustion (by a factor of 10-20% depending on conditions) and fewer exhaust emissions than the conventional ICE. It combines many of the better features of the Otto and diesel cycles: it has less rigid fuel requirements than either of them, greater fuel economy than the first, and a lighter, cheaper structure than the second. On the other hand, it still requires spark ignition, and would probably weigh and cost more than the conventional ICE.

A variation on the stratified charge approach is embodied in a system developed by the Azure Blue Corporation, of El Dorado, California. This system consists of a precombustion chamber built into a special spark plug, and a dual-ratio carburetor, which provides a lean air-fuel ratio (16 : 1 to 18 : 1) to the main chamber, and a rich mixture (6 : 1) to the precombustion chamber. During compression, some of the lean mixture is forced into the precombustion chamber, diluting the mixture. Ignition occurs in the precombustion chamber at the optimal 13.7 : 1 ratio. Orifices then direct the flame into the main chamber at sonic velocities, and turbulence results in complete and rapid combustion of the lean mixture there.

The developers claim significant reductions in emission levels for engines using the system. Hydrocarbons, for example, are rarely more than 10 ppm,

---

[12]Octane number is the percentage ratio by volume of iso-octane (which has excellent anti-knock qualities) to ordinary heptane in the fuel, and thus is a measure of "anti-knock."

while nitrogen oxides range between 35 and 37 ppm. Carbon monoxide content is very low. In addition, the system is easily adapted to existing vehicles. The spark plug containing the precombustion chamber is designed with a standard screw base and can be inserted into any engine block. The dual-ratio carburetor and a dual-ratio induction manifold are built into a special cylinder head, which is readily fitted to any standard engine.[13]

Exhaust Control Systems

Of the various sources of automotive air pollution, crankcase blow-by is the easiest to cope with. By using closed systems with no vent to the atmosphere, this problem can be eliminated completely. The gases are simply returned to the engine air intake system and recycled. This approach, which is called Positive Crankcase Ventilation (PCV), forms the basis for the numerous systems certified for use in new and used cars in California. In fact, all post-1968 model U.S. cars incorporate essentially complete control of blow-by gases at minor cost in terms of complexity and extra maintenance.

It is also possible to remove unwanted components from the exhaust after they have left the combustion chamber. Injecting air into the exhaust manifold to permit more complete oxidation is the most popular approach to date because it is relatively inexpensive and requires the least basic redesign and retooling. The air injection technique forces a constant flow of air into each exhaust port at the exhaust valve. Oxygen in this flow reacts to burn excess hydrocarbons and oxidizes most of the carbon monoxide into carbon dioxide. Extensive tests with simple air injection systems reveal emission values well below the 1966 California and 1968 federal standards of 275 ppm HC and 1.5% CO, although no appreciable improvement in nitrogen oxide emissions has been noted.

A more ambitious air injection approach utilizes a large exhaust manifold, called a "reactor," which allows the exhaust gases additional residence time for more complete reaction. In the laboratory, such devices have demonstrated low emission levels: DuPont reports experimental results of 54 ppm HC and 0.6% CO, while GM claims 27 ppm HC and 0.75% CO.[14] One such system, in operation on a vehicle for 50,000 miles, still displayed HC levels below 50 ppm and CO below 0.8%. There are some technical problems remaining, however: reactor temperatures must be maintained well above $1,000°F$, for one thing, and the bulkier system is difficult to fit into today's already crowded engine compartments. The high temperature creates an additional difficulty in this environment. However, use of a fuel-rich mixture in an engine with an exhaust manifold

---

[13]"Precombustion System Cuts Smog, Saves Gas," *Chemical and Engineering News*, September 22, 1969.

[14]N. K. Steinhagen, "Air Pollution Control Methods for the Gasoline Powered Internal Combustion Engine," Briefing for the Panel on Electrically Powered Vehicles, Office of the Secretary, U.S. Department of Commerce, 1967.

reactor has the advantage of inhibiting the formation of nitrogen oxides, which are notoriously difficult to remove, once formed. Although NO_x can also be reduced by recirculating the exhaust gas through the engine, this method does not combine well with some of the engine manifold techniques mentioned above, primarily because the cooler exhaust gas tends to reduce the temperature too much.[15]

Engine modification is another means of reducing emissions. This approach relies on carburetor and spark-timing adjustments and uses techniques such as fuel injection for optimizing the air-fuel ratio. The Chrysler Cleaner Air Package employs the engine and operation modification techniques, as did the 1968 GM engines, with the exception of the Cadillac, which utilizes an air injection method. By 1970, most U.S. cars had adopted this basic technique.

The exhaust problem can be attacked still farther downstream, so to speak, by replacing the muffler or tailpipe with a conversion device. Such devices, usually called afterburners, are of two basic types: catalytic and direct flame. Both types are aimed at changing the hydrocarbons and carbon monoxide into something else, preferably just carbon dioxide and water. Unfortunately, this is somewhat more difficult than it sounds. As the gases are much less concentrated and cooler in the exhaust pipe than they are in the engine combustion chamber, heat must be added to induce oxidation, and then the excess heat must be rapidly removed. Direct-flame afterburners therefore require expensive heat-resistant or heat-exchanging materials. For catalytic afterburners, the principal problem has been the high cost of the platinum or palladium catalysts, which have an extremely limited life (a few months at most) largely due to the "poisoning" caused by the lead in anti-knock gasolines. (Lead also tends to interfere with the performance of fuel-rich thermal reactors, by causing the hot inner core material to deteriorate.)[16]

Lead is responsible for a variety of problems. Some 70–80% of the lead additives are emitted with the exhaust as particulate matter.[17] Also, lead deposits in the combustion chamber are apparently responsible for marked increases in exhaust hydrocarbons, possibly due to the increased surface area (and increased quenching) created by the deposits. In tests with new catalytic mufflers, concentrations of HC and CO in exhaust gases dropped more than 90% initially, but efficiency decreased rapidly with miles driven.[18] At the moment,

---

[15]U.S. Department of Health, Education, and Welfare, *Control Techniques for Carbon Monoxide, Nitrogen Oxide, and Hydrocarbon Emissions from Mobile Sources*, NAPCA Publication No. AP–66, March 1970.

[16]U.S. Department of Commerce, *The Implications of Lead Removal from Automotive Fuel: An Interim Report of the Commerce Technical Advisory Board on Automotive Fuels and Air Pollution*, June 1970.

[17]Arthur C. Stern, ed., *Air Pollution*, Vol. 2 (New York: Academic Press, 1962).

[18]Steinhagen, "Air Pollution Control Methods for the Gasoline Powered Internal Combustion Engine."

the petroleum industry is reluctantly committed to removing the lead from gasoline even though this means revised refining methods and higher costs (estimated variously at 0.5¢ to 2¢ per gallon[19] and 0.5¢ to 1.5¢ per gallon).[20]

Tests run by the California Air Resources Board (CARB), using leaded gasoline, show catalytic converter efficiencies of 80% on HC and 90% on CO after 14,000 miles of operation.[21] And at least one company claims to have come up with an efficient "lead filter" to remove lead from the exhaust stream before it reaches the catalyst. However, the Interim Report of the Commerce Technical Advisory Board on Automotive Fuels and Air Pollution (1970) does not give much credence to the hope that lead will not continue to be a permanent hangup for catalytic convertors. While 1971 standards can probably be met with the "engine modification" approach alone, meeting the projected 1975 standards will probably require a reactor or catalytic convertor of some sort. Catalytic convertors promise at least one additional advantage not shared by emission control methods other than the fuel-rich manifold reactor system: it is possible, in principle at least, to reduce the level of nitrogen oxide emissions. One suggested technique is to use part of the CO emitted (probably with a rich air-fuel mixture) to reduce the nitrogen oxides with a catalyst, then add air and a second catalyst to oxidize the remaining CO.

Evaporative Losses

Evaporative losses occur from two sources: the "hot-soak" losses from the carburetor when the engine has been shut off, and the losses from the gasoline tank. In principle, carburetor losses can be eliminated by: eliminating the carburetor bowl; preventing evaporation from the gasoline surface; absorption of the vapor emitted; or removal of the gasoline from the bowl when the engine is shut off. It is, of course, also possible to eliminate carburetor losses by eliminating the carburetor itself. Thus fuel injection systems are attractive, not only for the reasons already mentioned, but also as additional methods of eliminating evaporative losses from carburetors.

Evaporative losses must be controlled on 1971 model-year autos sold in the United States, and several systems are now available that promise significant reduction in this category.[22] Ford and GM utilize a carbon absorption canister. Gasoline vapor collects in the top of the fuel tank. A smaller tank mounted above it separates the vapor from the liquid. This vapor, stored in the carbon canister when the engine is off, is extracted and burned with heated air drawn in through the canister when the engine is running. In the system advanced by

---

[19]Bonner and Moore Associates, *U.S. Motor Gasoline Economics* (New York: American Petroleum Institute, June 1967).

[20]U.S. Department of Commerce, *The Implications of Lead Removal from Automotive Fuel*.

[21]"Coming: Catalysts for Cars," *Chemical Week*, December 7, 1968.

[22]*Machine Design*, September 4, 1969, p. 40.

Chrysler the vapors are extracted from the fuel tank and carburetor through a ventilation valve, and then delivered to the engine and burned.

Future Prospects for Emissions Control

The 1968 and 1970 air pollution standards appear to have been met without undue strain on the part of the industry or cost to the public, although projected fuel savings to the car owner have not materialized. The 1970 federal standards are tantamount to about a two-thirds reduction in emissions of unburned hydrocarbons and carbon monoxide for new vehicles (since 1966) but—because of the diminished control effectiveness in older vehicles, the large existing population of uncontrolled vehicles, and the rapid increase in the number of vehicles on the road—the decrease in total emissions would never be more than 50% (from present levels), as shown very clearly in chapter 2 in figures 2-13 and 2-14.[23] Moreover, the lowest point would not occur until after the initial application of national standards (1980). In the case of Los Angeles, this would only be equivalent to "turning back the clock" to about 1953 or so, and in subsequent years the level would begin to rise once again. To reduce the pollution level to that of 1940—when the phenomenon of smog first became noticeable—would require a very considerable further tightening of standards, as was recognized by 1968 California legislation, and by the federal standards proposed by the National Air Pollution Control Administration (now Office of Air Programs). (See table 2-8, chapter 2.) The two sets of standards are not identical, but each calls for further cuts in emissions to about the 90% level of control by 1975.

In 1968 the California legislature set even stricter standards for automobiles to take effect with the 1970 model cars and become progressively tighter until 1974. The essence of these limits is as follows:[24]

Cars and Light Trucks:
    1970 models—2.2 grams of HC/mile (or, for a 3,000-lb car, 180 ppm HC instead of the earlier 275 ppm).
    1971 models—4 grams/mile of NO_x (the first limit imposed on NO_x).
    1972 models—1.5 grams HC, 23 grams CO, 3 grams NO_x.
    1974 models—1.3 grams NO_x.

Heavy-duty vehicles (over 6,000 lb):
    1970 models—275 ppm HC and 1.5% CO.
    1972 models—180 ppm HC and 1% CO.

Levels of 63 ppm HC, 0.3% CO, and 175 ppm NO_x were the objective of an Inter-Industry Emission Control (IIEC) Program, organized in 1967 as a joint

---

[23]NAPCA, *Summary of Emissions in the United States*, 1970 ed., May 1970.

[24]*Progress in the Prevention and Control of Air Pollution*, 2nd Report of the Secretary of HEW to Congress, January 1969.

effort of the Ford Motor Company and Mobil Oil and subsequently expanded to include five other U.S. oil companies (American, Atlantic Richfield, Sun Oil, Marathon, and Standard of Ohio) and four foreign car manufacturers (Fiat of Italy, Mitsubishi Heavy Industries, Nissan Motor Company, and Toyo Kogyo Limited, all of Japan). A considerable research effort was mounted by this group during the first eighteen months of activity. The various devices and concepts developed were embodied in a series of modified Ford and Mercury automobiles as part of an experimental test program. Typical of the features in these vehicles were: a pressurized gas tank and a modified fuel to minimize evaporation; a programmed by-pass computer for controlling exhaust flow; catalytic converters for nitrogen oxides, and for hydrocarbons and carbon monoxide; and air injection into the exhaust before entering the second (HC-CO) catalytic converter.[25]

An immediate aim of the IIEC group has been to develop a catalytic system for unleaded gasolines good for 50,000 miles, and an inexpensive disposable converter for use with leaded gasolines for shorter mileages.

In table 6-1, the principal emission standards are listed by year, along with the probable ranges of capital costs to the car buyer for each level of control.

Table 6-1. Emission Control Methods and Estimated Cost per Vehicle

Year	Standard	Control device	Approx. combined cost
1966	Control of crankcase emissions only	PCV	$5-$15
1968	Exhaust standard 3.2 gm/mile HC, 35 gm/mile CO, plus crankcase control	PCV plus engine modification	$25-$40
1970	2.2 gm/mile HC, 23 gm/mile CO, plus crankcase control	PCV plus engine modification	$40-$60
1971	Above plus evaporation control (6 gm NO_x/test)	Above plus evaporation control	$75-$100
1972-73	Above plus 3.0 gm/mile NO_x (California standard)	Above plus exhaust recycling system or catalytic converter	$125-$200

Sources: Compiled by the authors from U.S. Department of Health, Education, and Welfare, *Control Techniques for Carbon Monoxide, Nitrogen Oxide, and Hydrocarbon Emissions from Mobile Sources,* NAPCA Publication No. AP-66, March 1970; U.S. Department of Commerce, *The Implications of Lead Removal from Automotive Fuel: An Interim Report of the Commerce Technical Advisory Board on Automotive Fuels and Air Pollution,* June 1970; and *Cost of Clean Air,* 1st Report of the Secretary of HEW to Congress, in compliance with P.L. 90-148, June 1969.

---

[25]"The Search for a Low Emission Vehicle," Staff Report for the Senate Committee on Commerce, 1969.

The cost figures, which are taken from various (admittedly often conflicting) estimates in the literature, are also shown in figure 6-8.

To achieve a further improvement beyond the 1970 California standards it would appear to be necessary to: (1) eliminate carburetors completely in favor of fuel-injection or even stratified charge, (2) provide more sensitive means of fuel inlet control (perhaps through electronic sensors and circuitry), (3) provide catalytic or direct-flame afterburners for manifolds or exhaust pipes, and (4) provide some effective means of preventing vapor losses from gas tanks. The first two items, at least, promise to be expensive to the purchaser, and the third one may also necessitate eliminating tetraethyl lead from gasoline.

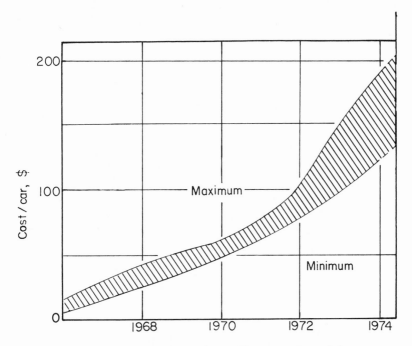

Figure 6-8. Estimated costs of controlling automotive emissions.

There has already been a noticeable sacrifice in engine "liveliness," since the system must be designed to prevent extra-rich mixtures from reaching the combustion chamber. Moreover, engine tuning and preventive maintenance have already become more precise and difficult, which eliminates one of the major advantages of the large understressed (overpowered) engines used in U.S. cars. In the future, these trends are likely to continue.

The effects of control measures on performance are important, for they point up the possibility of a serious revolt by automobile owners. In California in the early years many owners deliberately disconnected the PCV valves and other anti-smog devices, and, in some instances, dealers and used-car lots apparently cooperated.[26] Inspection and enforcement may be capable of minimizing this scofflaw problem, but only at substantial additional cost in terms of special test equipment. California and New Jersey are the only states that have taken steps to ensure that some level of effectiveness is maintained in automotive emission control devices. New Jersey, for example, has adopted legislation that requires each motor vehicle manufactured under federal emission control standards to pass an annual test of its emission control system.[27] A simple test cycle called ACID (for Acceleration, Cruise, Idle, Deceleration), which apparently simulates more ambitious test cycles, has been instrumented. The fact that the state already owns and operates a chain of safety inspection stations has facilitated the implementation of the plan. Even with this head start, so to speak, preliminary indications are that installing emission control equipment will cost approximately $15,000 per inspection lane. Constructing new facilities designed only for emission inspection would cost considerably more.

Cost and effectiveness per car for the New Jersey ACID system are outlined in table 6-2.

Table 6-2. New Jersey ACID System: Cost and Effectiveness per Car

Annual out-of-pocket cost per car	$4-$7.50[a]
Exhaust HC reduction[b]	13%-19%
Exhaust CO reduction[b]	7%-14%
$ per car/% HC removed	$0.31-$0.39
$ per car/% CO removed	$0.57-$0.54

Source: U.S. Department of Health, Education, and Welfare, *Control Techniques for Carbon Monoxide, Nitrogen Oxide, and Hydrocarbon Emissions from Mobile Sources,* NAPCA Publication No. AP-66, March 1970, pp. 4-10.

[a]Estimated; large-scale, full-utilization of the system.

[b]Based on 22% equipped, 70% not equipped with exhaust controls.

The projected future situation in Los Angeles is often cited by representatives of the automobile and petroleum industries as an indication that the problem is under control and well on the way toward being solved. However, it should be recognized that California was many years ahead of the rest of the country in recognizing the smog problem and setting up machinery to deal with it. Furthermore, the projections (see figures 2-12 and 2-13 for California, and 2-14 and 2-15 for the nation) assume that control devices will continue to be effective for perhaps 50,000 miles or so—probably an overly optimistic assumption. Actual emissions will almost certainly exceed the projected values by some margin, perhaps a considerable one.

It has been remarked more than once by industry advocates that the internal combustion engine is a "moving target," the implication being that by the time

---

[26]*Wall Street Journal,* August 17, 1967.

[27]HEW, *Control Techniques for Carbon Monoxide, Nitrogen Oxide, and Hydrocarbon Emissions from Mobile Sources.*

battery-electrics or steam engines are able to compete with today's ICEs, the latter will have substantially overcome its current drawbacks. So why worry?

To a skeptical non-Detroiter, this argument appears to rest on shaky ground. Indeed, it seems likely that in the effort to reduce air pollution even to the 1970 California standards, the ICE will become substantially more expensive and complicated, and somewhat less responsive and reliable, than it is today. On the other hand, the air quality standards are likely to get tougher and tougher to meet, necessitating ever more elaborate means of adapting this engine to a role for which it is not inherently well-suited. Indeed, the director of research for one major European manufacturer of small cars privately stated to one of the authors that in his opinion the (then) proposed 1970 standards could not be achieved by any of the current easy-going approaches (at least without a catalytic afterburner and the concomitant changes in fuel additives and refining practice which that would entail). While he probably overstated his difficulties, this pessimistic conclusion is substantially similar to the one reached by the Division of Air Pollution, U.S. Public Health Service, in its report to the National Commission on Technology, Automation, and Economic Progress (February 1966):

> It is not anticipated that either new add-on devices or minor modifications in prevailing engine design, however helpful they may be, will solve this problem satisfactorily, even apart from cost considerations; increasing attention, therefore, may have to be given to more radical changes in gasoline engines and even to alternative power sources for motor vehicles.[28]

Recent developments have not contradicted this statement.

Alternate Fuels

The uninitiated often assume that the pollution is somehow inherent in the fuel, not the engine, whereas some semi-sophisticates make the opposite mistake of assuming that the fuel does not matter. Actually it matters to the extent that various fuels have different burning characteristics, which affect the rate of ignition and the completeness of combustion. In principle, any substance capable of supporting rapid controlled oxidation could be used as a fuel in a combustion engine. In practice, however, the list of possibilities would be fairly concise, because the fuel must not only be energetic enough, but also inexpensive, available, and reasonably safe to handle—or at least as safe as gasoline.

Of the nonhydrocarbon fuels, one obvious possibility is ammonia (NH_3). From an air pollution standpoint, ammonia is probably acceptable because its exhaust products are mainly water and nitrogen gas, plus some nitrogen oxides. Anhydrous ammonia has been tested successfully at some length as a fuel for

---

[28]"Technological Change as it Relates to Air Pollution," A Report to the National Commission on Technology, Automation, and Economic Progress, Appendix, Vol. 5, February 1966.

conventional spark-ignition internal combustion engines.[29] The experimental results of these tests, however, indicate that the ammonia must be introduced as a vapor and partially dissociated into hydrogen and nitrogen; the latter requirement, of course, demands the expenditure of a certain amount of energy (2.9 hp were required to dissociate 30% of the ammonia needed in a 100-hp engine), which must be considered a negative attribute. In addition, power output per pound of fuel was only about 70-75% of the output of an equivalent gasoline engine. Also, the smaller air-fuel ratios required result in decreased volumetric efficiency. Neither disadvantage is crucial, but it is clear that an ammonia-fueled engine would not perform as well as a gasoline-powered engine of the same size and weight.

Hydrazine (N_2H_4), which is somewhat more energetic than ammonia, is being used in some experimental fuel cells. Its immediate drawbacks are toxicity and high cost ($8.50 per gallon at present). The first problem might possibly be mitigated by appropriate handling procedures, while the second may be only a temporary hurdle. Knowledgeable people in the chemical industry feel that hydrazine could be synthesized from natural gas and atmospheric nitrogen, or by hydrogenating ammonia. In either case, prices should ultimately not be too different from ammonia (5¢ per pound). The problems that might arise from burning hydrazine in an ICE have not been thoroughly investigated.

Among the various hydrocarbon derivatives of petroleum, the most likely candidates are light paraffins, such as propane or butane, and alcohols, such as ethanol or methanol. In particular, a considerable amount of work has been done in the use of liquid propane (LP) in internal combustion engines. The results are quite encouraging. LP offers better fuel economy than gasoline and a notably reduced emission level.[30] Improvements in fuel economy have ranged from 4-5% to about 15% (depending on engine speed and load). The paraffins are also intrinsically less harmful than olefins and aromatics. These values, furthermore, were obtained from a basic engine, without the benefit of the various exhaust decontamination techniques discussed earlier. Results are even more impressive in reducing carbon monoxide content. Values for the volumetric percentage of CO from propane exhausts ranged from highs of 0.4% and 0.7% during idle at 1,050 rpm and 400 rpm, respectively, to zero at 1,400 rpm and 1,800 rpm under full load.

A considerable variation in the spark advance was noted during these tests. Thus, the principal engine modification requirement for use with LP gas would probably be the redesign of the distributor. In general, propane offers performance comparable to gasoline with better fuel economy and reduced exhaust

---

[29]E. S. Starkman et al., "Ammonia as a Spark Ignition Engine Fuel: Theory and Application," SAE Paper No. 660155, Automotive Engineering Congress, Detroit, January 1966, *SAE Transactions*, Vol. 75 (1967), pp. 765-84.

[30]Milton C. Baker, "LP Gas–a Superior Motor Fuel," SAE Automotive Engineering Congress, Detroit, January 1967.

pollution. It is certainly a potential alternative fuel—the more so because it is both readily available and relatively inexpensive (20 cents per gallon, exclusive of tax).

It is estimated that about half a million vehicles in the United States have been converted to LP gas. Most of these are fleet cars, such as the 1,500-odd buses of the Chicago Transit Authority, and the 1,000 vehicles of General Telephone in Florida.

Parenthetically, General Telephone has noted a decrease in operating and maintenance costs from $2.30 per 100 miles using gasoline to $1.85 per 100 miles with propane. This is due in part to the lower price of the propane and in part to lower maintenance requirements.

Conversion of an internal combustion chamber to LP gas requires a pressurized fuel tank, an electrically operated fuel valve with filter, a converter to reduce pressure and to vaporize fuel, and a special carburetor. Present cost of such a conversion for a standard car ranges between $200 and $300, although Chevrolet is offering an engine specifically modified for use with propane.[31]

Similarly, internal combustion engines can also be converted to burn natural gas (which is largely methane). This has been done on occasion. A demonstration vehicle (a modified 1968 Ford) has been developed by the San Diego (California) Gas and Electric Company, and a dual-fuel car that can switch from gasoline to natural gas has been operated by Pacific Lighting Corporation in Los Angeles. Both companies reported that pollutant emissions were significantly lower than those of a comparable gasoline drive. Pacific Lighting, for example, claims CO emissions down 93%, hydrocarbons down 45%, and NO_x down 87%.[32]

Other fuel possibilities include butane, methanol, ethanol, and liquid hydrogen. Alcohol fuels would presumably burn relatively cleanly; they are often used in racing-car engines for that reason. Ethanol can be produced for about 30 cents per gallon, and methanol for 25 cents per gallon, but they do not provide the same power per pound, or per dollar, that gasoline does. Moreover, although they do not mix with gasoline, they do mix with water, which may be a problem.

In the long run, however, hopes of converting any sizable fraction of the immense automotive market to the use of natural gas or any of the natural gas liquids would seem to be doomed by problems of supply. Projections of present and future demands vs. probable reserves indicate that natural gas will be in short supply by the end of the twentieth century, even with no greater load than the anticipated growth of present (i.e., nonpropulsive) uses.[33] Future reserve discoveries may, of course, change this, but these can scarcely be counted on.

---

[31]*IDEAS*, IR&T Newsletter, Vol. 2, No. 1 (Washington, D.C., October 1969).

[32]Ibid.

[33]Hans H. Landsberg, Leonard L. Fischman, and Joseph L. Fisher, *Resources in America's Future* (Baltimore: The Johns Hopkins Press for Resources for the Future, Inc., 1963).

Hydrogen, on the other hand, would not only be clean and energetic, but also cheap (since it can be produced very inexpensively from natural gas, at the wellhead). Although liquid hydrogen must be shipped in pressurized cryogenic containers, and therefore offers some minimal risk, the problems associated with bulk handling have been largely overcome through technology developed as a by-product of the NASA space program. (All the present generation of multi-staged rockets use liquid hydrogen and liquid oxygen in their upper stages.) Retail distribution would pose some fairly difficult technical problems, however. It is somewhat hard to envision burning hydrogen in a standard, unmodified engine, but the possibility of using hydrogen as a fuel for automobiles should at least be considered. However, we cannot evaluate this possibility at the present time.

Chapter Seven

Applications of Gas Turbines to Automotive Vehicles

The principle of the turbine as an energy conversion system has been known over a very long period, dating back to a crude form of reaction turbine, known as the aeolipile of Hero, of about 130 B.C., and the simple impulse turbine of Giovanni Branca of 1629. Later, the possibility of utilizing the combustion gases from burning hydrocarbon fuel with air in a turbine stimulated a good deal of inventive activity. Old literature is filled with references to machines of one type or another that could be classified as gas turbines.

In the seventeenth century a gas turbine known as the "smokejack" was invented to operate a spit. The name of the inventor is unknown, although the idea, in fact, had been anticipated by Leonardo da Vinci, who in 1550 made a sketch of a similar device. The principle of the smokejack is quite similar to that of the windmill. The smokejack was placed in the chimney, and rotation was induced by the passage of the hot chimney gases upward through the bladed wheel. Although it was inefficient and feeble, it successfully performed certain limited functions such as turning a spit for roasting meat.

The earliest patent on a gas turbine was that of John Barber (Nuneaton, England) in 1791. This patent (Patent Spec. No. 1833) was the earliest practicable engine and the original turbine scheme. Although crude as to form, Barber's gas turbine included a compressor, combustion chamber, impulse turbine, and even water-injection to prevent exposing the turbine blades to excessive temperatures. Barber's invention was a constant pressure gas turbine, with continuous combustion.

John Dunbell (England) invented the first constant-volume (or "explosion") type of gas turbine in 1808. Products of combustion from a coal fire passed through several rows of moving blades attached to a single rotor. Since stationary guide blades were not utilized, the advantages of multistaging were not achieved.

In 1884, Sir Charles Parsons (England) patented his reaction steam turbine, which specified the essential mechanical features of the modern gas turbine.

Parsons' steam turbines were adopted very quickly for marine propulsion and stationary electric-power generating plants.

The next important step in the development of the gas turbine *per se* appears to have been taken by René Armengaud and Charles Lemale, two French inventors who actually made a 500-hp engine working at 5,000 rpm, which utilized a liquid fuel atomizer, electric ignition, and water-spraying jets to cool the turbine nozzles and blades. The unit consisted of a two-row impulse turbine driving a Rateau multistaged centrifugal compressor. The compression absorbed about one-half of the total power developed by the turbine. Combustion took place in a combustion chamber by injecting and igniting a liquid fuel in the compressed-air stream on its journey from compressor to turbine. This gas turbine was said to have worked satisfactorily over a period of some years. Thermal efficiencies slightly below 3% were reported.

Mention should be made here of another early version, designed by F. Stolze (Germany) about 1872, but constructed and tested between 1900 and 1904. The Stolze gas turbine is very similar in many ways to modern units. Atmospheric air was heated in an externally-fired combustion chamber and finally expanded in a reaction turbine directly coupled to the compressor. Special features of this unit included a multistage turboexpander, a multistage axial-type air compressor and a heat exchanger. Despite its ingenious conception and the incorporation of some of the basic features of later gas turbines, the Stolze turbine was not a success.

S. A. Moss (United States) operated a De Laval turbine wheel with the aid of combustion products at Cornell University in 1902, and from about this time experimented with exhaust-driven turbines. This work subsequently led to the development of the Sherbondy and later turbochargers for aircraft engines.

A considerable amount of experimental work was undertaken by Hans H. Holzwarth (Germany) on the constant-volume (or "explosion") gas turbine, beginning in 1905, and (except for the World War II period) has been carried on by his followers to this day. The explosion gas turbine, as conceived by Holzwarth, provides for intermittent combustion of a liquid fuel in a combustion chamber at constant volume at elevated pressure. The high pressure in the combustion chamber is obtained by supplying combustion air under pressure. In one stationary plant, steam-turbine-driven centrifugal compressors are used to compress the combustion air, the steam being supplied by a waste-heat boiler located in the exhaust hood of the gas turbine. Several Holzwarth gas turbines were built by the German firm of Thyssen between 1914 and 1927, but only one of these appears to have been put into commercial operation. In 1920, a Holzwarth gas turbine, built by Thyssen, was supplied to the Prussian State Railway for an electricity generating station. Another Holzwarth gas turbine was built in 1928, by Brown, Boveri and Co. It operated using two combustion chambers on a two-cycle principle, and, when completed in 1933, was installed in a Ge•man steel manufacturing works, where it was in operation on blast furnace gas. Although Holzwarth gas turbines have been built in Germany and Switzerland, they have not met with great favor because of their complicated design, intermittent combustion, bulkiness, and generally low efficiency. The highest thermal efficiency ever obtained with a Holzwarth unit is reportedly about 13%.

Dr. Alfred Buchi (Switzerland) in 1908 investigated the use of waste exhaust gases from diesel engines to operate air compressors for supercharging these engines and later was responsible for the well-known Buchi-turbochargers, which have been made over a long period by Brown, Boveri and Co.

A. Rateau (France) was responsible from about 1917 for the first aircraft piston engine turbochargers which were used to maintain engine power up to the higher flying altitudes of that period.

Frank Whittle (England) applied in 1930 for a patent covering both the basic gas-turbine as it is known today and the propulsive duct, or "athodyd," system. The publication of Whittle's earlier patent with its drawings is believed by some to have been an inspiration to German aircraft engine manufacturers, since the experimental work on aircraft gas turbines was taken up by the B.M.W., Junker, Heinkel-Hirth, and Daimler-Benz companies from about 1937 on.

General Electric (United States) commenced research work upon gas turbines in 1903, and in the following year a number of gas turbine models were operated at Schenectady and Lynn, under the guidance of Elihu Thomson and R. H. Rice. These turbines each had a combustion chamber, a heat exchanger nozzle, and a single-stage impulse-turbine wheel. All of the earlier GE gas turbine work was carried out with the highest operating temperatures for the available metals. Combustion temperatures were, accordingly, reduced by injecting excess air or water, but the resulting efficiencies were relatively low. Somewhat later (about 1907) research work on the gas turbine was postponed, but the development of the centrifugal air compressor was actively pursued. With the successful application of the gas turbine in other fields, notably in aircraft and in electric power generation for marine and locomotive uses, much thought and attention have been given to the possibility of employing small gas turbines for commercial vehicles and passenger cars.

Among the companies with sizable investments in automotive gas turbine development are Chrysler, Ford, General Motors, American Motors (joint development with an independent firm), Caterpillar Tractor Co., International Harvester Co., Rover Co. (now a division of Leyland Motors), Daimler-Benz, Citroen, Fiat, and a joint project of six Japanese automobile manufacturers. In addition, a considerable development effort has been expended on small gas turbines for non-automotive applications by such organizations as Boeing, Solar, Garrett Corporation (a subsidiary of Signal Gas), Lycoming Division of AVCO Corp., and Continental.

The work at Chrysler Corporation, one of the earliest in the field began in 1939, but was interrupted during World War II and did not reach a serious level of development until early in the 1950s. In 1953, after an initial design and analytical phase, Chrysler's first gas turbine-powered automobile (a 1954 Plymouth) was test driven on the streets of Detroit. During the succeeding decade, a continuing development effort evolved the gas turbine through several "generations." In 1963, Chrysler made 50 experimental turbine cars using 130-hp free turbine regenerative units for the purpose of gathering data on performance and public reaction. These vehicles were loaned to various randomly selected qualified drivers for periods of three months each. In all, during the ensuing 30-month program, 203 people in 48 states and the District of Columbia used these vehicles. User reactions were favorable. However, it was felt that further improvement was required in low-speed fuel consumption, noise level, acceleration lag, and engine-braking characteristics, although behavior in all these areas was at least tolerable.[1] The company is now at work on a "sixth-generation" turbine, but neither its status nor any plan for its application has been discussed recently in any public forum.

The Rover Company, Ltd., a British automobile producer, also began to explore gas turbines after World War II, with a view to their eventual automotive use. Their first engine, a 100-bhp device, was tested as a plain jet in 1947, but a practical turbine for automotive use was not tested until 1949, when the Twin-Shaft T.8 engine was actually installed in a mobile test bed. This was, in fact, the world's first gas-turbine–powered car. In a series of tests conducted by the Royal Automobile Club, it accelerated from 0 to 60 mph in 14 seconds, and reached speeds exceeding 85 mph.

Rover continued its development efforts, and in 1953, set up a subsidiary company—Rover Gas Turbines—to manufacture and sell gas turbines. Rover's 1S/60 gas turbine (60 hp) has been used in many applications—for example, as an auxiliary airborne power plant in the RAF Vulcan Bomber.

Later gas turbine models were installed in various automobiles to gather data on such applications. For instance, in 1956, a 2S/100 (100 hp) turbine was installed in a special fiberglass sports car called the "T.3," which would accelerate from 0 to 60 mph in 10.5 seconds, displaying a fuel consumption of 14 mpg at 60 mph. Ten years later, this vehicle was still in everyday use on the British public roads.

The current 2S/140 gas turbine, a twin-shaft 140-hp engine, has been installed in the "T.4" automobile, a four- or five-seat car designated by Rover as a gas turbine prototype. Eventual production was the objective, but the project has been de-emphasized in recent years.[2]

---

[1] Statement of George Huebner, Jr., "Economic Concentration," Hearings before the Senate Subcommittee on the Judiciary, 90 Cong., 2 sess., 1968, pp. 2757 ff. J. A. Hoess et al., "Study of Unconventional Thermal, Mechanical, and Nuclear Low-Pollution Potential Power Sources for Urban Vehicles," Summary Report to National Air Pollution Control Administration, Battelle Memorial Institute, Columbus Laboratories, March 15, 1968.

[2] Noel Penny, "Rover Case History of Small Gas Turbines," Paper No. 634A, *SAE Transactions*, Vol. 72 (1964).

As might be expected, the other major automotive companies have also been working on the gas turbine for some time. The General Motors GT-309 unit, for example, the latest in a series which began development in the early 1950s, is a 280-hp regenerative system designed for installation in trucks and other commercial vehicles. A unique variable-slip coupling between the gas generator and power-turbine sections of this engine, provides both improved part-load fuel economy and engine-braking characteristics.[3]

Field tests by GM indicate that at a speed of about 65 mph the turbine delivers approximately the same fuel mileage as an equivalent diesel engine, but at low speeds the diesel is more efficient. The turbine's major advantage seems to be its compactness and light weight (which results in increased payload) and reliability. Production versions of the GT-309 are reportedly in the works.

Ford's gas turbine effort is also directed largely at developing a unit for trucks and commercial vehicles. The 375-hp Model 707, a low-pressure regenerative design with a variable geometry nozzle, has been shipped to various industrial concerns for the purpose of developing applications. A somewhat smaller second-generation version (200-335 hp and 1,000-1,200 lb), will eventually be produced in volume.[4] Although the main area of application is expected to be in heavy-duty trucks, other possible applications are buses, generator sets, "total energy" systems, marine equipment, and so on.[5]

The general status of a number of turbine entries of recent years is summarized in table 7-1.

Most of the development work on the gas turbine has been done with goals other than automotive power in mind. Since 1950, serious efforts have been made to adapt the technology for passenger vehicle use, but as yet there is no commitment to even limited production for this purpose. Of the major automobile manufacturers involved, none now seems to be aiming at passenger vehicles; both Ford and GM, in particular, are bending their efforts almost exclusively toward the truck and bus market. They are thus competing more with the diesel engine than with the spark ignition (gasoline) engine.

In some respects, the gas turbine appears to be an ideal vehicular power plant. It is long-lived and relatively lightweight, produces a smooth and vibrationless rotary output, and is apparently low in pollutant emissions. There are problems, however, and their solutions add significantly to the complexity and cost of the gas turbine.

In the application of the gas turbine to automotive vehicles, it is impractical to drive the compressor directly from the turbine shaft because the torque from the single turbine is very low at low shaft speeds, which is precisely when high

---

[3] Hoess et al., "Study of Unconventional . . . Power Sources."

[4] "Ford Selling Turbines Now . . . for '71 Delivery," *Gas Turbine International*, July-August 1970.

[5] Huebner, "Economic Concentration."

Table 7-1. Vehicular Gas–Turbine Parameters

Characteristic	Chrysler (A-831)	Ford (704)	GM (GT-305)	Rover (2S/140)	Volvo	Fiat
Rated output, (bhp)	130	300	225	150	250	200
Compressor ratio	4:1	4:1	3.5:1	3.9:1	4.25:1	4.5:1
Compressor efficiency at full load (%)	80	80	78	79	82	n.a.
Compressor-turbine efficiency (%)	87	83	84	86	86	n.a.
2nd stage turbine efficiency (%)	84	86	81	86	86	n.a.
Regenerator or heat exchanger effectiveness (%)	90	75	86	78	85	none
Compressor turbine speed (max rpm)	44,600	91,500	33,000	65,000	43,000	29,000
2nd stage turbine speed (rpm)	45,700	37,500	27,000	36,000	43,000	22,000
Inlet temp. (°F)	1,707	1,697	1,597	1,538	1,562	1,472
Airflow (lb/sec)	2.2	2.71	3.5	2.15	3.01	n.a.
Specific fuel consumption (lb/hp-hr)	0.51	0.566	0.535	0.55	0.401	0.948
Weight (lb)	397	651	596	470	805	574
Specific weight (lb/hp)	3.08	2.19	2.68	3.17	3.26	2.90

Source: R. U. Ayres and Roy Renner, "Automotive Emission Control: Alternatives to the Internal Combustion Engine," Paper given at Fifth Technical Meeting, Air Pollution Control Association, West Coast Section, San Francisco, October 8–9, 1970.

n.a. = not available.

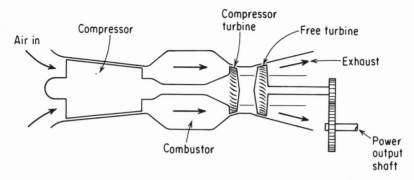

Figure 7-1. Schematic of split-shaft "free" turbine.

torque is most desired. A possible solution is to install a second turbine rotor on a "split-shaft" and have it driven by the exhaust gases from the compressor turbine (see figure 7-1). The first turbine runs at top speed and simply acts as a fan; it creates a high-speed gas stream to drive the second turbine; this so-called free turbine provides the output shaft power, which may be geared as necessary. When a simple turbine slows down, so also does its compressor; the air-fuel charge decreases and therefore the torque drops also. With the free turbine, however, this does not happen. Even when the drive shaft stops, the compressor turbine shaft is still spinning full tilt, hence the maximum exhaust stream is hitting the free turbine blades. In fact, since the effective force acting on the blades decreases with the rotational speed, the torque must be maximum when the free turbine is standing still. The dual arrangement thus solves a number of problems:

1. Both turbine units can operate at their respective optimum speeds.
2. The system can be started more easily, since the starter has only to turn over the compressor-turbine unit.
3. Output torque-speed characteristics are good for many automotive applications, particularly for vehicular propulsion as can be seen in figure 7-2, which compares output torques from the free turbine, the simple gas turbine, and the reciprocating-type internal combustion engine.[6] In this respect, the free turbine is comparable with a steam engine or a series-wound electric motor (see chapter 10).
4. The lack of any mechanical coupling between power turbine and compressor allows more rapid acceleration of the free turbine unit to its maximum speed.
5. Output shaft speed can be readily varied while maintaining a constant compressor speed.

---

[6] B. G. A. Skrotski, *Basic Thermodynamics* (New York: McGraw-Hill, 1963).

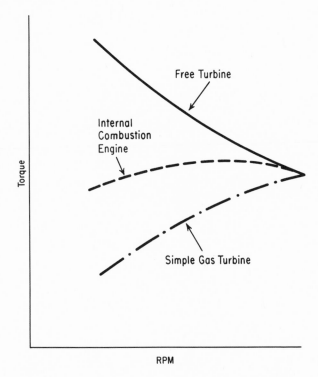

Figure 7-2. Torque characteristics of simple and free turbines.

Disadvantages remain, however. For an automotive application requiring many starts and stops and many partial loads, the engine is often running at top speed even when the power is not needed. Efficiency thus drops sharply. Obviously there are many possible design variations in the various turbine system components. Compressors may utilize either an axial flow of gas or a centrifugal one, for example; the turbine rotor itself may also operate on an axial flow or a radial inward flow. In general, a centrifugal flow pattern is preferred in smaller gas turbines, because of cheaper construction, greater structural ruggedness, and less liability to damage from dust or foreign matter in the air flow. Axial flow systems, however, are usually more efficient and result in lighter and somewhat smaller diameter engines.

Possibly the greatest disadvantage of the gas turbine—simple or compound—is its extremely low efficiency under zero load or under partial loads, which constitute a considerable fraction of actual usage in any vehicular application. The free turbine, which does produce maximum torque at zero speed, still provides relatively low power at low speeds; the power available for compressor acceleration is then low, yet specific fuel consumption is high. This problem can be reduced to some degree by adding a transmission or torque-corrector so that the compressor (input) fan and turbine (output) fan are interconnected by means of a differential gear (see figure 7-3). The compressor shaft is connected to a central sun gear, the turbine to an annular or ring gear, and the power output shaft to the intermediate planetary gears. In this arrangement, the compressor and the turbine rotate in opposite directions.

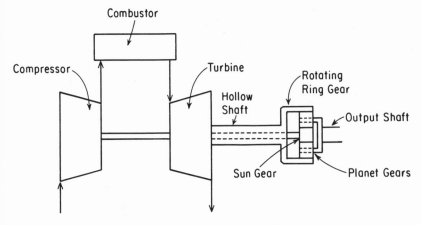

Figure 7-3. Schematic of differential turbine.

When the output shaft of such a differential turbine system is held stationary, the planetary gears function as idlers. The turbine then drives the compressor (on the inner concentric shaft) at maximum speed, the turbine itself (on the hollow outer shaft) rotating in the opposite direction at about one-half the compressor speed. If the turbine speed is increased to the compressor speed, the planetary gears (or roller bearings)—whose centers move around the central axis—drive the output shaft in the direction of the turbine at perhaps one-third of its speed. The net effect is to maintain the compressor speed constant while varying the turbine speed over a range from zero to several thousand rpm. Because of the differential gear, half of the turbine torque is exerted on the compressor and half on the output shaft. As output speed is reduced to zero, a progressively larger portion of the turbine's power—up to 100%— is transmitted to the compressor. The differential gas turbine can provide up to a 65% reduction in fuel flow with decreasing output speed while still providing maximum torque on the output shaft.[7] Typical turbine "machine efficiency" might vary from 84% at optimum speed to 77% at zero speed, while for an ordinary free (split-cycle) turbine of equal power, efficiency would have dropped to zero at stall.

The differential principle can also be applied with the free power-turbine attached to the rotating ring gear via a hollow outer shaft and the inner

---

[7] Arthur W. Judge, *Small Gas Turbines* (New York: Macmillan Co., 1960).

compressor-turbine shaft fixed to the sun gear. Furthermore, by a suitable blow-off valve inserted in the flow between the "captive" and "free" turbines, a negative torque can be provided for vehicular braking purposes. Opening this valve reduces the back pressure (i.e., load) of the captive turbine, increasing its torque and also decreasing the torque necessary to drive the compressor. The result is a negative torque at zero output shaft speed. The free turbine slows down to about one-third speed in this condition, which tends to drive the output shaft in the reverse direction, which, in turn, further decreases the speed of the free turbine. When the free turbine speed reaches zero, the compressor turbine exerts large negative output torques and reverse output speeds.

The gas turbine has many attractive features as a portable heat engine. It is mechanically relatively simple, as compared with the reciprocating engine, for example, and it exhibits a high order of mechanical efficiency (as high as 96% to 98%). The specific weight of small gas turbines ranges from about 0.6 lb/hp to 3–4 lb/hp (including gearing and accessories but excluding transmission); the larger figures generally correspond to the smallest practical power levels for turbines.

Turbines can operate satisfactorily on many fuels—including kerosene, diesel oil, coal-gas, natural gas, methane, and others. Since the usual gas turbine compressor supplies three or four times the amount of air theoretically required for complete combustion of the fuel, the exhaust gases are relatively free from smoke and from carbon monoxide. (See table 7–2, which lists the volumetric emissions from the Rover 2S/140 gas turbine (140 bhp) for various load conditions.)

In summary, the gas turbine is an efficient lightweight power plant which can be (and has been) successfully applied to vehicles, although its performance tends to suffer somewhat at low speed and partial loads, which, of course, occur frequently in the automotive driving cycle. It is currently manufactured in relatively small quantities (i.e., a few thousand) for aircraft, power generators, and other specialized applications. Thus there is insufficient information available to evaluate the probable cost of such devices in mass production. At least one major automotive executive is on record with an assertion—probably no more than a pious hope—that vehicular-sized turbines (circa 150–200 hp) might be made in large numbers for roughly the same cost as internal combustion engines (probably $2–$3 per hp) and in sizes as small as 30 hp for $3.00–$4.50 per hp, but this is still very much open to question.[8]

The question of minimum size itself is still unresolved. The lowest-powered turbine listed in table 7–1 is the Chrysler engine, rated at 130 hp. However, the Williams Research Corporation has developed an efficient regenerative turbine of 70 hp, which was installed in a boat, and it has also operated a 30-hp device,

---

[8] Hoess et al., "Study of Unconventional . . . Power Sources."

Table 7-2. Emissions from the Rover 2S/140 Gas Turbine

	Zero load (ppm)	1/3 load (ppm)	2/3 load (ppm)	Full load (ppm)	Average over variable load (ppm)	Total emissions (lb per lb of fuel)
CO	310	440	260	255	340	0.02
NO_x	5	6	8	5	5.5	0.0003
SO_2	40	80	120	140	60	0.002
Formaldehyde	11	12.5	7.5	1.5	11	n.a.

Source: Noel Penny, "Rover Case History of Small Gas Turbines," Paper No. 634A, *SAE Transactions*, Vol. 72, 1964.

Note: Conditions: 63 lb air/lb of fuel burned. Fuel consumption under full load = 0.5 lb/bhp-hr. Fuel consumption under variable load = 0.7 lb/bhp-hr.

although at some loss in efficiency. It is probable that 50 hp to 100 hp represents a practical minimum size range for gas turbines.

The advantage of the gas turbine in such areas as low exhaust emissions and low specific weight is not quite so clear-cut as it once seemed to be. Even with the much greater mass flows in the turbine (4 to 10 times those of the ICE), the hydrocarbon and CO content of a turbine exhaust is only about 4% to 5% of a typical ICE exhaust, but the picture is somewhat clouded for emissions of nitrogen oxides. Data from General Motors, for example, show that their GT-309 produces approximately 1.75 times the total mass of NO_x produced by a comparable piston engine. Conversely, both Ford and Chrysler claim that NO_x emission levels from present turbines are below projected federal standards.[9]

In addition, the specific weight of automotive gas turbines is not likely to improve to any significant degree during the next decade. Any improvement made possible by new technology will probably be offset by the use of heavier but less costly materials in an effort toward cost-reduction—perhaps the most desired area for improvement. A specific weight of 3 lb/hp including air cleaner has been projected for a turbine in the 100-200 hp class and 4 lb/hp for a 30-50 hp machine, which is not startlingly better than today's values.[10]

It has been estimated that the costs of small gas turbines of 100 hp or less would range from $3.00-$4.50 per hp if they were produced in large quantities for passenger-car use. High-performance, longer-lived turbines for truck or bus applications would probably be produced in smaller quantities and cost considerably more.[11]

---

[9]Ibid.

[10]Ibid.

[11]Ibid.; and Robert Kirk and David Dawson, "Low-Pollution Engines: Government Perspectives on Unconventional Engines for Vehicles," ASME Paper No. 69-WA/APC-5, 1969.

Chapter Eight

Applications of Rankine-Cycle Engines to Automotive Vehicles

By the first decade of the twentieth century when the automotive revolution had begun in earnest, steam engines, internal combustion engines, and battery-powered electrics were in neck and neck competition. The best-remembered early steamers—Locomobile, White, and Stanley—were generally superior to their ICE rivals in speed, acceleration, and smoothness, but they had severe drawbacks. The early boilers were potentially dangerous; and they were heavy and slow to develop a suitable head of steam. Also, they consumed large amounts of water, which imposed severe limits on range: the Stanley, with a 30-hp engine, consumed 25 lb of water per hp-hr; yet it was not until after 1914 that the Stanley was equipped with a condenser for recycling. Meanwhile, the ICE was becoming more and more reliable and convenient, especially after the invention of the self-starter by Kettering, and, perhaps more important, it was being mass produced. Even the early ICE auto plants produced as many cars in a day as a typical steam manufacturer made in a year; in 1906, the heyday of steam, the maximum output of Stanley Bros. was 650 cars per year.

By the 1920s steam was virtually confined to railroads, except for the Doble, which is a legend among classic car (as well as steam car) enthusiasts.[1]

The Doble featured a flash-boiler and a completely closed system, could reach full steam pressure in less than 60 seconds, and could smoothly and quickly accelerate to a top speed of 90 miles an hour. It cost around $10,000, however, and only a few dozen cars were built over an eight-year period. When Doble went bankrupt in the early 1930s its assets were acquired by the Besler Corporation of Emeryville, California.

The efforts of Besler and other developers are examined here at some length before we consider the steam engine in principle.

---

[1] G. Borgeson and E. Jaderquist, *Sports and Classic Cars* (Englewood Cliffs, N.J.: Prentice-Hall, 1955).

DEVELOPMENT HISTORY

Besler

In April 1933, a Travelair airplane powered by a Besler steam system was test flown. For its horsepower, this condensing power plant is believed to be the lightest ever built.[2] One feature of the aircraft was its remarkable silence; the pilot was able to converse in shouts with observers on the ground. A reciprocating compound-expansion engine was employed, developing 150 horsepower at 1,650 rpm. Water rates as low as 10 lb per horsepower-hour were demonstrated.

High-pressure steam was also applied to a two-car commuter train, operated by the New Haven Railroad. This train was operated for almost a million miles over a seven-year period during the late 1930s and early 1940s. Utilizing steam at 1,500 psig and 750°F, the power plant delivered 700 horsepower. Because this was a condensing system, the train could be run about 500 miles before it was necessary to replenish the water supply. Overall thermal efficiency at the wheels was on the order of 20%, which was two-to-four times that of conventional steam locomotives of that period. While fuel consumption was higher than for diesel-electric equipment, the space saved with the steam system was the equivalent of twenty additional passenger seats over the space required by a diesel engine of the same power. Thus, even at two cents per passenger-mile, the additional revenue potential would have been more than sufficient to pay the entire fuel bill.

Most, if not all, of Besler's steam generators have been of the once-through (monotube) type. Packaged, automatically controlled steam generators have been marketed principally for industrial use; over 3,000 units were delivered to the U.S. Navy alone. Besler steam generators have been employed for conditions up to 3,000 psig and 1,200°F. Heat release rates up to 3 million Btu per hour per cubic foot of combustion space have been demonstrated.

Other noteworthy applications of Besler systems have included a "motor" launch for the U.S. Navy[3] and a steam-converted Chevrolet automobile for General Motors (1969).[4]

Sentinel Truck

During the early years of the twentieth century, several dozen English manufacturers produced steam-driven commercial vehicles. Most of these trucks and buses were fueled by coke or coal, an important advantage in the British Isles.

---

[2]The complete 1933 power system, less fuel and water, weighed 4.5 lb per horsepower. Subsequent work on a very similar system by Besler in 1956 reduced this even further.

[3]"Steam Tries a Comeback," *Machine Design*, March 17, 1960, pp. 26-27.

[4]"The GM SE-124 Steam Car with Besler Engine," Press Release, May 7, 1969, General Motors Progress of Power Show, Warren, Mich.

Rising labor costs and discriminatory vehicle taxation were factors causing the demise of steamers in that region.

One of the better-known makes was the Sentinel, produced in Shrewsbury from 1906 to 1952. The last production series was the Model S, introduced around 1930. The Model S was mounted on pneumatic tires and was shaft-driven from a four-cylinder single-acting engine. Steam was supplied from a manually fired water-tube boiler at 255 psig and up to 800°F. Up to 120 horsepower could be drawn for overload periods of short duration. Water rates of 14 lb/hp-hr were quoted.[5] No condenser was used, and water was consumed at the rate of over 4 gallons per mile at full load. The Sentinel Model S was built in two, three, and four axle configurations. With a trailer, gross combined weights were rated at up to 32 tons.[6]

During the mid-1930s, Abner Doble was retained as a consultant by Sentinel. Under Doble's direction, a prototype truck with a condensing, compound-expansion power plant was developed. Based on this work, it was concluded that truck power plants could be built to travel 45 to 50 miles per long hundred-weight (112 lb) of coke, instead of the 18 to 25 miles per hundredweight expected from a noncondensing, simple-engined truck of 14 tons gross weight.[7]

Henschel Bus

Shortages of high-grade motor fuels in Germany led to experiments with steam-powered commercial vehicles, and from 1932 to 1934 Warren Doble was retained as a consulting engineer by the Henschel Works at Kassel. Doble steam power units were installed in buses, trucks, railcars, and motor launches during the mid-1930s. High torque characteristics at the rear axle provided smooth, rapid acceleration. Another advantage for a bus was the ease of operation; the absence of a transmission eliminated the need for approximately 4,000 gear changes per day in city traffic.[8] Engines up to 150 hp were fitted, and condensers were used on all road and rail applications. While Doble cars produced in America in the 1920s burned gasoline or kerosene, fuel oils derived from coal were preferred in these German vehicles.

With the approach of World War II, steam vehicle developments were suspended. Henschel did, however, continue to supply Doble-type automatic steam generators for industrial applications until at least 1952.

McCulloch "Paxton" Engine

The McCulloch Corporation of Los Angeles, which is well-known for its successful line of chain saws and lightweight gasoline engines, had an active interest

---

[5] Ronald H. Clark, *The Development of the English Steam Wagon* (Norwich: Goose and Son, 1963).

[6] R. D. Cater, "Sentinel S4T Steamer," *Commercial Motor*, December 20, 1968.

[7] J. N. Walton, "Light Steam Power," *Doble Steam Cars, Buses, Lorries, and Railcars*, Isle of Man, 1965.

[8] Ibid.

in steam automotive development from 1951 until 1954. A superior engine was being sought for their proposed Paxton automobile, and steam seemed to fulfill the requirements for quietness and good performance.

Doble principles were clearly evident in the one prototype chassis developed. Abner Doble had been retained as advisor for this project; several of his innovations were the result of twenty years' reflection since the demise of the Doble car. The Doble-type monotube steam generator (mounted in front, under the hood) was smaller, lighter, and more efficient than earlier designs. Maximum capacity of the generator was 900 lb of steam per hour; the operating pressure was 2,000 psi—the highest pressure ever used in an automobile, so far as is known.

A rear-mounted engine drove the independently suspended wheels at a fixed gear ratio of 1.6 : 1. Utilizing compound expansion, each of three high-pressure cylinders was mounted in tandem with a low-pressure cylinder—a total of six cylinders. A maximum expansion ratio of 24 : 1 (higher than direct pressure) resulted in a very high overall thermal efficiency, up to 23%.

The total weight of the power system was 953 pounds—about the same as the present-day ICE system including the necessary starter, radiator, transmission, mufflers, etc. Power output was 120 bhp (continuous), and 150 bhp (intermittent maximum).[9] The engine was tested in a converted Ford V-8, which recorded 32 miles per gallon of fuel.

A design objective was to operate with a steam temperature of 1,200°F; this was later dropped to 900°F to avoid the decomposition of cylinder lubricants.

Development was discontinued in 1954 (for nontechnical reasons), and the power system was sold to the U.S. Navy for their further evaluation.

Yuba Steam Tractor

Immediately following World War II, there was an increasing demand for large tractors suitable for earthmoving, logging, and other off-road activities. Sensing this growing market, the Yuba Manufacturing Company (San Francisco) developed a steam-powered, pneumatic-tired, all-wheel-drive prime mover during the years from 1946 to 1951. While production plans did not materialize, the prototype tractor represented a significant advance in this branch of engineering.

The use of an internal combustion engine was initially considered for this prime mover. After a study of alternative transmissions (mechanical, hydraulic, electrical), the decision was to bypass the problems of a complex transmission by installing steam power instead.

In the steam-driven prototype, high maneuverability was achieved by independently driving each of the four wheels by its own steam engine. Since the front wheels could be cramped 90° either side of center, an extremely short turning radius was attainable. Two-speed gearing in the final drive of each engine

---

[9] J. L. Dooley and A. F. Bell, "Description of a Modern Automotive Steam Powerplant," SAE Paper S338, January 22, 1962.

allowed a choice of two modes of operation: a low-speed with a drawbar pull up to 32,000 lb, and highway travel with speeds to 40 mph. The tractor had a rating of 200 drawbar horsepower.

Steam was generated in a monotube boiler, and delivered at 1,500 psig and 900°F. Power plant and vehicle auxiliaries were driven by a small gasoline engine, at constant speed. Each steam engine was single-acting, four-cylinder, opposed-piston, with cam-operated poppet valves. Departing from conventional steam engine practice, lubrication of the cylinder walls was by splash from the crankcase (rather than by injecting oil into the feed steam). It was claimed that this method avoided oil contamination of the condenser and steam generator. Driver's controls were so simple that the wives of the development engineers were able to operate the tractor with minimal instruction. Based on the successful prototype, Yuba conducted subsequent design studies on steam power for military tanks.[10]

The Keen Steam Car

The late Charles F. Keen, of Madison, Wisconsin, was the fifth generation of the Keen family to be involved in steam engine development. His last steam car, developed during the 1950s, was a sports roadster known as the Keen Steamliner.

As have many other developers, Keen adopted the Doble monotube design for his steam generator. Located under the hood, this generator was fired by a pressure-atomizing burner with an electrically driven forced-draft blower. Automatic steam generator controls were also of the Doble variety, sensing both pressure and temperature. Maximum steam pressures of 1,000–1,200 psi were used. Control of steam pressure was by automatic on-off switching of the fire and the basic feedwater circuit. Maximum steam temperatures of 750–800°F were controlled by a thermostatic sensor and "normalizer" arrangement, in which water from a secondary feed pump was sprayed into the superheater section of the boiler.

Keen's engine was a V-4, single-acting engine mounted in the rear compartment of the car. Bore and stroke were 3¼" X 3'½". Uniflow expansion was used, with cutoff positions adjustable at 75%, 28%, and 18%. An automatically variable compression ratio was employed to avoid rough running at low speeds. No clutch or transmission was used; the final drive ratio was 1.25 engine revolutions per axle revolution. Upper cylinder lubrication was by injection of oil into the steam line. A "dry-sump" system lubricated the crankshaft; blow-by water was separated from the oil by centrifugal force and by a suction blower. Nonemulsifying oils were specified.

A road test of this car in 1968 showed it to be capable of steady speeds of 60–70 mph. Fuel consumption (kerosene) was said to be 12 miles per gallon. In

---

[10]R. L. Harris, R. E. Hulbert, and M. Lathrop, "Steam Power Package for Military Vehicles," Yuba Manufacturing Co., San Francisco, Calif., 1963.

late 1967, this car was sold to the Thermal Kinetics Corporation, of Rochester, New York.

Thermo Electron (TECo)

Thermo Electron Corporation (TECo) of Waltham, Massachusetts, is a small science-based company specializing in research and development on advanced energy conversion systems (notably thermionics). The company has also entered the steam power field with a completely new, hermetically sealed, closed-cycle, 4-cylinder (in-line), single-acting, uniflow engine operating at high pressures and temperatures. Small versions have been built for the U.S. Army for operating silent electric power generators in remote areas (e.g., for radio transmitters).[11] Details are shown in table 8-1. The principal engineering advancement in these

Table 8-1. Summary of Characteristics for Three TECo Steam Engines

	(1)	(2)	(3)
Rated shaft power (hp)	1/4	2/3	3
Shaft speed (rpm)	1,800	1,800	3,600
Bore (in)	1.063	1.315	1.315
Stroke (in)	0.500	0.625	0.625
Number of cylinders	1	1	3
Total displacement (cu in)	0.445	0.84	2.52
Type of lubrication	Splash	Splash	Forced
Design inlet pressure (psia)	300	700	700
Design inlet temperature (°F)	800	800	850
Weight (lb)	4.6	6.1	17
Overall efficiency (%)	50-60	55-65	60-70
Longest uninterrupted period of operation (hr)	300	600	600
Total accumulated running time (hr)	1,500	900	3,000

Source: S. S. Kitrilakis and E. F. Doyle, "The Development of Portable Reciprocating Engine Rankine Cycle Generating Sets," SAE Paper No. 690046, Automotive Engineering Congress and Exposition, Annual Meeting, Detroit, January 1969.

engines is a complete separation of the steam-side from the lubrication side, and a successful upgrading of operating temperature to 900°F and 600 psia. Both working fluid and lubricants are sealed in for the life of the engine. Valves are actuated internally by a new (patented) technique. Exclusion of contaminants and noncondensibles results in improved heat transfer in the steam generator and improved condenser performance, as compared with prior technology. A prototype simplified 5-hp engine using an organic working fluid (thiophene, with the chemical formula C_4H_4S) is currently being tested.[12] In April 1968, Thermo

---

[11] S. S. Kitrilakis and E. F. Doyle, "The Development of Portable Reciprocating Engine Rankine Cycle Generating Sets," SAE Paper No. 690046, Automotive Engineering Congress and Exposition, Annual Meeting, Detroit, January 1969.

[12] D. T. Morgan, E. F. Doyle, and S. S. Kitrilakis, "Organic Rankine Cycle with Reciprocating Engine," Intersociety Energy Conversion Engineering Conference, Washington, D.C., September 1969.

Electron and Ford Motor Company jointly embarked on a development program to explore commercial applications, possibly including vehicles. Ford also acquired a substantial block of TECo stock.

In July 1969, Thermo Electron was chosen by the National Air Pollution Control Administration for a design study of a Rankine-cycle propulsion system for an automobile. Parameters determined from this study are given in table 8-2. In May 1970, TECo signed an agreement with the Ford Motor Company, embarking on a five-year, $6-million, joint-development project with Ford.

Williams Brothers

The Williams family (Calvin C. Williams, Sr., Calvin, Jr., and Charles) of Ambler, Pennsylvania, spans two eras. The elder Williams began working on steam development in the 1920s; he was joined by his two sons in 1937. Their "final" design was patented in 1946,[13] although important refinements were added later. The basis for the 1946 patent was a recompression feature called the Williams Cycle to which the Williams attribute their engine's impressive performance, although its value is generally doubted by engineers.

The basic Williams design is a 4-cylinder (in-line), single-acting, uniflow steam engine, which has been built in several size ranges. Pertinent data are given in table 8-2. The present version incorporates several engineering innovations in addition to the controversial recompression feature. One is an automatic "bleeder" valve which allows condensed liquid to escape from the cylinder and thus permits the engine to operate 'cold' on water pressure alone if necessary.[14] In other words, it is not necessary to warm up the cylinder walls by running steam through the engine block for several minutes before the engine will run, as is the case with most older designs. The second major innovation is a second valve, built into the engine, which automatically matches engine compression to feed-steam pressure.[15] Without such a mechanism a uniflow steam engine cannot operate smoothly under rapid variations in load conditions: the engine tends to buck or stall whenever the engine compression exceeds the inlet pressure.[16]

A Williams engine with 25,000 miles on it, tested by the Mobil Oil Co. over the so-called California driving cycle, emitted 0.05% (560 ppm) carbon monoxide, 20 ppm of unburned hydrocarbons, and 70 ppm of nitrogen oxides.[17] It is claimed that newer models will reduce these figures by 50% or more. Even so, these figures are far below the 1970 California maximum emission standards (1.0% CO, 180 ppm HC).

---

[13] C. C. Williams, U.S. Patent No. 2,402,699 (June 1946).

[14] C. C. Williams, U.S. Patent No. 2,513,982 (1950).

[15] C. C. Williams, U.S. Patent No. 2,943,608 (1960).

[16] This problem particularly vexed Abner Doble, who actually attributed his company's failure to not having solved it.

[17] *Road and Track*, September 1, 1967.

Table 8-2. Modern Reciprocating Steam or Vapor Engines

Characteristic	Keen		Williams		McCulloch	TECo	Pritchard
Piston displacement (cu in)	100	56	105	262	23-116[a]	184	17.6[b]
Cylinders	4	4	4	4	6	4	2
Inlet pressure (psi)	850-1,200	1,000	1,000	1,000	2,000	500	1,000
Inlet temperature (°F)	750-850	980	1,000	1,000	900	550	870
Engine weight (lb)	285	200	300	500	285	265	125
Total power plant wt. (lb)	800	650	800	1,000	953	957	450
Rated power (bhp)	55	63	150	199	120	103	35-40
Rated rpm	1,700	2,500	3,300	3,000	1,200	2,000	1,400
Overload capability (%)	50	100	100	100	20		20-80
Stall-torque (lb-ft)	500	500	1,105	2,734	900	800	331
Maximum thermal efficiency (%)[c]	17-19	26	26	26	23	18.5	
Starting time (sec)	60	30-60	30-60	30-60	30	30-45	60-80
Vapor/bhp-hr (lb)		7.8	7.8	7.8	9.0	71.5	
Earliest normal cutoff (%)[d]	18	10	10	10	10	14	14

Source: R. U. Ayres and Roy Renner, "Automotive Emission Control: Alternatives to the Internal Combustion Engine," Paper presented to the Fifth Technical Meeting, West Coast Section, Air Pollution Control Association, San Francisco, October 8-9, 1970. McCulloch data from J. N. Walton, "Light Steam Power," *Doble Steam Cars, Buses, Lorries, and Railcars,* Isle of Man, 1965. TECo data from D. T. Morgan and R. J. Raymond, Conceptual Design. Rankine-Cycle Power System with Organic Working Fluid and Reciprocal Engine for Passenger Vehicles, Report No. TE4121-133-70, Thermo Electron Corp., NAPCA, June 1970.

Note: All use steam as the working fluid except TECo, which is designed to use thiophene. TECo system is a design study only.

[a]Double-acting compound. Displacements are for high- and low-pressure cylinders.

[b]Double-acting.

[c]Engine only, without auxiliary loads such as fan or feed pump. Efficiency figures given by Williams conflict with the more conservative efficiencies calculated from Williams thermodynamic data. The latter figures are given here.

[d]Cutoff (hence expansion ratio) is variable. A maximum of 60% to 80% is typically used for starting under load.

Pritchard

Pritchard Steam Power Proprietary Ltd. of Melbourne, Australia, is one of the few groups working on steam engines outside the United States. In the 1950s and early 1960s the Pritchards (father and son) designed and built an engine for a 5-ton truck.[18] More recently the Pritchards designed and built a small steam engine, which was installed for road testing purposes in a 1963 Ford Falcon. The engine generates up to 80 hp in a very compact, double-acting V-2 configuration with a displacement of only 17.6 cubic inches (see table 8-2).[19] The steam generator and burner is quite advanced, being only 15″ in diameter by 16½″ high. Both engine and boiler incorporate a number of novel features which greatly improve their heat transfer characteristics.

Other Developers

Steam Engine Systems Corporation, formerly Energy Systems, Inc., is a new firm, formed early in 1968. The firm proposed to develop auxiliary equipment to permit the adaptation of existing 2-cycle Otto-cycle or diesel-cycle engines. The benefits of such a conversion are illustrated by figure 8-1, which compares the power and torque characteristics of a GM 6V-71 diesel engine with those of a hypothetical steam conversion of the same basic unit, assuming inlet steam pressure of 100 psi and temperature of 1,000°F.[20]

Lear Motors Corporation of Reno, Nevada, is another new venture launched in 1968 with considerable fanfare. By mid-1969 plans for a reciprocating engine had been dropped and the company was concentrating its efforts on developing a closed-cycle vapor turbine. However, in June 1970, Lear contracted to build and install a steam engine using a novel screw-type expander for a bus to be tested in the San Francisco Bay area. More recently, Lear has returned to the vapor turbine concept, with which he claims to have achieved promising results. Two other bus engines are being built by firms not previously engaged in steam engine development: Wm. A. Brobeck Associates of Berkeley, California and Steam Power Systems, Inc., of San Diego, California. Both of these engines will have fairly conventional reciprocating designs.

Less conventional designs also show some promise. The "Elliptocline" engine, invented by Thomas Hosick and R. A. Gibbs of Winston-Salem and Greensboro, North Carolina, is a hybrid barrel-type, rotary-reciprocating, single-acting uni-

---

[18] *Truck and Bus Transportation* (Sydney), August 1963.

[19] E. Pritchard, "The Steam Engine as a Feasible Alternative to the Petrol Engine," *Automobile Steam Engine and Other External Combustion Engines*, Joint Hearings before the Senate Committee on Commerce and the Subcommittee on Air and Water Pollution of the Senate Committee on Public Works, May 27 and 28, 1968, 90 Cong., 2 sess. (1968). Also, E. Pritchard, personal communication.

[20] Richard S. Morse, Joint Hearings before the Senate Committee on Commerce and the Subcommittee on Air and Water Pollution of the Senate Committee on Public Works, May 28, 1968.

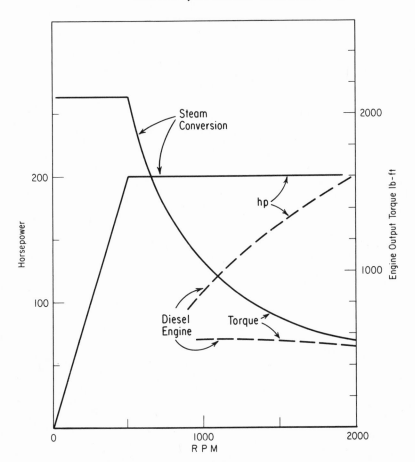

Figure 8-1. Torque-speed characteristics of Rankine cycle (steam) and diesel engines using the same engine block. (Statement of Richard S. Morse, in *Automobile Steam Engine and Other External Combustion Engines*, Joint Hearings before the Senate Committee on Commerce and the Subcommittee on Air and Water Pollution of the Committee on Public Works, 90 Cong., 2d sess., May 27-28, 1968, pp. 92 ff.)

flow steam engine.[21] The action of its pistons is reciprocating, but the output motion of the engine is rotary. A central axial steam input channel is surrounded by 9 parallel steam-piston cylinders arranged like revolver chambers which are secured by slip-rings at both ends to toroidal collars fixed to rotating tilted swash plates mounted as shown in figure 8-2. Steam is admitted from a channel in the central rotor to the cylinder whose pistons are closest to the channel. The resulting expansion of the piston forces the rotor to turn. Eventually, at what-

---

[21] R. A. Gibbs and Thomas A. Hosick, in *The Steam Automobile*, Vol. 7, No. 4, Winter 1965; Thomas A. Hosick, in *The Steam Automobile*, Vol. 8, No. 4, Winter 1966; and S. S. Miner, in *Popular Science*, Vol. 188, February 1966.

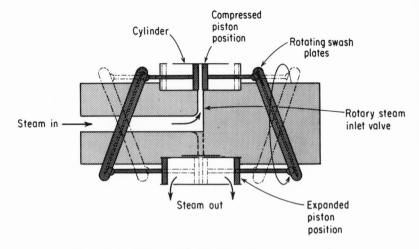

Figure 8-2.   Gibbs and Hosick elliptocline steam expander.

ever angle is desired, the rotary motion automatically cuts off the steam and—at the point of maximum expansion—uncovers the exhaust ports. The continuing eccentric rotary motion then forces the twin pistons back towards each other in position for another steam charge. The engine is inherently balanced and essentially vibration-free. A single valve mechanism services all the cylinders (as contrasted to three valves *per* cylinder for the Williams design). No cams or connecting rods are needed. Because of these simplifications, R. A. Gibbs estimates that manufacturing cost would be substantially less than competitive ICEs, even if the engine is produced in relatively small quantities.

An engine developed by D. E. Johnson of General Steam Corporation (Newport Beach, California) also has a revolver-like arrangement of eight cylinders. Only one piston per cylinder is used, and the piston positions are controlled by a single helical camshaft which serves much the same function as the toroidal "collars" in the Gibbs-Hosick engine. However, the engine block is stationary, unlike the G–H version. Details have not been published, but the design is apparently very similar to the so-called Herman "cam" engine, built and tested (in an internal combustion configuration) during the 1930s.

A vapor engine was built by Douglas Paxton of Ventura, California in 1957, but the early prototypes no longer exist.[22] It relied on a closed, regenerative Rankine-cycle, utilizing an undisclosed organic working fluid, possibly a commercial solvent such as ethylene trichloride. A small 4-cylinder reciprocating model with a displacement of 50 cubic inches was built, with a total system weight of 212 lb. Cutoff was indefinitely variable, controlled by a cam. The engine apparently utilizes a catalytic combustor, the details of which are not

---

[22]*Road Test*, October 1967.

known. Horsepower, torque, and thermal efficiency figures are unverifiable. In 1968 a new company (Paxve, Inc.) was formed to exploit this technology.

Interesting results have also been claimed for a vapor engine designed and built by Wallace Minto of Kinetics, Inc., Sarasota, Florida. The original Sarasota engine is a rebuilt 2-cylinder, double-acting Stanley engine using Freon F-11 as a working medium. (Freon F-11 is the DuPont trade name for an industrial refrigerant with the chemical formula CCl_3F.) This fluid freezes at $-168°F$, boils at $+75°F$ (at 1 atmosphere), and is apparently chemically stable up to about $300°F$. Freon F-11 is also a good lubricant for steel, is nontoxic, and costs about 22 cents per pound in bulk. The maximum theoretical efficiency of a Rankine cycle engine using F-11 as a working fluid between $90°F$ and $300°F$ is 22%. Volume flow per hp-hr is relatively low (compared to steam), which suggests that a more compact condenser design may be feasible. A demonstration Freon engine is also being installed in a Datsun automobile: the Nissan Motor Co. (which makes the Datsun) is reported to be interested in the possibility of producing such a vehicle—presumably depending upon the outcome of these early tests. The major problem with using F-11 (or any of the other Freons) is the low temperature of thermal decomposition; possible decomposition products include hydrochloric acid, hydrofluoric acid, and phosgene gas—all highly toxic and corrosive.

EFFICIENCY AND "WATER RATE"

In chapter 4, the thermodynamics of the Rankine cycle were discussed briefly. In general, it can be stated that efficiency can be improved by (1) raising the *average* temperature at which heat is added to the working fluid, (2) reducing the *average* temperature at which heat is rejected by the fully expanded vapor, and (3) eliminating irreversible losses.

Translated into specific terms, one of the key design objectives is to increase the temperature at which vaporization (boiling) occurs since most of the heat is added during this phase. This involves operating at the highest possible pressures. It is also desirable to superheat the vapor before expansion (in the case of steam). Partial expansion and reheating makes it possible to raise the average temperature at which heat is added without resorting to excessive peak temperatures.

In addition to high feed temperatures, thermal efficiency also benefits from a low exhaust or "back" pressure, which permits a low temperature for the rejection of waste heat. The exhaust temperature must, of course, be above the condensation temperature, but the latter can be cut by reducing the back pressure at the exhaust valve—below one atmosphere if desired. Low back pressures thus have a marked effect on thermal efficiency, although the magnitude of this effect also varies somewhat with the engine inlet (i.e., throttle) pressure. For example, at a constant inlet temperature of $1,000°F$ and a throttle pressure of 200 psi, the theoretical thermal efficiency of a Rankine engine (using steam) can be increased from 23.3% to 36.3% by decreasing the back pressure from 30 inches of mercury absolute (1 atmosphere) to 1 inch (1/30 atm.). With a throttle pressure of 5,000 psi and the same inlet temperature and back pressure drop, the thermal cycle efficiency increases from 37.7% to 46%. Thus, in this case the higher throttle pressure permits only a smaller—although still impressive—relative increase in efficiency. In small vehicle applications, both very high and very low pressures are impractical because there is an unfavorable trade-off between increased thermal efficiency and increased mass of the structures required to withstand pressures above 1,000 psi or so, on the one hand, and the increased condenser volume needed to accommodate the very low pressure (less than 15 psi) steam, on the other. Much the same objections (extra weight and complexity) also tend to argue against the use of reheat or multiple expansion cycles in small sizes despite possible gains in efficiency.

To hold irreversible losses to a minimum, it is important to prevent steam from condensing on the walls of the cylinder during expansion. Apart from mechanical problems such as "water locks," this is crucial because condensation greatly increases the irreversible heat losses during expansion. Dry steam is a relatively poor conductor, but condensation involves the release of large quantities of heat at the cylinder walls, at temperatures and pressures considerably above the condenser temperatures.

A frequently used index of steam engine efficiency is the "water rate," or the quantity of steam required for one hp-hr output. In practice, this will range from 5 or 6 pounds to about 30 pounds of steam per indicated hp-hr, depending on the type of engine and the power size. For unsophisticated low temperature and pressure engines, it will be at the upper end of the range; for large triple-condensing engines, or for extra high inlet temperatures and pressures, it can be below 7 lb/hp-hr. Table 8–3 shows the effects of some of the above artifices on the performance of a typical steam power system.

A more elaborate approach involves the use of the *binary vapor* cycle, in which two vapors condensing at different temperatures (such as mercury and steam) are combined, yielding a high overall system efficiency. The technique is analogous to the multiple-expansion method. In the mercury-steam case, heat is used to produce mercury vapor, which is then expanded in a mercury turbine and condensed in a heat exchanger. The latter serves as the boiler for the steam cycle, the steam receiving energy from the condensing mercury and then being expanded in a steam turbine and finally condensed in a steam condenser. Actual operating thermal efficiencies of 37% have been reported using this cycle in large units.[23] A combination of steam and Freon or some other dense fluid might be more appropriate at lower temperatures (Freon boils at approximately room temperature).

---

[23] C. F. Werner, *Thermodynamics Fundamentals for Engineers* (Ames, Iowa: Littlefield, Adams and Co., 1957).

Table 8-3. Effect of Various Approaches to Increasing Steam Engine Efficiency

Parameter	Saturated steam	Superheat	Vacuum condensation	Further increase feed pressure	Further superheat	Reheat
Feed steam pressure (psi)	300[a]	300	300	2,000	2,000	2,000
Feed steam temperature (°F)	417	750	750	750	1,050	1,050
Back pressure (psi)	14.7	14.7	6	6	6	6
Exhaust moisture content (%)	17	4.1	8.3	25.7	18.8	0
Theoretical cycle efficiency[b] (%)	17.8	23.3	27.3	35.8	38.6	40.7
Engine conversion efficiency (%)	50	70	65	45	50	85
Engine water rate (lb/hp–hr)	23.4	12.8	11.4	13.8	9.6	4.4
Overall thermal efficiency −85% boiler efficiency, 10% steam for auxiliaries (%)	6.8	12.5	13.5	12.3	14.8	26.5

Source: T. A. Hosick, in *Steam Automobile*, Vol. 5, No. 1 (1962).

[a]Psi absolute.

[b]The calculation is quite simple: heat converted to work is the difference between the enthalpy in the "feed" and "exhaust" states, or 337.2 Btu/lb, while total heat consumed is the above plus the enthalpy difference between feedwater (180 Btu/lb) and saturated steam (1,150.4 Btu/lb) or 1,308 Btu/lb. The efficiency is the ratio of the two. The enthalpy difference between exhaust steam and saturated vapor is assumed to be 100% recovered by a regenerator or preheater.

Various forms of binary Rankine cycle systems have been developed for space applications in the past few years using mercury or alkali metals in the mercury Rankine cycle and usually a hydrocarbon in the lower temperature cycle. Thermal efficiencies of 33% have been developed with these systems and overall power/weight ratios of 25 watts/lb. The weight includes the weight of all electrical generating equipment because, as a space power unit, the system was designed to produce electrical energy; it also includes the weight of various components such as radiators, which would normally only be required for a space application.[24] Presumably, as portable terrestrial power plants, higher power/weight ratios would be possible, but, in general, extra weight is one of the penalties to be paid for higher efficiency.

OPERATIONAL CHARACTERISTICS

Serious thermal losses occur in a steam engine if steam condenses on the interior surfaces of the engine cylinder, which it tends to do if these surfaces are substantially cooler than the feed steam. In most of the older engine designs,

---

[24]V. P. Kovacik, "Dynamic Energy Conversion," *Astronautics and Aerospace Engineering*, May 1963.

condensation was inevitable because the ports and surfaces were exposed first to hot feed steam, and then to cool exhaust steam.

One practical solution to this problem was the uniflow engine introduced in 1903 by Johannes Strumpf of Berlin. Strumpf's engine featured inlet valves in or near the cylinder head; the expanded steam was exhausted through ports in the cylinder wall, which are exposed when the piston is at the end of the stroke. Thus, the hottest steam was always in contact with the hottest end of the cylinder, and exhaust steam was discharged only from the cooler end of the cylinder. See figure 8-3.

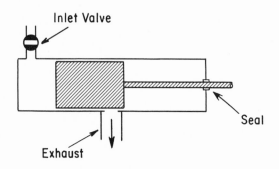

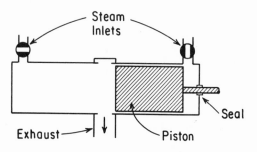

Figure 8-3.   Single-acting uniflow engine.

As originally conceived, the uniflow engine was intended to be operated with an evacuated condenser. Unless the uniflow engine is suitably modified, a high compression is imparted to any steam remaining in the cylinder after the early closure of the exhaust ports. (By contrast, in the older "counterflow" engine, the exhaust valve remains open for almost the entire return stroke of the piston.)

Early attempts to adapt the uniflow engine to automotive propulsion were unsuccessful. Doble models (Doble-Detroit, circa 1917) used a uniflow engine; the results were discouraging because of rough running. Abner Doble turned thereafter to compound-expansion engines, which reduce condensation by limiting the temperature drop in each expansion stage, each stage having its own cylinder.[25]

More recently, a number of prototype steam engines have embodied the uniflow idea. Among these are the Keen, Williams, Pritchard, Smith, Lear (Delta engine, since abandoned), Thermo Electron, and General Motors (SE-101).

Little or no data have been published to confirm the assumed advantage of using uniflow engines in road vehicles. Some of the general characteristics, however, are known:

Torque variation. Because of the high compression, torque variations per piston cycle are larger than those of a counterflow engine. The network of the cycle would also appear to be lower in the case of the uniflow engine, for the same displacement.

High speed. For the above reasons, automotive uniflow engines are generally operated at high rotational speeds.

Low speed application. Since the basic uniflow principle would result in rough running when the engine is delivering high torque at low speeds, various remedies are applied, such as compression-relief valves, automatically variable compression ratios, auxiliary exhaust valves, and variable-ratio transmissions.

Compression reheating. Drawing from the analogies of elasticity and hysteresis losses, one would never expect to recover all of the energy expended in compressing a gas. However, some value is derived from the heat of compression, since it raises the temperature of the upper cylinder and thus could reduce losses from initial condensation of the feed steam.

Incomplete expansion. As usually designed, uniflow engine exhaust ports are open during the last 10% or so of piston travel. Some expansive energy is thus lost to the system.

Rotary engines. The uniflow principle can be applied to other forms of positive-displacement expanders. One of these, for example, is the vane-type rotary engine. With these engines, recompression of the residual exhaust steam can be largely eliminated, if desired.

Efficient energy conversion in a heat engine requires a high ratio of expansion of its working fluid.[26] High expansion ratios can be obtained by closing the steam inlet valve early in the piston stroke (early "cutoff"). If one ignores the clearance volume above the piston at the top of the stroke, and if exhaust commences at the bottom of the stroke, the expansion ratio is computed as the reciprocal of the cutoff. For example, if the engine admits steam into the

---

[25] Multi-staging of expansion, in the limit, is the equivalent of the uniflow principle.

[26] "Expansion ratio" is defined as the ratio of final to initial volumes of a given mass of working fluid in the engine cylinder. With turbines, on the other hand, it is usually more convenient to refer to the "pressure ratio" of the expansion process.

cylinder for the first 25% of the stroke (25% cutoff), the expansion ratio would be 4 : 1. However, in order to obtain the maximum effort (torque) from a given engine, one might choose to employ a low expansion ratio (late cutoff) even though this means a sacrifice in efficiency. In this latter case, high torque results from the high mean pressure during a piston stroke.

Automotive steam engines are commonly fitted with a means of adjusting the cutoff. In this way, a relatively small engine can be made to serve widely varying loads without the need for a variable-ratio transmission. Late cutoff would be employed for heavy starting loads and for climbing steep hills; early cutoff would be selected for cruising speeds.

Even a very small engine with a displacement of, say, 50 cubic inches, can produce several hundred horsepower at any given instant (if it is designed to withstand the stresses) by operating with the steam inlet valve in a late cutoff condition. Within these constraints, the system can thus trade greater average horsepower for reduced efficiency, as shown in table 8-4. Thus (assuming a variable cutoff), the displacement of the motor would be determined by the desired efficiency at its *average* power level. This is usually fixed at somewhere near 10% cutoff, and corresponds to a mean effective pressure (MEP) throughout the entire stroke of about 10% of the feed steam pressure or an MEP of 100 psi for a maximum pressure of 1,000 psi.

Table 8-4. Performance Trade-offs for Reciprocating Steam Engine

Cutoff (% of stroke) (%)	Mean effective pressure (psi)	Indicated engine power[a] (ihp)	Indicated thermal efficiency[b] (%)	Water rate (lb/hp-hr)
5	50	99.4	29.2	6.77
10	100	198.8	25.8	7.74
15	150	298.2	22.9	8.95
20	200	397.6	20.8	10.0
30	300	596.4	16.8	13.0
40	400	795.4	13.65	16.5
70	700	1,391.8	6	39

Source: Data from Williams Engine Company, Ambler, Pa., for a 4-cylinder, 262-cubic-inch, 200-hp (nominal) engine. Figures for theoretical thermal efficiency and water rate based on authors' calculations.

[a]Refers to the engine alone; the steam generator efficiency is usually about 85%. Frictional and other irreversible losses are also not included. Figures correspond to Williams' engine.

[b]Actual efficiency would be reduced by the mechanical efficiency of the engine (about 85%), whence brake hp = 0.85 indicated hp.

In view of the clear advantages to be gained by utilizing temperatures above 700-750°F and pressures above 500-600 psi, why did earlier designers not do so? In part, the explanation lies in the quality of steel that was available, which required greater weights for a given stress. In part, earlier engines were designed and built by mechanics and engineers who designed largely by rule of thumb. There was very little "systems" analysis of the steam power plant as a whole to see where the maximum advantage could be obtained. Most important, however, there was—and is—a serious lubrication problem at high temperatures. This problem has been solved, at least in some instances, by improvements from two sources: better high-temperature lubricating oils and innovations in design. It is even within the realm of possibility to use steam itself as the sole lubricant for the piston when it is in the very hot part of the cylinder, although steam is not a good "wetting agent" for steel. During the expanded phase, when the steam is cooler, oil can be used in the cylinder. Oil is then scavenged by the exhaust steam and can be recovered in the condenser and recycled. This is the approach successfully taken by the Williams brothers.[27] In a hermetically sealed closed-cycle system, however, oil cannot be used within the cylinder at all, because it would eventually deteriorate and carbonize. The problem of maintaining a tight seal between the steam chamber and the crankshaft is doubly critical if steam is to be the sole lubricant and the engine is to be permanently sealed. Indeed, engine maintenance in general is likely to pose a very serious problem for a hermetically sealed closed-cycle system. Partly for these reasons, a number of firms are currently investigating alternative working fluids with superior lubricating properties.

Steam engine designers have, over the past two centuries or more, invented many ingenious mechanisms for securing variable cutoff. Most of these devices also serve to reverse the direction of crankshaft rotation; indeed, many of these "valve gears" can provide infinitely variable cutoff in either forward or reverse rotations.

There are other trade-offs to consider in varying the cutoff, besides the basic one of high efficiency vs. high torque. Short cutoff results in a high cyclic variation of torque output per cylinder. This may dictate the use of many cylinders, or high rotational speeds, or a heavy flywheel, or some combination of these to avoid output pulsations and vibration.

Another related problem has to do with the self-starting ability of the engine. With sufficiently late cutoff, as few as three or four single-acting cylinders can provide enough overlapping of steam admission to guarantee self-starting. With early cutoff, there may be significant "dead spots" in initial torque generation.

Still another situation stems from the short period of inlet valve opening, with early cutoff. With some valve gears, the lift of the valve is curtailed also, resulting in serious flow losses. If a high lift is to be maintained over the short period, stresses and noise levels in the valve apparatus may be high.

Because of the problems cited above, valve gears of the past were often compromised in favor of quietness, smoothness of torque delivery, and durability. The Model-E Doble engines, for example, gave the driver a choice of

---

[27]Reported by C. C. Williams, Jr.

83%, 65%, and 45% cutoffs. Although the corresponding expansion ratio was enhanced by compounding expansion in high-pressure and low-pressure cylinders, the overall maximum expansion ratio would scarcely have been more than about 5 : 1.[28]

Recent workers have advocated the use of considerably higher expansion ratios, ranging from 10 to 20 or more (Williams, McCulloch) while others (Pritchard) have advised moderation for the sake of operational advantages.[29]

The steam generator must, of course, produce enough steam to maintain at least the desired *average* power level. A steam engine system can be designed to have an overload capacity of 500% or more above its nominal horsepower rating by drawing upon stored heat. This is desirable because peak power demands for automobiles typically exceed average power consumption by roughly that margin. When steam consumption exceeds the capacity of the condenser the excess may be simply vented, but the lost water will have to be replenished. If the reserve supply of steam or heat is used up, of course, the power will drop to the level that can be maintained continuously by the steam generator.

Some indication of the current status of steam generator technology and of what may be anticipated in the near future may be gained from table 8-5, which gives the vital statistics of a Doble Standard "F" generator, two devices from Steam Engine Systems Corporation of Newton, Massachusetts, and a McCulloch generator. Note that the heat flux ("heat absorption rate" in the table) produced in the Doble generator is exceeded by that produced in the SES 200-hp proposed design, which is much smaller in volume and weight and which is claimed to be feasible with current technology.

A short calculation shows the usefulness of the energy storage concept in a realistic application. Assume a highly efficient steam engine at a temperature of 1,000°F and 1,000 psia and a condenser temperature of 200°F. The mass flow will be approximately 6 lb/hp-hr. If a surge of horsepower is wanted, the temperature of the steam will drop—the amount of the drop being dependent on overload capacity of the boiler. Assuming a small drop of 25°F or approximately 15 Btu/lb in the enthalpy/lb available for work, the unavailable enthalpy for a 100 hp system will be approximately 9,000 Btu/hr.

If the surge of power is to take place over a short increment such as 20 seconds, then the Btu needed to replace that lost in the temperature drop of 25°F will be about 49 Btu. Since most molten salts have a heat of fusion between 50-100 Btu/lb, only ½-1 lb of material would be needed. However,

---

[28] Assuming equal cutoffs in both high-pressure and low-pressure cylinders, the expansion ratio in each would not exceed about 2.2; since the overall expansion ratio in a multiple-expansion engine is the product of the individual ratios, two sequential expansions in this case would yield an overall ratio of around 4.9 (ignoring initial clearance volumes, which would reduce this number).

[29] *Automobile Steam Engine and Other External Combustion Engines.* Joint Hearings before the Senate Committee on Commerce and Subcommittee on Air and Water Pollution of the Senate Committee on Public Works, 90 Cong., 2 sess. (May 1968).

Table 8-5. Comparison of Four Steam Generators

Characteristic	Doble Standard "F" 100-hp (at 2.12 × 10⁶ Btu/hr)	SES 10-hp (at 300,000 Btu/hr), tests	SES 200-hp (at 2.8 × 10⁶ Btu/hr), design	McCulloch
Pressure (psig)	980	300	1,000	2,000
Steam flow (lb/hr)	1,532	228	2,170	900
Furnace volume (ft³)	2.5	0.68	0.4	1.1
Furnace intensity (Btu/hr-ft³)	1.105×10^6	4.4×10^6	8×10^6	1.22×10^6
Boiler efficiency (%)	77 at 74°F feedwater	95 at 80°F feedwater	89 at 250°F feedwater	90
Heat absorption rate (Btu/ft²-hr)	26,420	125,000	152,000	21,000
Total length of tube (ft)	558	47	115	215
Dimensions (in)	22 × 40¾ (overall)	6.62 × 7	12 × 17	18 × 27
Weight (lb)	484 (incl. burner)	20	150	223 (incl. burner)

Sources: Doble data from Doble Papers, University of California Bancroft Library, Berkeley, California. SES data supplied by Steam Engine Systems, Inc., Newton, Mass. McCulloch data from J. L. Dooley and A. F. Bell, "Description of a Modern Automotive Steam Powerplant," SAE Paper No. S338, January 22, 1962.

an estimate[30] of the heat transfer surface needed to bring the steam back up to 995°F (5°F below what it was before the power surge) implies approximately 41 square feet of surface area, or about 12 feet of 1-inch-diameter standard pipe.

The fact that such a large amount of tubing is required to transfer only 49 Btu suggests that the limiting factor in applying the heat storage method is the rate of heat transfer or, inversely, the surface area required. Another phenomenon slowing down the rate of heat transfer will be the lag time in solidifying a molten salt,[31] which is due to the necessary nucleation of a liquid material before freezing. However, this probably can be avoided by appropriate seeding techniques.

---

[30]If an overall heat transfer coefficient of 125 Btu/hr-ft²-°F is assumed, most of the heat transfer resistance will be on the steam side ~ 160 Btu/hr-ft²-°F. Heat transfer coefficients on the molten salt side will be in the vicinity of 250 Btu/hr-ft²-°F. (A. W. D. Hills and M. R. Moore, "Use of Integral Profile Methods to Treat Heat Transfer during Solidification," *Mass Transfer in Process Metallurgy*, Institute of Mining and Metallurgy, London, 1967, p. 141).

[31]In "molten salt" we include binary as well as ternary eutectic mixtures of a salt.

In order to keep the cost of the heat exchanger reasonable, iron rather than stainless steel should be used. To obviate corrosion problems in an iron-molten salt system the salts must be extremely pure both from the viewpoint of chemical impurities and moisture.[32] Moisture is particularly insidious as it is a synergistic agent in the corrosion process. Precautions to ensure extreme purity might raise the cost of a particular salt by 50-100%, but this would be cheaper than using a stainless steel system.

There has been no open discussion in the literature of the appropriate molten salt to use at a particular temperature for heat storage purposes. General Motors has examined the applicability of aluminum oxide[33] and lithium hydride[34] thermal storage systems for integration with the Stirling cycle. GM has also built a modified Corvair powered by a Stirling engine using aluminum oxide (melting point, 3700°F) as the heat storage material (regenerator). Work is continuing at GM on a second generation concept for a vehicular thermal energy storage power system using lithium fluoride (melting point, 1588°F).

The major industrial suppliers of molten salts sell only about four mixtures that are used as heat transfer agents in annealing processes. Thus, actual corrosion tests will have to be made before choosing a particular molten salt system. However, it is possible to eliminate many salts of potential interest[35] by considering only those mixtures that are stable and reasonable in cost. On such considerations the following unstable chloride salts may be eliminated: zinc, tin, cuprous, magnesium, lead, and cadmium. The eutectics that are eliminated because they are too expensive contain the salts of gold, silver, cesium, beryllium, rubidium, thallium, etc.

An innovation proposed by a Massachusetts engineering firm, Comstock and Westcott, is to reduce the quantity of working fluid (e.g., steam) required by utilizing a second fluid to perform one of its major functions—heat storage. The length of time an engine can be overloaded is limited by the quantity of heat stored in the vapor and, to some extent, in the vapor generator tubing walls. The heat capacity of the system as a whole can be increased by surrounding a length of tubing by a suitable material that can store surplus heat in the form of heat-of-fusion and release it automatically whenever the temperature of the vapor drops much below its design operating level. An ideal thermal storage

---

[32] H. Shimotake and J. C. Hesson, "Corrosion by Fused Salts and Heavy Liquid Metals— A Survey," *Regenerating Fuel Cells, Regenerative EMF Cells* (Washington, D.C.: American Chemical Society, 1967), p. 149.

[33] G. Flynn, Jr., W. H. Percival, and M. Tson, "Power from Thermal Energy Storage Systems," SAE Paper No. 608B, 1962.

[34] "Investigation of a 3 KW Stirling Cycle Solar Power System, Vol. VI: Energy Storage Analysis and Experimental Research," Report to Flight Accessories Laboratory, Aeronautical Systems Division, Air Force Systems Command Report No. WADD–TR–61–122, February 1962.

[35] G. J. Janz, *Molten Salts Handbook* (New York: Academic Press, 1967).

medium for a steam engine generating at 1000°F would be a light material with a melting point a few degrees lower.[36]

THE FLUID CIRCULATION SYSTEM

The fluid circulation system consists of a vapor generator, an expander (reciprocator or turbine), a condenser, and a pump. The general arrangement is shown in figure 8-4. The two critical elements are the vapor generator and the condenser, which are the high and low temperature heat exchangers for the system.

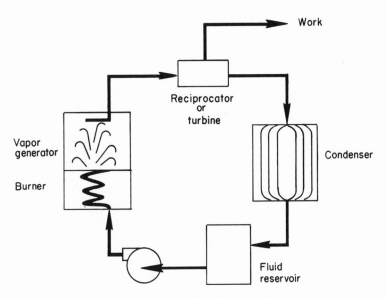

Figure 8-4. Fluid cycle schematic.

One of the major weaknesses of early steam automobiles was the cumbersome boiler. The shift from a simple boiler to a coil of heavy iron pipe represented a major improvement early in the century. Abner Doble first introduced the more refined monotube "flash" generator in 1921, and modifications of this design are used today. The present-day Williams monotube steam generator, utilizing stainless steel to withstand the high temperatures and pressures (1,000°F and 1,000 psi) achieves acceptable performance (pressure in 30 seconds from a cold

---

[36] To facilitate fast starts without waiting for the thermal storage medium to reach its full "charge" of heat, it will probably be necessary to introduce a bypass valve permitting vapor to proceed directly to the engine and "bleeding" hot vapor into the heat storage unit only as surplus heat is available.

start), but it weighs 250 pounds and occupies too much space for a truly practical power plant.

The materials conventionally used are steel and/or stainless steel; however, at temperatures below 800°F aluminum or copper tubing might be satisfactory. Richard Smith and Karl Peterson of Los Angeles have successfully used 3/8″ copper coils at temperatures of 900°F, even though the yield strength of copper is reduced by 75% at this temperature. The Smith steam generator, capable of producing 500 pounds of steam per hour, weighs less than 80 pounds complete.[37] The crucial element in the Smith system is a reliable automatic self-regulating device which prevents the temperature from rising beyond the 900°F point. These generators are being produced and sold on a custom basis.

Details of condenser and vapor generator design are determined by operating temperatures, although generally speaking the goal is to maximize the efficiency of heat transfer while minimizing mass volume and the use of costly materials. Thus a small sacrifice of efficiency might be well worthwhile if, in exchange, it were possible to cut the weight and cost of the vapor generator (and possibly avoid difficult lubrication problems as well).

The same arguments also suggest the value of utilizing every possible means of improving heat transfer at both ends of the fluid circuit, i.e., between the flame and the vaporizing fluid, and between the condensing vapor and the external coolant (ambient air).

At the low temperature end there is an important trade-off between operating temperature and condenser bulk which can be resolved only when a particular application has been selected. To maximize thermal efficiency, the heat rejection temperature should be as low as possible, but since the heat sink for an automotive vehicle is ambient air, a low condensation temperature implies a small temperature difference (ΔT) across the condenser walls, and therefore a very large condenser volume. Volume can be sharply reduced by increasing the temperature gradient. Moreover, at a higher temperature the saturated vapor is at a higher pressure and density; and thus the volume of saturated vapor that the condenser must handle will be markedly decreased. Thus, small condenser size tends to correspond to higher condenser pressure. Hence one of the salient working fluid parameters is the temperature-vapor pressure curve: other things being equal, the higher the vapor pressure (at low temperatures) the better.

A critically important factor in determining condenser size where steam is the working fluid is the fraction of air in the exhaust steam. It has been discovered that if the partial pressure of air in the steam rises above about one psi, condenser efficiency drops drastically. But, if air is *absolutely* excluded from the system, the minimum condenser size necessary to handle a given quantity of steam decreases by a factor of 2 or 3. This consideration was a major factor in the decision of Thermo Electron Corporation to design a hermetically sealed

---

[37]*Road Test*, October 1967.

steam engine in which the initial supply of distilled water lasts the life of the engine.[38]

Table 8-6 summarizes the pertinent information for all the fluids that have been seriously considered for Rankine cycle application, and indicates the cycle efficiency, the volume flow of vapor through the condenser per hp-hr, and the heat removed by the condenser (heat flow) per hp-hr. This information provides the basis for estimating the volume and surface area needed for the condenser. Heat transfer coefficients for the organic vapor-liquid interface are quite constant, independent to a first approximation of the specific fluid. The coefficient of heat transfer for the steam-water interface is about seven to ten times larger than for the organics. However, the bottleneck in the heat transfer process for an air-cooled condenser is on the air interface side; for this type of condenser the surface area required for a given heat load would be independent (to within 10% or so) of the fluid that is being condensed. For a water-cooled condenser the surface area required to dissipate a specified heat load would be fairly sensitive to the fluid and would be much smaller for the condenser of a cycle using steam as the power fluid.

The mass flow per hp-hr, which is the amount of fluid that must go through the cycle for one shaft horsepower to be produced for one hour, gives an indication of the rate of flow of the fluid and hence of the pressure drops and the pumping requirements that are needed in the various components of the system. Obviously, the lowest possible mass flow is desired.

Since the efficiency measurements are made at different temperatures and amounts of superheat, these parameters are noted as well as the amount of regenerated desuperheat. Maximum temperatures assumed were based on published data on thermal stability.

For 90°F condenser conditions without superheating, CP-28 (hexafluorobenzene) appears to be among the best of the organic compounds. It has an efficiency of 38.2%, with the upper temperature of the cycle being only 450°F. The only drawback of hexafluorobenzene is that the volume flow rate through the condenser is quite high at 1,040 cubic feet per hp-hr as is the mass flow at 45.4 lb/hp-hr. However, a few percentage points could be sacrificed from the efficiency by raising the condenser temperature and reducing the volume flow.

Thiophene and benzene have an efficiency several points below hexafluorobenzene but do have a smaller volume and mass flow. Benzene might be a good compromise except for its flammability and its high freezing point of 42°F.

It is not clear where the Freon products might be useful, if anywhere. Even though they have low volume flow, their maximum efficiencies are 50% below that of the organics mentioned above. This is due mainly to their low critical temperature and their poor thermal stability. The "G" products of Allied Chemical have better thermal stability than the Freons and therefore show up a little

---

[38] R. Harvey, TECo, personal communication.

Table 8–6. Summary Table for Rankine-Cycle Systems with 90°F, 200°F, and 400°F Condensers

Fluid	Expansion from— (°F)	Super-heat (Δ°F)	Regenerated desuperheat ΔH$_d$ (Btu/lb)	$\dot\eta$ (%)			Vapor volume flow (ft³/hp–hr)	Heat removal rate (Btu/hp–hr)	Mass flow (lb/hp–hr)
				ΔH$_d$ = 0	0.5ΔH$_d$	0.8ΔH$_d$			
				90°F condenser					
P–1D	700	270	120		20.9	29.6	5,950	3,650	101
P–100	495	115	45		27.2	30.6	416	4,950	86.7
G133aB$_1$	350		20		24.5	25.1	223	8,860	111
G133a	400	120	17.3		23.7		105	7,290	86.6
Freon C–318	300	75	23	20.6	17.8	19.5	138	8,360	207
Freon 12	300	75		20.5			62.5	10,000	170
Freon 21	320	60					162	9,850	100
Freon 114	340	60	15		23.2	24.6	117	7,570	140
Freon 11	300	20		22.0			248	9,120	118.5
Freon 113	300		8.5		22.1	22.7	370	8,690	130
CP–17 (benzene)	600	100	48		34.4	36.5	670	4,180	22.8
CP–25 (toluene)	700	150	106		35.5	38.8	1,590	3,560	20.4
CP–25 (toluene)	550		45		33.9	35.5	1,950	4,350	25.0
CP–28 (hexafluoro-benzene)	450		77		37.9	38.2	1,040	4,100	45.4
CP–32	600		19		36.2	36.7	2,500	4,300	20.2
CP–34 (thiophene)	550		9		35.4	35.8	811	4,530	28.9
Superheated steam									
400 psia	600	152		35.3			2,100	4,770	5.80
	6??	??		37.8			1,900	4,210	5.57

P-1D	430	270	108	14.3	21.3	612	5,710	178
Dowtherm E	600		34	28.7	30.1	1,580	5,610	41.5
CP-17 (benzene)	600	100	67	24.3	26.3	159	6,630	53.6
CP-25 (toluene)	550		55	25.0	26.8	350	6,540	40.6
CP-25 (toluene)	700	150	106	26.8	30.0	269	5,040	31.3
CP-28 (hexafluoro-benzene)	450		85	26.3	27.1	189	6,720	84.2
CP-32	600		28	28.3	29.1	344	5,740	28.9
CP-34 (thiophene)	550		14	26.4	26.8	185	6,570	43.8
Superheated steam								
400 psia	600	152		25.0		262	7,630	8.96
800 psia	600	80		28.4		220	6,430	8.14
1,000 psia	1,000	455	10	32.8		180	5,220	5.83
400°F condenser								
CP-9 (MIPB)	750		111	20.0	23.4	1,630	7,100	55.7
CP-63	750		76	20.4	23.4	1,910	7,580	55.1
Dowtherm A	775		112	20.4	23.8	726	7,000	50.6

Source: Joel Jacknow, *Analysis of Alternate Working Fluids for Rankine Cycle Engines*, IRT-N-62, February 1969.

better in comparison. For example, GI33aB$_1$ is close to being an isentropic fluid and has a cycle efficiency of 24.5% upon expansion from 350°F to 90°F. If Freon 11 could be safely heated to 350°F, it would have a comparable efficiency to FI33aB$_1$, and a similar volume flow of about 200–220 cubic feet per hp-hr. Since the heat flow rate in the condenser for the "G" fluids and the Freons is quite large, approximately 9,000 Btu/hp-hr (more than twice as great as for CP-32), the surface area rather than the volume flow will most likely define the size of the condenser. This means that one could not take full advantage of the low volume flow of these fluids when sizing this important part of the system.

At a condenser temperature of 200°F, toluene (superheated to 700°F) and Dowtherm E have efficiencies of 30% or better. However, Dowtherm E has a much larger volume flow. Among the fluids whose thermodynamic properties have been published, CP-32 might be the fluid of choice with an efficiency of 29.1% if superheating is too costly or otherwise undesirable. Both toluene and CP-32 under the conditions given in the table have comparable volume, heat, and mass flow values.

At a condenser temperature of 400°F, the choice of a fluid would be fairly clear. With the efficiencies of CP-9, CP-63, and Dowtherm A almost identical, Dowtherm A would be chosen on the basis of the lowest volume flow through the condenser, and the lowest mass flow through the system.

Comparing steam to the more efficient organic fluids is a very tricky task. One can always superheat the steam to a high enough temperature and pressure so that the efficiency of a steam cycle is better than would be obtainable from an organic cycle. This is noted in figure 8–5, where the cycle efficiency of steam (condenser temperature of 150°F) is plotted as a function of the temperature and pressure of the steam before expansion.

However, given the same inlet and exhaust temperatures a few organics actually show a more favorable cycle efficiency than superheated steam. For example, CP-32 shows a cycle efficiency of 28.3% and 29.1% (for 50% and 80% desuperheat regeneration) at 600°F and 567 psia with a 200°F condenser, while steam must be heated to 600°F and 800 psia to give a cycle efficiency of 28.4%. Between 1,000°F and 1,000 psia, and the same condenser temperature of 200°F, steam has a cycle efficiency of 32.8%. This is only four points better than CP-32 with 80% regeneration, with a much lower maximum temperature and pressure. All of the more efficient organics are "drying fluids" and have a desuperheat which must be recovered for maximum efficiency. On the other hand, as previously noted, steam is a "wetting fluid" and must be substantially superheated before its cycle efficiency will surpass the better organic fluids.

As noted elsewhere, one of the major drawbacks of using steam at high superheat in a reciprocator is the lubrication problem. (No such problem would exist in a turbine, however.) The thermal stability of the lubricant would probably limit the maximum allowable superheat temperature.

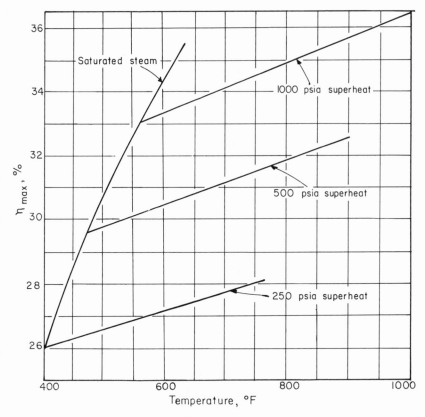

Figure 8-5. Efficiency of steam Rankine cycle.

RANKINE-CYCLE TURBINES

It is well-known that the steam turbine provides the bulk of the utility electric power produced today. In principle, at least, the vapor turbine, which operates on more or less the same general principles as the gas turbine, is much simpler than a reciprocating vapor engine. In large fixed installations, overall operating efficiencies above 40% are currently achievable. It is rather difficult, however, to translate such large stationary designs into small portable installations of comparable efficiency. High efficiency is achieved in large units by utilizing very high steam-jet speeds (steam velocities of 3,000–4,000 feet per second are easily attainable with expansion from quite moderate pressures), high temperatures and pressures, and multiple expansion. Performance is quite sensitive to the so-called velocity ratio (the ratio of turbine blade tip speed to vapor velocity). Thus, the smaller the turbine wheel, the higher its rotational speed must be to keep the velocity ratio constant.

Since the advent of the space program, considerable interest has been manifested in the development of various types of Rankine turbines for space purposes. All of these systems, because of their intended use in space, have been sealed closed-cycle systems and most of them have been designed to utilize solar or nuclear energy rather than chemical fuels. Power to weight ratios tend to be extremely poor. For example, SNAP-8, a 35 kw mercury vapor system, has a specific weight of 323 lb/kw at present.[39] It must be remembered, however, that these estimates apply to complete systems including the nuclear reactor plus shielding and the alternator, with various radiating and condensing components that would not be necessary in "down to earth" versions.

The bulk of the early astronautical development work has concentrated on Rankine systems using mercury or alkali metals such as potassium or rubidium or even lithium.[40] At high temperatures and speeds, serious engineering problems, such as erosion of turbine blades, have slowed development in this field. For vehicular applications toxicity and high reactivity would pose further difficulties. An interesting concept of more recent origin is the use of an organic working fluid. An approximate rejection temperature of 375°F for space applications limits the choice of fluids to those having high molecular weights, e.g., biphenyl $(C_{12}H_{10})$ and mono-isopropyl biphenyl $(C_{15}H_{16})$, which is better known as MIPB. Basic characteristics are shown in table 8-7.

When a saturated vapor of biphenyl or MIPB is expanded isentropically, it becomes increasingly superheated, i.e., the condensation temperature drops faster with decreasing pressure than the temperature of the vapor itself. Thus the vapor becomes drier as it expands. This absolutely eliminates any possibility of condensation in the turboexpander and, incidentally, the risk of the turbine blades being eroded by droplets. The fact that the fully expanded vapor is

Table 8-7. Possible Organic Working Fluids for Vapor Turbines

	Biphenyl	MIPB
Molecular weight	154.2	196.3
Melting point	156°F	-65°F
Boiling point	491°F	563°F

Source: H. D. Linhardt and G. P. Carver, "Development Progress of Organic Rankine Cycle Power Systems," Intersociety Energy Conversion Engineering Conference, 1967, p. 103.

[39] J. N. Hodgson, R. G. Geimer, and A. H. Kreeger, "SNAP-8: A Technical Assessment," Intersociety Energy Conversion Engineering Conference, 1967, p. 135. Also J. N. Hodgson and R. P. Macosko, "A SNAP-8 Breadboard System—Operating Experience," Intersociety Energy Conversion Engineering Conference, Vol. 1, 1968, p. 338.

[40] G. Szego, "Space Power Systems, State of the Art," *Journal of Spacecraft and Rockets*, Vol. 2 (1965), No. 5; V. Kovacik, "Dynamic Energy Conversion," *Astronautics and Aerospace Engineering*, May 1963; D. R. Snoke and G. L. Mrava, "Silent Mercury Rankine Cycle Power System," SAE Paper No. 883D, 1964; J. P. Davis and B. M. Kikin, "Lithium-Boiling Potassium Rankine Cycle Test Loop Operating Experience," Intersociety Energy Conversion Engineering Conference, 1967, p. 87.

superheated makes it imperative to incorporate a regenerative heat exchanger to transfer waste heat from the turbine exhaust to the boiler feed, thus increasing overall efficiency.

A major difficulty with organic substances of the biphenyl or MIPB family is that at high temperatures they tend to break down, but at low temperatures their vapor pressures are too low. Practical operating temperatures tend to be in the 650°-750°F range at the upper limit down to 350°F or so at the lower end. This obviously limits the thermal efficiency that can be achieved.

A way of getting around this limitation is to introduce a compound or hybrid cycle. One possibility, alluded to earlier, is the binary vapor cycle, where the exhaust from a high temperature (mercury or other fluid) vapor turbine is used as a heat source for a lower temperature organic vapor turbine. Overall efficiencies of 33% and specific weights of the order of 25 watts/lb for a 20 kw output level have been projected for space power units.[41] Another possibility of considerable interest—at least for space applications—is a combined Brayton-Rankine cycle, wherein the hot exhaust gas from a gas turbine (Brayton cycle) becomes the heat source for an organic Rankine cycle turbine. Figure 8-6 shows the thermal efficiencies that seem to be achievable for these conceptual systems. The simple Brayton cycle curve is based on a turbine efficiency of 85%, a compressor efficiency of 80%, a regenerator efficiency of 80%, and the optimum pressure ratio for each temperature.[42]

In this connection, it must be noted that, while Brayton cycle (gas turbine) systems are ordinarily *open cycle* internal combustion engines, it is possible to utilize the cycle with an external heat source. A small (3-kw) closed Brayton cycle gas turbine operating at 64,000 rpm and utilizing argon as a working fluid has been built by the Air Research Division of Garrett Corporation.[43]

More recently Mechanical Technology, Inc. (MTI) has been studying the application of turbines to a Rankine cycle using a variety of working fluids, e.g., FC-75 (a fluorinated fluid of the 3M Company), monochlorobenzene (CP-27), CP-63, and water. The MTI group asserts that closed-cycle turbines operate at power levels from a fraction of a horsepower and up, at overall efficiencies up to 35% with peak temperatures of 1,000°F and (surprisingly) efficiencies in the 10% range even with peak temperatures as low as 160°F. MTI has also considered the application of turbomachinery to Rankine cycles with a power output suitable to automotive applications.[44]

---

[41] V. Kovacik, "Dynamic Energy Conversion."

[42] S. Luchter, "Current Status of the Technology of Organic Rankine Cycle Power Plants," Intersociety Energy Conversion Engineering Conference, 1967, p. 117.

[43] J. E. McCormick and T. E. Redding, "3 Kilowatt Recuperated Closed Brayton Cycle Electrical Power System," Intersociety Energy Conversion Engineering Conference, 1967, p. 1.

[44] J. W. Bjerklie and S. Luchter, "Rankine Cycle Working Fluid Selection and Specification Rationale," Paper No. 690063, SAE Annual Meeting, Detroit, January 1969.

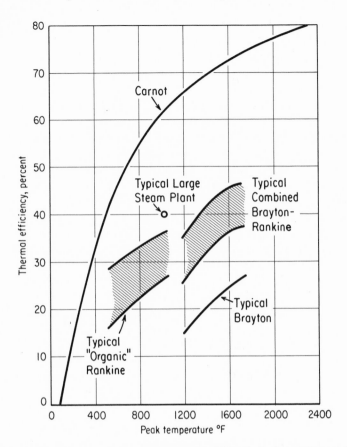

Figure 8-6. Efficiency of various turbine systems. (From S. Luchter, "Current Status of the Technology of Organic Rankine Cycle Power Plants," *Advances in Energy Conversion Engineering*, Intersociety Energy Conversion Engineering Conference, 1967.)

In a comparison of Otto and Rankine engines for automotive use[45] Bjerklie and Luchter have outlined the designs of 150-hp vapor engines (comparable to 250-hp Otto-cycle engine) using steam and CP-27. For CP-27 a single stage turbine was envisioned with a 9″ diameter and operated at 30,000 rpm. Inlet conditions were assumed to be 548 psia and 650°F with 190°F condensing temperature. Turbomachinery is compromised by using steam at these power levels. Top speed limitation restricts the efficiency to 70%, and torque characteristics require a two-stage turbine for steam. The turbine will be less than 8″ in diameter and have a speed of over 40,000 rpm. Top speed for both turbines will be on the order of 1,000 feet per second. Thus the steam turbine will tend to be

---

[45] J. W. Bjerklie and B. Sternlicht, "Critical Comparison of Low Emission Otto and Rankine Engine for Automotive Use," SAE Annual Meeting, Detroit, January 1969.

less efficient, multi-staged, smaller in diameter, and operating at a higher rpm as compared to the organic fluid turbine.

Bjerklie and Sternlicht[46] also expect that turboengines will be cheaper than reciprocating engines (if new capital costs for mass production equipment are ignored). Because both top speed and temperatures are low, inexpensive materials can be used. Also, current technology indicates that the shaft bearing assembly can be made relatively inexpensively.

The probable costs of steam engines in general are rather difficult to pinpoint without any mass-production data. Somewhat conflicting estimates appear in the literature. One source envisions a specific cost of $4-$6/hp,[47] while another projects $8-$13/hp.[48] A recent estimate of approximately $9/hp for the complete steam power plant is probably close to the likely mark for present systems.[49] However, because the steam engine has highly favorable torque characteristics that enable it to provide a given acceleration with lesser power rating, it can probably be expected to be competitive in cost with the internal combustion engine for a given vehicle, if large-scale production is assumed.

[46]Ibid.

[47]J. A. Hoess et al., "Study of Unconventional Thermal, Mechanical, and Nuclear Low-Pollution-Potential Power Sources for Urban Vehicles." Summary Report to National Air Pollution Control Administration, March 15, 1969.

[48]Robert Kirk and David Dawson, "Low-Pollution Engines: Government Perspectives on Unconventional Engines for Vehicles," ASME Paper No. 69-WA/APC-5, 1969.

[49]R. U. Ayres and Roy Renner, "Automotive Emission Control: Alternatives to the Internal Combustion Engine," Paper presented to the Fifth Technical Meeting, West Coast Section, Air Pollution Control Association, San Francisco, October 8-9, 1970.

Chapter Nine

Application of Noncondensing Gas Cycle-External Combustion Engines to Auto Use

The first use of working fluid other than steam dates back to the "hot air" engine patented by Robert Stirling in 1816 and developed by him over the next thirty years before being abandoned.[1] The "Stirling Cycle" concept was revived by N. V. Philips Research Laboratories in the Netherlands in the late 1930s and has been brought to a fairly advanced state of development. Thermal efficiencies greater than 40% have been achieved with helium or hydrogen used as the working fluid.

A related cycle, also based on the use of a light gas as a working fluid—originally hot air—is the Ericsson cycle, named after a versatile nineteenth century sea captain-engineer-promoter. Thousands of Ericsson hot air engines were built during the nineteenth century (mostly to run household appliances) despite their low efficiency and considerable bulk in relation to power output. There is no known contemporary version, despite its basic simplicity. In operation, the Ericsson engine would be somewhat similar to the steam engine (Rankine cycle) except that no condenser is needed because the exhaust air is simply returned to the atmosphere, which is a constant temperature and pressure reservoir.

We also consider, briefly, the possibilities inherent in an external-combustion Brayton or diesel cycle.

THE STIRLING CYCLE

The Stirling cycle was discussed from the theoretical point of view in chapter 4. Two of the steps in the cycle involve (respectively) heating and cooling the working fluid at constant volume. To accomplish this, the space containing the working fluid (gas) could be alternately heated and cooled—which would involve heating and cooling much more than just the gas itself—or the fluid could be physically transferred from a "hot space" to a "cold space." The latter is obviously the

---

[1] R. A. J. O. van Witteveen, "The Philips Stirling Engines, Present and Future," Philips Research Laboratories, August 31, 1966.

most economical procedure, but it can be seen that this implies that the working fluid be as light and have as little viscosity as possible to minimize the work done in moving it from one place to another. It must also have a very high thermal conductivity in order to come rapidly to equilibrium with its environment during the isothermal stage in the cycle. For this reason the developers have been led to use highly compressed (100 atmospheres) helium or hydrogen. The best results are obtained with hydrogen.

Another basic feature of the Stirling cycle engine is its heavy dependence on thermal regeneration. If all the waste heat in the hot gas were carried to the "cold chamber" and dissipated there (and vice versa), thermal efficiency would be extremely low. In practice, the hot gas, after expansion, travels through a regenerator where it gives up most of its heat en route to the cold chamber. Then, after it is compressed, it passes back through the regenerator and regains much of the heat on its way to the heater. Stirling cycle engines of today gain much of their high performance from a remarkably efficient (approximately 95%) heat storage unit (regenerator) developed by Philips Research Laboratories.

To transfer the working fluid from the "hot space" to the "cold space" and back, a double piston arrangement is used in each cylinder (see figure 9-1). In addition to the power piston, there is a "displacer" whose motion is out of phase with the power piston, and whose function is to move the helium or hydrogen back and forth.

At the beginning of the cycle the displacer is at the top of the cylinder, the power piston is at the bottom of its range, and the gas is in the cold space, at its maximum volume and lowest temperature. In the first stage, the power piston moves up, compressing and heating the gas in the cold space. Next, the displacer moves down while the working piston remains in place, and the gas is transferred (at constant volume) through the regenerator and heater into the hot space, being heated en route. The third stage is the power stroke. The displacer and power piston move down together to their lowest points, as the hot gas expands isothermally (since it is still gaining heat from the heater). In the last stage the gas is transferred from the hot space back to the cold space at constant volume as the displacer moves back to the top of the cylinder while the power piston remains in place, and heat is given up to the regenerator.

The piston rod is hollow, and the displacer rod runs through its center. Both shafts are linked to a unique phasing and drive mechanism called the rhombic drive, illustrated in the lower part of figure 9-1.[2] The drive is perfectly balanced so that even a single-cylinder engine runs smoothly and free of vibration.

One of the most serious problems that had to be faced in the engineering development of the engine was to seal off the working chamber from the crankshaft, to prevent any escape of pressurized gas into the oil bath or vice versa.

---

[2] R. J. Meijer, "The Philips Hot Gas Engine with Rhombic Drive Mechanism," *Philips Technical Review*, Vol. 20 (1958/59), pp. 245–62.

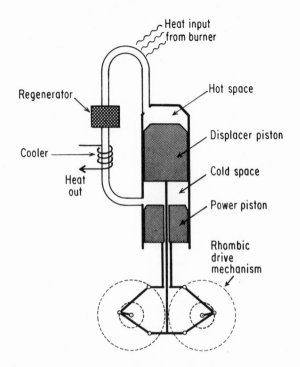

Figure 9-1. Schematic of Philips Stirling engine. (From R. J. Meijer, "The Philips Hot Gas Engine with Rhombic Drive Mechanism," *Philips Technical Review*, vol. 20, 1958/59, pp. 245-62.)

(This problem was mentioned earlier in connection with double-acting vapor engines.) An adequate solution has apparently now been found. A flexible diaphragm, called a rollsock, is permanently affixed to both the moving piston rod and to the fixed wall of the housing. It is supported on the crankcase side by oil pressure, which can be adjusted to any desired level. With this seal the only possible source of leakage is diffusion through the diaphragm. The rollsock seal permits engine operating lifetimes of more than 10,000 hours with little or no maintenance.

The Stirling engine is inherently rather costly. It has a complex drive mechanism and piston-displacer combination, and because it operates at high temperatures and pressures it requires rather expensive "space-age" materials. Doubling the mean operating pressure to 200 atmospheres and raising the burner temperature are being considered as a means of increasing efficiencies still more, but this would also increase the materials cost. The advantages to be gained from going to higher temperatures and pressures are shown in figure 9-2. In terms of size, weight, efficiency, and torque-speed characteristics, the Stirling engine most

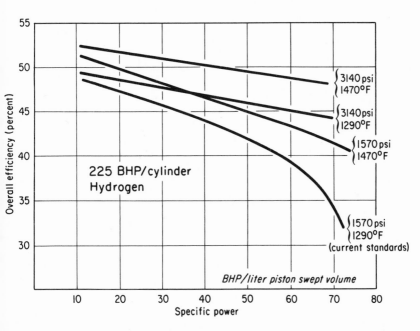

Figure 9-2. Efficiency versus specific power for various temperature-pressure conditions. (From R. A. J. O. van Witteveen, "The Philips Stirling Engine, Present and Future," Philips Research Laboratories, Eindhoven, Netherlands, August 31, 1966.)

nearly resembles the diesel engine. For equal continuous power outputs (say 900 bhp), the Stirling engine would simultaneously offer greater efficiency and smaller bulk (but higher cost) than the diesel, although there are some trade-offs between size and efficiency for a given power level. Alternatively, if a given size (bulk) Stirling engine is specified, one may trade greater efficiency for smaller horsepower and vice versa. These relationships are implicit in figure 9-3.

Power to weight ratios appear to be comparatively poor for the Stirling engine, though still comparable to the diesel (at least in larger sizes). A prototype engine producing 40 bhp weighed 400 lb, suggesting a specific weight of about 10 lb/hp. However, in larger sizes and with reduced weight as a design objective (which it has not been, to date), this figure can certainly be reduced somewhat, although probably not to the level required for private automobiles (∿3 lb/hp). For buses and large trucks the Stirling engine offers real possibilities.

Torque-speed relationships are somewhat more favorable than they are for diesel engines, i.e., the relationships are relatively flat. However, torque is typically at a maximum at around 1,000 rpm and drops sharply below 250 rpm. Hence, although the conventional gearbox for buses or trucks might be dispensed with, a clutch would probably still be needed.

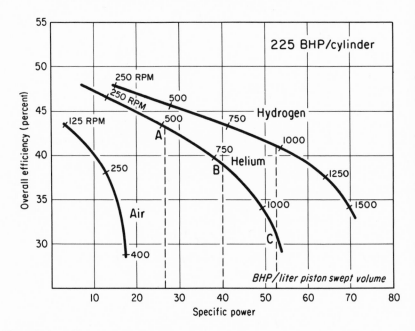

Figure 9–3. Efficiency versus specific power for various working fluids. (From R. A. J. O. van Witteveen, "The Philips Stirling Engine, Present and Future," Philips Research Laboratories, Eindhoven, Netherlands, August 31, 1966.)

Reliable cost estimates for Stirling engines are difficult to make at this stage of development. One source cites $5–$7 per hp as a probable specific cost,[3] while another one estimates $13 per hp.[4]

Residuals from fuel combustion are extremely low for a Stirling engine, as they are for steam engines. In volumetric terms, the Stirling engine would produce approximately 70 ppm carbon monoxide (compared to 1,000 ppm for a diesel), and only 1 or 2 ppm unburned hydrocarbons (compared to 200 ppm to 4,000 ppm for a diesel). Production of oxides of nitrogen (NO_x) would also be low (100 ppm to 200 ppm), compared to 400 ppm up to 2,000 ppm for a diesel. In addition, the engine would emit no smoke at all, whereas diesel engines tend to emit a rather unpleasant blue smoke, particularly when they are in need of tuning. (Emissions data for the Stirling and other engines are compared in table 14–3 in chapter 14 in terms of weight of residuals per mile instead of in terms of volume. Volumetric measurements, such as those cited above, are somewhat

---

[3] J. A. Hoess et al., "Study of Unconventional Thermal, Mechanical, and Nuclear Low-Pollution-Potential Power Sources for Urban Vehicles." Summary report to National Air Pollution Control Administration, Battelle Memorial Institute, Columbus Laboratories, March 15, 1968.

[4] Robert Kirk, and David Dawson, "Low-Pollution Engines: Government Perspectives on Unconventional Engines for Vehicles," ASME Paper No. 69–WA/APC-5, 1969.

misleading in comparisons because different engines use different amounts of excess air.)

Data on several prototype Stirling engines are summarized in table 9-1.

A MODIFIED STIRLING ENGINE

The phases of the Stirling cycle have also been embodied in a somewhat different engine by Vannevar Bush, the former president of Carnegie Institution and the wartime head of the Office of Scientific Research and Development.[5] The Bush engine differs from the conventional Stirling machine physically and, to some degree, conceptually. In essence, it pairs two simple Stirling engines, and in doing so, eliminates the displacer piston from each. Furthermore, a more or less conventional crank and differential mechanism replaces the special rhombic drive typical of the other Stirling engines.

Table 9-1. Data on Four Experimental Philips Stirling-Cycle Engines

Engine Characteristics	(1)	(2)	(3)	(4)
Piston displacement (in^3)	~20	~20	~20	~1,000
Cylinders (number)	1	1	4	4
Inlet pressure (psi)	~2,000	~2,000	~2,000	~2,000
Inlet temperature ($^\circ$F)	~1,300	~1,300	~1,300	~1,300
Working fluid	. . . . Hydrogen			
Total power plant wt. (lb)	440	–	–	–
Total volume (ft^3)	–	–	–	17.3
Rated bhp	40	90	360	900
at rpm	2,500	1,500	1,500	1,000
Overload (%)	~50%	~50%	~50%	~50%
Starting torque (lb–ft)	81			
Torque at 1,000 rpm (lb–ft)	100			
Torque at 2,500 rpm (lb–ft)	32			
Thermal efficiency (%)	38% (at 1,300 rpm)			42% (at 1,000 rpm)
Starting time (sec)	~75	~75	~75	~75

Source: Compiled by the authors from Philips' literature.

The Bush Stirling system consists of two pairs of double-ended hot and cold pistons; the hot piston ends are enclosed in a suitable furnace or burner and the cold encased in a radiator or refrigerator. Each hot piston is connected to its matching cold piston by a regenerator of unique design. The "lower" hot-cold system is linked by piston rods to the crankshafts and differentials.

---

[5]Vannevar Bush, U.S. Patent No. 3,457,722, July 29, 1969.

Operation of the linked hot-cold pistons is 90° out of phase—i.e., the cold piston follows the hot one in position by one-quarter cycle—while the upper and lower systems themselves are 180° out of phase.

As indicated in figure 9-4, which is a schematic representation of the Bush scheme, a pair of bypass coils is situated between each regenerator and its associated hot and cold pistons. Hot gas or cold gas, trapped in these coils by the operation of suitable valves during one phase of the cycle, can be released into either the regenerator or the piston at some other phase to improve the heat flow or pressure characteristics. The net effect (at least theoretically) is to improve the operation of the Stirling cycle engine.

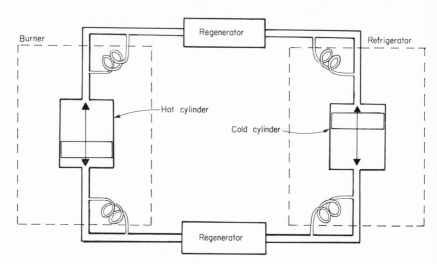

Figure 9-4. The modified Stirling engine of Vannevar Bush in schematic form. (Adapted from U.S. Patent No. 3,456,722, July 1969.)

THE BRAYTON CYCLE HOT-AIR ENGINE

This concept is an adaptation of the Brayton cycle to a positive expansion system. A number of implementations are possible. For example, one can use a rotary expander of the vane type similar to the Dawes steam engine[6] or the Mallory internal combustion engine.[7] The system we consider here, however, is an open-cycle external combustion system using air as the working fluid.[8] Essentially, it consists of a rotor mounted off center in a housing. The crescent-shaped space between the rotor and the housing is divided by vanes into a number of

---

[6] Bailey P. Dawes, U.S. Patent No. 2,463,155, March 1, 1949.

[7] W. L. Linn and G. A. Dotto, U.S. Patent No. 3,301,233.

[8] Theodore B. Taylor, U.S. Patent applied for.

chambers that expand and contract in volume as the rotor turns. In contrast to the simple vane pumps familiar to most engineers, the vanes are held to a minimum clearance from the housing walls by a control ring mechanism that is concentric with the center of the housing. This permits relatively high speeds and low wear.

In operation, air is introduced into the engine approximately at 1. It is then compressed by the compressor stage and led out of the engine at 2, from which point it is conducted through an external heat source (burner), heated to its maximum temperature, and reintroduced into the engine at 3. The subsequent expansion of the hot compressed air exerts a net force on the vanes, which manifests itself as a torque on the shaft. The expanded gases are then exhausted from the system at 4. Since this engine receives heat at roughly constant pressure, it presumably operates in a cycle somewhere between the Brayton and the diesel, with incomplete expansion, as shown in figure 9–5. It is similar to a gas turbine in some ways, but both the compressor and expander are positive displacement devices.

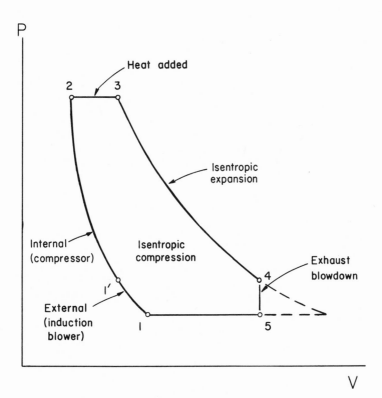

Figure 9–5. Idealized air engine cycle.

Initial development on a form of this engine has been carried out by International Research and Technology Corporation, Washington, D.C.[9]

In principle, the rotary air engine has many attractive aspects. It delivers a smooth torque with very little vibration. As an external combustion system, it inherently has both a multi-fuel capability and low pollutant emission. Preliminary investigations indicate ratios of weight (including the burner and all necessary auxiliary equipment but not including such components as transmissions that might be required in various applications) to power of about 2.5 lb/hp, even in fairly small sizes. Careful design in larger sizes should reduce this ratio still further.

The system has some of the disadvantages of the internal combustion engine types (including the gas turbine). As an open-cycle air engine, it has no starting torque at all, and would therefore need a starter motor (or a source of compressed air). However, initial indications are that even fairly high leakage rates have only a very slight effect on power and torque at speeds above several thousand rpm, and, in general, leakage effects decrease markedly with increasing engine size. Power and torque curves for a typical rotary air engine are shown graphically in figure 9-6, assuming a maximum temperature of 800°F and an air leakage rate for the system of approximately 25 lb/hr.

As an automotive power plant, the rotary air engine appears to have some attractive possibilities. For one thing, it seems to be generally lighter in weight for an equivalent power output than the internal combustion engine which typically weighs in at 3-5 lb/hp not including transmission, fuel, or generator. In

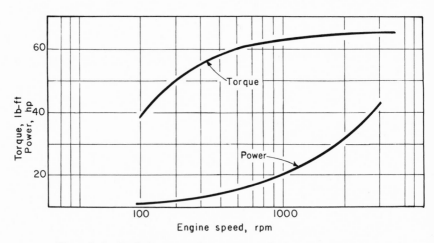

Figure 9-6.   Typical power and torque curves for rotary hot-air engines.

[9] Richard P. McKenna and Roy Renner, "Performance Analysis of Rotary Air Engine Concept," International Research and Technology Corporation Paper No. IRT-N-80, Washington, D.C., August 15, 1969.

addition, the rotary air engine concept is mechanically rather simple, and, in principle, easy to maintain. Its torque characteristics at very low speeds are similar to those of the standard automotive gasoline engine. Like the ICE, it therefore would need a starter, clutch, and variable-speed transmission.

Because the rotary air engine appears to have favorable performance characteristics at higher speed (3,000 rpm), it might well serve as the prime mover in a hybrid-electric system. In such an application, the engine would operate constantly at optimum conditions (i.e., under constant load at high speed). Its leakage losses, and other shortcomings, would thus be minimized, and its low torque at low speeds would be unimportant.

Chapter Ten

Electric Propulsion Systems for Automobiles

TORQUE-PRODUCING (ROTARY) MOTORS

The choice of an electric propulsion unit for a vehicle is more complicated than at first appears, especially if we consider some of the relatively undeveloped alternatives. In most discussions of electric cars, only the familiar series-wound DC motor and the "squirrel cage" AC induction motor are mentioned. The latter is simpler, more reliable, lighter in weight, and lower in cost, but its speed can be controlled only by altering the frequency of the AC power supply, which requires a rather sophisticated solid-state electronic control system (inverter). So far, at least, this immediately reintroduces problems of weight, high cost, cooling, and unreliability. On balance, if one or the other had to be chosen for mass production at this time, one would probably choose the DC motor in a low-cost, low-performance urban "runabout," and the AC induction motor for a high-performance electric car designed to compete with ICE-powered vehicles over a wider range of driving situations.

However, if the constraint of immediacy is lifted, a somewhat broader field of choice exists and a correspondingly greater range of trade-offs must be considered. A good starting point is the observation that all electric motors rely on the interaction between an electric circuit, confined to an electrical conductor, and a magnetic circuit, usually confined to a magnetically conductive (or low reluctance) material such as soft iron or silicon steel. Most motors rely on magnetic field reversal—or sequential activation of pole pairs in an n-pole arrangement of windings—to "rotate" the magnetic field, thus producing torque on the rotor. However, the simplest of all configurations is the homopolar motor (or Faraday disk), which utilizes a fixed axial magnetic field and obtains an azimuthal force (torque) by causing electric current to flow in the radial direction, viz., from the axis to the rim of the disk or vice versa.

The axial magnetic field may be provided by a permanent magnet or by a winding which could, in principle, be connected either in series or in parallel

190

with the rotor circuit. Moreover, although the homopolar configuration is always thought of as a DC motor, there is no reason why both the rotor and stator could not be fed by alternating current; this would rapidly reverse the direction of both the radial current flow and the axial field, but would leave the direction of the torque on the rotor unchanged. Thus an AC homopolar motor seems to be a theoretical possibility. Torque-speed relations would be comparable to DC rather than AC motors, however.

The next simplest configuration is probably the familiar DC motor with a winding on the rotor that causes it to line up (N pole to S pole, etc.) with respect to an external pair of magnetic poles. The rotor circuit reverses itself (by means of a commutator) each time it lines up, so it is continuously pulled or pushed from one pair of fixed poles to the next. The stator may have any even number of poles: 2, 4, ... , $2n$. Again, these may be permanent magnets or they may be electromagnets; if the latter, they may be connected either in series or in parallel with the rotor circuit. The torque-speed characteristics naturally depend upon the specific arrangement. The parallel (shunt) wound motor approaches a definite limiting speed, whereas the series wound motor is limited in speed only by the external load imposed. Compound (series parallel) motors have inter- mediate characteristics as might be expected. Figure 10-1 indicates the differ- ences.

All of the above basic configurations involve an electrically active rotor in conjunction with a variety of possible stators. The basic "squirrel cage" AC induction motor, in contrast, utilizes a passive rotor whose interaction with the external electrical circuit is strictly inductive or magnetic. The torque on the rotor is produced by the interaction of induced currents and a rotating multipole magnetic field created by a multiphase alternating current supplied to the stator windings. A variant utilizes windings on the rotor linked via a commutator to an external variable resistor, which makes it possible to adjust the speed at which maximum torque occurs. In the above cases, the rotor lags somewhat behind the rotating field under load. If the rotor itself is permanently magnetized, however, with the rotating field imposed on the stator windings by the polyphase current, its speed is independent of load and determined only by the frequency of the multiphase current supplied to the stator. Such constant speed motors are called "synchronous," for obvious reasons.

Figure 10-2 shows the schematics and torque characteristics of AC motors. A summary of the basic morphological possibilities is shown in table 10-1. Note the matrix is essentially symmetric across the diagonal, if rotor and stator are functionally interchanged. Most of the "conjugate" motors are of limited practi- cal importance, since they would involve rather cumbersome and pointless mechanical rearrangements. It is interesting to observe, however, that the AC synchronous motors are essentially symmetrical conjugates of the DC com- mutator motors. These types are so different in practice that one would be ill-advised to dismiss all conjugates as *ipso facto* uninteresting permutations of

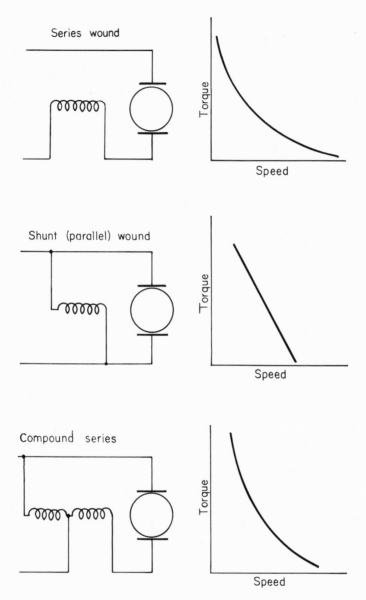

Figure 10-1. DC motor types.

existing types. The majority of the matrix squares are blank because they do not correspond to torque-producing modes of interaction.

From the standpoint of vehicular operational requirements, torque should be a maximum at zero speed, with a monotonic decrease to zero at higher speeds.

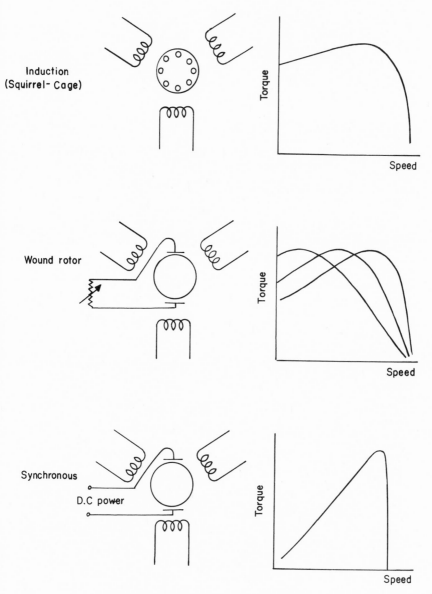

Figure 10-2. Polyphase AC motor types.

In this respect the DC series wound motor is nearly ideal (see figures 10-1 and 10-3), and the homopolar motors would be similar. On the other hand, under conditions of varying load, motor speed will fluctuate unless compensated for by positive voltage (and current) control. An AC inductive motor's speed can be

Table 10-1. Morphological Summary of Torque-Producing Devices

Rotor Stator	Passive	Permanent Magnet (PM) 2-pole	Electromagnetic (EM)			
			DC 1-pole	DC 2n-pole	AC 1-pole	AC 2n-pole
Passive						Inductive conjugate
(PM) 2-pole			Homopolar			DC (PM) (with commutator)
(EM) DC 1-pole		Homopolar conjugate (PM)		Homopolar conjugate (EM)		
(EM) DC 2n-pole			Homopolar (EM)			DC series/ shunt (with commutator)
(EM) AC 1-pole						AC homopolar conjugate (?)
(EM) AC 2n-pole	Inductive (squirrel cage)	Inductive synchro (PM)		Inductive synchro (EM) (slip rings)	AC Homopolar	

maintained (in principle) by controlling the frequency only. Other comparisons worthy of note are as follows:

1. The efficiencies of both AC and DC motors can be approximately the same.

2. The torque characteristics of both AC and DC motors can be the same, although the starting torque of the series DC motor is normally higher. Torque in an AC induction motor depends on the "slip," the lag between the "rotating" stator fields and the induced fields in the rotor. However, by adjusting input voltage with speed sensitive equipment deriving information from the motor shaft, the slip frequency can also be adjusted.

3. AC induction motors and DC shunt motors have a built-in maximum synchronous no-load speed. The speed of a series DC motor under no-load conditions, on the other hand, can become dangerously high, tending to a runaway condition.

Cost and weight cannot meaningfully be computed for motors alone; voltage, current, and frequency controls must obviously be included in the package, especially in view of the considerable differences in requirements among the four remaining possibilities: DC homopolar (series, parallel, or permanent magnet); AC homopolar (series or parallel); DC multipolar (series or parallel); AC inductive.

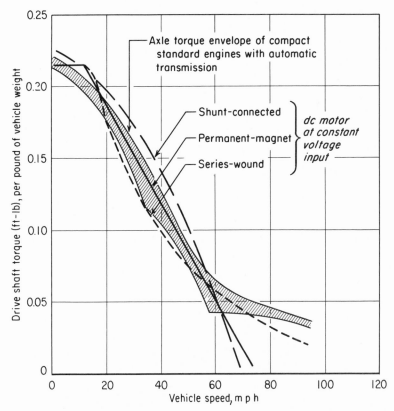

Figure 10-3. Torque characteristics of various DC electric motors and ICE with automatic transmission. (From George Hoffman, *Electric Motor Cars*, Rand Corporation, RM-3298-FF, March 1963.)

The DC homopolar motor has long been known to be capable of superior performance in terms of horsepower per pound, mechanical efficiency and ease of cooling (resulting in very high overload capacity) compared with conventional electrical machinery. Its drawbacks have been, primarily, that it is inherently a high current, low voltage device (up to 100,000 amperes at only a few volts), with the associated problem of supplying a very heavy current to the moving rotor via slip rings or sliding metal contacts. Heavy bus-bars are normally required to handle such currents, and switching—and therefore control—problems are acute. Nevertheless, Ford's Research Laboratories are investigating a homopolar motor with liquid metal (mercury) contacts, for a possible automotive transmission application.[1] The results of a preliminary study showed that

---

[1]F. L. Zeislar, "A High Power Density Electric Machine Element," First National Conference on Automobile Electrical and Electronics Engineering, AEEE, Detroit, September 1964.

power densities of several horsepower per pound can be expected at "useful" (~90%) efficiencies.

An AC homopolar device would also have to operate on relatively high currents and low voltages, and at fairly low frequencies to minimize eddy-current, hysteresis, and skin-effect losses. Inefficiencies might thus be too great to tolerate, but the obvious advantage over DC homopolar motors would be ease of current and voltage control. The control problem would be qualitatively similar to that of a DC multipolar device because no frequency variation is required. Whether an AC homopolar motor would have any weight advantages remains to be determined, but high speeds should pose about the same problems as in the DC case because slip-ring contacts would be inadequate above a few thousand rpm. In the case of the AC multipolar inductive motor, the control problem is, again, acute because of the need to vary the frequency (or conceivably, the magnetic flux). Possible methods of doing this are discussed later.

However, the conventional DC motor has a serious problem of its own in the commutator, which is a mechanical device that makes and breaks a current thousands of times per minute. At high speeds commutators simply do not work well; in small sizes the upper limit is 3,000–5,000 rpm, depending on whether the motor has interpoles.[2] But the top speed of a motor determines its specific power capability; a 25 hp DC motor at 4,000 rpm might have a weight to power ratio of 8 lb/hp, whereas at 16,000 rpm, the same motor would produce 100 hp or 2 lb/hp. The generally superior performance of AC inductive motors, at the present time, is largely due to the fact that they can operate safely at much higher speeds. Even allowing for the gearbox, inverter, and extra cooling requirements, 3.5 lb/hp is currently attainable.

This figure of 3.5 lb/hp tends to be borne out by what little actual experimental work has been done with high-performance electric cars. The GM Electrovair,[3] for example, which is based on a Corvair chassis, uses an induction motor that weighs 130 lb and produces about 100 "peak" horsepower; with a 235-lb modulating inverter the ratio of overall weight to power is 3.6 : 1.

As implied in an earlier paragraph, the peak horsepower capability may be quite a bit larger than the (1-hour) rated hp, because electric motors can generally be overloaded for short periods. The allowable overload is a function of weight, design, and cooling arrangements.

Typical safe overload time ratings for some commercially available electric motors today are: 1 hour at 165%, ½ hour at 190%, ¼ hour at 210%, and shorter times at 250%. It should be pointed out that these ratings are for continuous use throughout the specified period. In a vehicle, the motor would be subjected to peak overloads for only a few seconds at a time during acceleration. Such brief

---

[2] J. W. March, "Motors for Electric Cars," Symposium on Electric Vehicles, Institution of Electrical Engineers, London, April 1967.

[3] H. Wilcox, "Electric Vehicle Research," GMC Research Publication, GMR–645, March 1967.

overloading should have no effect on the life of the motor in an electric car. (However, even if overloading were to cut the expected life of the motor in half, the remaining life expectancy would still be better than that of most internal combustion engines—or automobiles in general, for that matter—because industrial electric motors are often designed for continuous operation for twenty years or more.)

To get around the commutator problem that limits DC motor output, increasing attention has been given in recent years to so-called "brushless" motors. Much work has been directed toward providing solid state electronic switching, triggered by devices that sense the instantaneous position of the rotor with respect to the stator. Photoelectric, Hall-effect, and reluctance switches have been tried by NASA for space applications.[4] A less exotic and possibly more practical arrangement consists of a double commutator, with brushes linked to a set of four thyristors (silicon-controlled rectifiers) and arranged so that as the brush reaches a commutator contact the current is switched on and remains on while the brush is entirely within the contact, but is switched off again, by the thyristors, before shaft rotation brings the next contact under the brush. The brushes need carry only a light current.[5] In effect, they act as triggers for the thyristors, but carry no heavy currents themselves. Similar designs have been reported elsewhere.[6]

Indeed the use of sophisticated high-power electronic-switching circuitry is beginning to blur the classical distinctions between motor types. For instance, a "synchronous" motor, utilizing a permanent magnet rotor can be operated at variable loads and speeds by controlling frequency and voltage. One prototype of this type recently reported by Richard D. Thornton at MIT is expected to achieve 30 kw for a weight of 30 kg, with 85% efficiency at 6,000 rpm.[7]

Just as electronic alternatives to the commutator may improve the outlook for DC motors, so there may be nonelectronic solutions to the problem of speed control for AC motors. As noted previously, the stator for such a motor typically involves $2n$ windings (poles). As multiphase alternating current of a fixed frequency is supplied, the rotor speed is governed by the number of poles. If the number of *effective* poles in the stator can be changed, so will the motor speed. This is tantamount to varying the amount of magnetic flux intercepted by the rotor as it turns. A number of ingenious attacks on this problem have been made in England since 1955, notably by G. H. Rawcliffe at Bristol University and

---

[4] "Power Systems Research at MSFC," Research Achievements Review, Series No. 14, NASA TM–X–53419, 1965.

[5] J. J. Bates, "Thyristor Assisted Sliding Contact Commutation," *Proceedings, Institution of Electrical Engineers*, February 1966, p. 113.

[6] A. Tustin, "Electric Motors in Which Commutation Is by Switching Devices Such as Controlled Silicon Rectifiers," *Proceedings, IEEE*, January 1964.

[7] R. D. Thornton, "High Frequency Motors for Electric Propulsion," Intersociety Energy Conversion Engineering Conference, Denver 1968.

F. C. Williams at Manchester University. Phase multiplication—both AM and FM—and phase mixing of a number of separate signals have been tried, as well as spherical rotors and even a variable transformer that was an electrical analog of a slide rule (called "logmotor"). None of these schemes need be described here in detail because they all suffer from obvious drawbacks.[8] Nevertheless, it is quite conceivable that, within the next decade or so, an invention in this area may succeed in overcoming the major difficulties.

The problem of controlling the speed of a DC motor is not necessarily congruent with the problem of improving its performance by eliminating or changing the function of the commutator brushes. Voltage variation for DC motors has usually been implemented by inserting varying amounts of resistance into the circuit. This can be done by means of a rheostat or, if the power is supplied by a battery, by manually switching varying numbers of individual battery cells into or out of the circuit in succession as different voltage levels are needed, or by switching into various series-parallel combinations of cells. In fact, virtually all of the electric cars of the past and present—with the exception of a few very recent experimental designs—have used one or the other of these two schemes.

For example, in the Henney Kilowatt, which is an electrified Renault with a 7-hp series DC motor (modified by the Eureka Williams division of the National Union Electric Corporation) the required variable voltage is produced by switching the array of lead-acid batteries from a parallel to a series arrangement through the use of heavy-duty contact relays and a field shunt resistor.[9] The accelerator pedal of the car simply operates a sequence of micro-switches by a mechanical linkage, cutting the necessary voltage levels in or out as dictated by the demands of the accelerator position. The power being drawn at any instant is read directly from an ammeter and a voltmeter on the dashboard. This system, which utilizes relatively little variable resistance, is not too inefficient at most speeds; most of the power from the battery goes into the motor, although there are substantial losses. However, even neglecting the inefficiencies, which necessarily detract from an already limited speed and range, the control system is somewhat less than ideal. Physical switching of the batteries causes rapid changes in motor power, and thus a tendency to discontinuity or "jerkiness" in acceleration. Also, the frequent openings and closings of heavy contactors produce a constant "clacking" noise, which tends to detract from the inherent quietness of electric propulsion.

Variable resistance and electromechanical techniques are practical up to a point but they all have certain disadvantages. For one thing, considerable heat energy losses occur if resistors are used to vary voltages, particularly when the currents are large (as much as 500 amperes must be switched on occasion in the

---

[8]E. R. Laithwaite, "New Forms of Electric Motor," *Science Journal*, February 1966.
[9]Henney Kilowatt Operation Manual.

Henney Kilowatt). If the voltage is varied by switching individual cells in and out of the load circuit, there is some disparity in the work load. Unless the cells are rotated or "randomized" by some means, those at one end of the battery tend to be used all the time while those at the other end are used only when the "throttle" is wide open. Furthermore, mechanical switching techniques tend to be inherently rather slow. Of course, none of the above is applicable to AC motors.

ELECTRONIC CONTROLS

Despite the various means of speed control noted, the most favored current approach for controlling both AC and DC motors is based on the use of solid-state switching devices—transistors, and silicon-controlled-rectifiers (SCRs) or thyristors. The device used for DC motors is a DC "chopper" in which a solid-state diode (switch) is "on" for relatively short times and "off" for relatively long times, with a very high rate of repetition. The result of this is a low *average* voltage to the motor. Voltage is increased by increasing the percentage of time during which the switch is "on" until it is on at all times; if the voltage is to be decreased, the "on" time is reduced until it is cut off entirely. No power is dissipated in such a system, and the theoretical efficiency is therefore high. Furthermore, very fine control is afforded, since the process is stepless and almost infinitely variable (see figure 10-4).

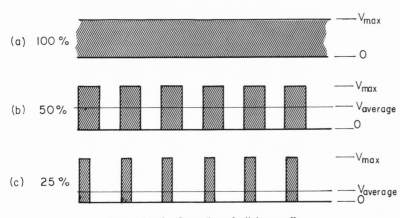

Figure 10-4. Operation of a "chopper."

A pulse control modulation (PCM) technique can also be used in the control of polyphase AC induction motors such as would be used in vehicles, by using a chopper to control the DC input voltage to a three-phase inverter; the variable frequency output from this inverter in turn controls a three-phase motor. The motor's rotational speed is controlled by timing a sequence of six transistor or thyristor switches in the inverter output to the motor, so that current is led into the motor in the right sequence. However, since the total switch current capacity for the AC motor is approximately three times that of the DC motor (because of the three-phase current required), the AC pulse control modulation system is more complex and therefore more costly. The question of costs is discussed later, but it seems safe to predict on the basis of the past record of solid-state component costs that, by the time high-speed AC motors are in common use for automobile propulsion, AC motors and control systems will have become competitive in price with DC systems of the same rated power.

As indicated previously, there are two principal types of solid-state switches—the transistor and the silicon-controlled rectifier (SCR) also known as a thyristor. The main operational difference between the two is that the transistor can switch either "on" or "off," i.e., it can both initiate and interrupt current, while the SCR can only switch "on." If the device is also to interrupt a current, a capacitor that is charged during the SCR's "on" time must be discharged via an auxiliary SCR that diverts current from the main SCR and gives it time to recover its blocking capacity. Typical simplified circuits embodying transistors and SCRs are shown in figure 10-5.[10] The diode across the motor is necessary to prevent any excessive rise in voltage (Back EMF) due to motor inductance when the switch opens; this diode is necessary also for DC motor circuits. Although the transistor circuit has fewer components, the SCR

(a)   Transistor Circuit

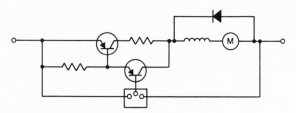

(b)   SCR Circuit

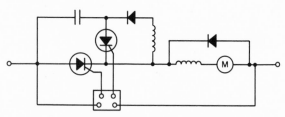

Figure 10-5. Basic electronic control circuit schematics.

[10]N. Mapham and J. Mungenast, "Semiconductor Power Controls for Electric Vehicles," National Electric Automobile Symposium, San Jose, Calif., February 1967.

controller is likely to be the cheaper of the two and therefore (all else being approximately equal) the more desirable.

Solid circuitry is also adaptable to control of regenerative braking. In deceleration, the motor is, in effect, being driven in reverse. Externally applied torque is acting to drive the rotor through a magnetic field, and the voltages are thus reversed and the motor becomes, temporarily, a generator. The excess kinetic energy of the car should be recoverable in part at least. The method by which this is implemented depends on the type of motor.

Whatever the motor type, if it is to function as a generator, current must be permitted to flow *out* of the terminals. This is ordinarily prevented by the diode across the motor, which must be bypassed in any case. Variable speed shunt motors can be made to regenerate by simply increasing their field strength. Series DC motors, on the other hand, may require either a reversal of the field direction or an additional shunt field coil that would allow current to flow from the armature through a diode into the battery. In an AC induction motor, the regeneration mechanism is basically simpler. Such a machine, if driven above synchronous speed, will automatically generate power with the same terminals and connections. Thus, control circuitry need merely sense the shaft speed at all times and ensure that the phase speed supplied to the stator is reduced when dynamic braking[11] or regeneration are desired. The sensing of the shaft speed so that the input voltage can be adjusted to ensure a "slip" at all speeds is one of the methods of producing high torque in such motors. When shaft speed exceeds synchronous speed a "reverse slip" is created. Active development projects in the area of "brushless DC motors" using solid-state control circuitry are or have been under way at the U.S. Army Engineer R&D Laboratories, Fort Belvoir; General Electric; General Motors (Delco-Remy Division); Fairchild Semi-Conductor; Gulton Industries; Linear Alpha, Inc.; Lansing-Bagnall (U.K.); C.A.V. Ltd. (subsidiary of Jos. Lucas); AEI Ltd.; Volkswagen (W. Germany); Fiat (Italy); and Toshiba (Japan).

WEIGHT AND COST

Since the power produced by a motor increases in direct proportion to its speed, the first and most obvious approach to increasing specific power is to operate at higher speeds. For DC multipolar motors, this means modifying or replacing the commutator, as remarked earlier. The next major constraint on

---

[11]While regenerative braking can slow down the car, it cannot by itself bring the vehicle to rest quickly. When the speed has reached a level (usually 15 mph or so) at which it is no longer possible or practical to maintain the regenerative requirements, mechanical brakes will have to be used. A certain degree of reorientation may be needed to accustom drivers to correlating their use of the conventional and electrical brakes. However, because kinetic energy is a quadratic function of speed, the use of mechanical brakes below 15 mph does not introduce much inefficiency. Reducing the vehicle speed to one-half, for example, will cut the kinetic energy by three-quarters, and most of this energy can (in principle) be recovered.

operating speed is imposed by cooling requirements and the need for gearing down by a large factor. It is probably within the state of the art to solve these problems economically for speeds of 20,000 rpm; with further intensive development, it may become possible to exploit motor speeds of 50,000–80,000 rpm in the next twenty years.

The main avenues to weight-reduction involve reducing (a) the mass of copper (or aluminum) wire required for the windings and (b) the mass of soft iron or silicon-steel cores required for the magnetic circuitry. Both approaches have already yielded significant results: "printed" disk-type motors have shown themselves to be lighter—and perhaps cheaper—for a given horsepower output than conventional wire-wound types, largely through a saving in electrical conductor weight. The weight could be reduced tremendously if metallurgical research results in lighter (but inexpensive) low-reluctance materials that could be used in place of iron (or steel). Topological ingenuity on the part of designers has recently been focused increasingly on the magnetic circuit, and several ways have been found to make motors more compact by reducing the length of flux paths and reducing flux leakage at air gaps. Some of the newer aircraft starter-generator motors, such as North American Aviation's "Nadyne," have incorporated very unusual rotor designs to increase efficiency and minimize weight.

Because of the diversity of electric motor types and sizes, it is impossible to assign a single weight/power ratio that is meaningful. In general, specific weights will be lower for larger motors than for smaller ones, owing to geometrical scaling relationships that tend to become more favorable with increase in size. In particular, power output tends to vary as the fourth power of diameter, while electrical losses vary as the third power. Hence larger motors can be far more efficient: as diameter increases from 3 to 14 inches, in a typical case, efficiency increases from 55% to 90%.

Another standard figure of merit, the ratio of peak or stall torque to motor weight, has been plotted for a number of electric motor types by George Hoffman (see figure 10-6).[12] The approximate limiting case, it will be noted, is defined by the envelope curve $T_s/W = 8$, where T_s is the stall torque in lb-ft, and W is the motor weight in lb. For most present-day motors of possible interest for vehicular propulsion (AC industrial motors, DC truck motors, etc.), T_s/W is much less than 8. For DC motors, the range is from 0.5 to about 1, for AC motors from roughly 0.75 to 1.5 or so. Stall torque represents perhaps 300–400% of the maximum torque at rated load for DC motors, and roughly 200% for AC motors. Thus, if rated torques for typical stall torque values on the graph are calculated for likely motor sizes of automotive interest (at, for example, 1,800 and 3,600 rpm), and using stall torque/weight ratios of 1 for DC and 1.5 for AC, estimates of motor weight per hp can be found. For DC motors

---

[12] George A. Hoffman, *Electric Motor Cars*, Rand Corporation Memorandum RM-3298-FF, March 1963.

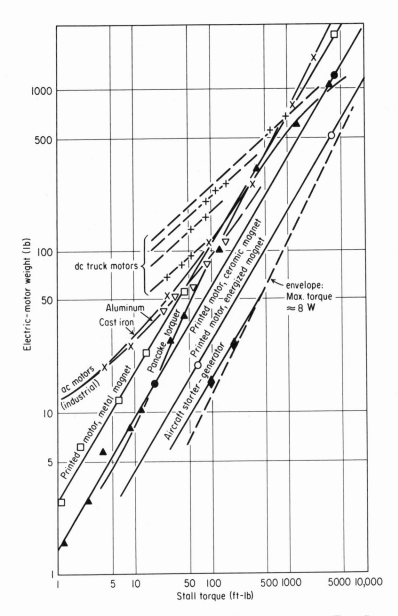

Figure 10-6. Stall torque versus weight for various electric motor types. (From George A. Hoffman, *Electric Motor Cars*, Rand Corporation, RM-3298-FF, March 1963.)

under these conditions, the figure is about 8 lb/hp at the lower speed, dropping to 4 lb/hp at 1,800 rpm for AC motors and 2 lb/hp at 3,600 rpm. These figures tend to agree rather well with actual cases.

With improved materials, higher operating speeds, etc., it should be possible to approach closer to the envelope stall-torque/weight ratio of 8. In this case, the weight/power ratio for electric motors could be reduced to as low as 0.5. The specific weights of various high-speed special purpose AC motors, in fact, already come close to this value.

It is equally difficult to assign an all-inclusive rule of thumb to the costs per unit weight or per unit power output of electric motors and controls. Assuming mass production of several million units per year, one 1960 study estimated an average cost of $3.30 per pound for wound DC motors, the actual values being somewhat higher for low power ratings and somewhat lower for high power ratings.[13] A figure in dollars per hp output would be better, in view of the sharp decrease in lb/hp looked for. With more sophisticated designs, a higher price per pound must be anticipated. The increase might, or might not, be offset by savings from mass production. Until a final motor design is reasonably well established, reliable estimates of manufacturing cost are impossible to make.

The cost of speed (voltage) control equipment is another important variable in the total picture. For conventional switching and variable-resistance types of control equipment, mass production prices have been estimated at less than 25% of the cost of mass-produced DC motors (of conventional design) for the same power rating.[14] Solid-state transistor or thyristor controls with the requisite power capabilities as of the late sixties were still quite expensive, since they are still in rather early stages of development. High-volume production costs for solid-state components can be assumed to be roughly proportional to the area of silicon required to handle a given current. Since the silicon area for transistors is approximately five times that for SCRs, the SCR circuits, despite a somewhat greater complexity, should prove cheaper to manufacture in very large numbers. In lots of thousands, the price for SCRs in 1967 was approximately $50 per square inch, while the standard automotive diodes, which are produced in very large quantities, were priced at $10 per square inch. (Prices of solid-state electronic devices have been dropping rapidly for a decade, however, so it is hard to make long-term predictions.) Thus, it is probable that in volume production, the costs of solid controls will be between $50 and $10 per square inch. Based on a compromise of $30 per square inch, the cost of a 30-kw variable pulse length DC controller in volume production would be $40, and a cost for an equivalent AC controller—approximately three times as complex—would be about $100.[15]

---

[13] D. R. Adams et al., *Fuel Cells—Power for the Future* (Willow Grove, Penna.: Fuel Cell Research Associates, 1960).

[14] Ibid.

[15] Mapham and Mungenast, "Semiconductor Power Controls for Electric Vehicles."

Estimates made privately to the authors by other representatives of the semi-conductor electronics industry have been even lower.

It has been projected by one source that solid-state chopper controls for DC motors should eventually exhibit manufacturing costs of $2 per hp, and that corresponding costs of solid-state inverters for AC motors will be $5 per hp.[16]

In the newer systems it is difficult to separate the costs of motors and controls. However, very rapid progress is being made in overall terms. According to Richard Thornton of MIT, new systems have reduced motor weights by nearly an order of magnitude in a decade with no increase in costs per unit weight.[17] This is an indication of spectacular technological progress that seems likely to continue for some time to come.

---

[16]J. H. B. George, L. J. Stratton, and R. G. Acton, *Prospects for Electric Vehicles—A Study of Low-Pollution-Potential Vehicles—Electric*, A. D. Little, Inc., Report for the Department of Health, Education and Welfare, May 15, 1968.

[17]Thornton, "High Frequency Motors for Electric Propulsion."

Chapter Eleven

The Electric Vehicle

The all-electric vehicle has appeared to many people to hold out the promise of truly modern and pollution-free transportation. This promise is somewhat chimerical. Freedom from pollution cannot be claimed for a vehicle powered by a fuel cell, since significant conversion inefficiencies do in fact exist. And in the case of a battery electric, the source of pollution is merely transferred to the central electric power plant. Residuals from such a plant may be different in kind, but are not necessarily less in magnitude or easier to deal with. As it seems unlikely that a fuel-cell vehicle will be really practical within the next decade or two, only the characteristics and requirements of a battery-powered car are examined in this chapter. (Fuel cells are described in Appendix A, and mentioned briefly in the discussions of current electric-vehicle efforts at the end of this chapter.)

GENERAL PRINCIPLES

Battery-powered electric vehicles are distinguished from fuel cell–battery hybrids in that "fuel" is—directly or indirectly—electricity itself. Apart from considerations of air pollution, the major attractions of an all-battery system would be simplicity, reliability, ease of recharging ("plug it in"), and the use of utility electricity, which is the cheapest available source of energy.

Actually, although the *option* to recharge electrically is a valuable one— without which regenerative brakes would be impossible, for example—the *necessity* to recharge an electric car electrically is a potential drawback, and the first limitation to be overcome, if possible. Even assuming that a large battery could be recharged in a very short time (small specially designed nickel-cadmium cells, for example, can accept ten or fifteen minute recharges), a fast recharge is inevitably as inefficient as a fast charge. Moreover, the problem of generating and distributing the enormous amount of electric power that this would entail remains. For example, a minimal 7-kwh lead-acid battery for a 2,000-pound vehicle, such as has been repeatedly suggested for urban use, would require an actual input of up to 15 kwh if a 25-minute charging rate is assumed. Even then, the 25 minutes required to charge is not comparable with the time spent in simply refilling a gasoline tank. Furthermore, a maximum input power of about 40 kw—or 3,300 amps at 12 volts (DC)—would be required. For a 70-kwh battery the peak power levels would obviously be ten times larger. The size and weight of various battery components would increase drastically with increase in charging current (decrease in charging time), with cables reaching bus-bar proportions if a large battery is to be charged in half an hour, for example. This, of course, does not mean that it could not be done; but recharging as fast as this is not at all a simple matter. Presumably a network of "charging stations" would be needed—much like today's network of filling stations—and the charging stations would have to be equipped to handle extra heavy electrical loads.

In 1967, John Campbell, then editor of *Analog* (science fiction magazine), graphically envisioned "a Howard Johnson restaurant on the New Jersey Garden State Parkway on a summer Saturday noon, with their big parking lot solidly filled with cars, each thirstily drinking in the juice at 1,000 to 1,500 amps apiece—and the dull roaring of the six 60-inch cooling fans trying to keep the boxcar-sized transformers and silicon rectifiers from frying in the 95° heat," and a 30-inch sodium-filled steel pipe for feeding the charge current to withstand the magnetic forces of 4 tons per running foot, and so on.[1]

While this sort of massive installation would be technically feasible, at least on major traffic arteries, it would hardly be practical to space them every few blocks, like neighborhood service stations. Thus electric car drivers would sacrifice some of the convenience and flexibility that makes the automobile an attractive means of personal transportation.

A rechargeable battery with optionally replaceable anodes (and electrolyte) would apparently be the ideal solution because it could be refueled in minutes on the road or recharged overnight at home as the occasion demanded. There are undoubtedly possibilities in this direction, although no practical system is yet in sight.

It has also been proposed that batteries might be recharged locally, either at home or in local charging stations, most of the time, but that for long trips discharged batteries could be *exchanged* for fully charged ones. In effect, one would be renting a battery for the trip. To make this possible, batteries would have to be designed in modular units with provision for rapid connecting and disconnecting. For instance, GM scientists have suggested that a high-energy battery of the lithium-chlorine type might be rented for a trip at $0.05 to $0.06 per mile, which is somewhat cheaper than renting a complete car. Alternatively, ICE or ECE powered "battery chargers" might be available on a rental (or purchase) basis. Such a unit could, for instance, be permanently mounted in a trailer, which would also provide additional luggage space as well as a long-range

---

[1] John W. Campbell, "Portable Power" (editorial), *Analog*, Vol. 77, No. 1 (March 1967).

energy source. (For campers, a unit of this sort would have the additional advantage of doubling as a power takeoff.)

The major competitive advantage of an electric car of small size and relatively modest performance capabilities would be maneuverability; possibly a dual-mode capability (i.e., able to travel on the road under its own power, or on a closed guideway using distributed electricity) as in the case of the Alden StaRRcar or the Cornell Aeronautical Laboratory "Urbmobile" concepts described later; long life; and very low operating cost.

The three elements of cost—amortization of capital costs including capital replacement over the lifetime of the vehicle; maintenance; and energy costs—are discussed below.

To estimate capital costs, one must make some assumptions about structural materials. Because weight is an important consideration, the choice for an electric car will probably be between aircraft-type aluminum and the new high-strength thermoplastics, such as ABS (Acrilonitrile-Butadiene-Styrene). With new molding techniques now highly perfected, a thermoplastic would probably be preferred on grounds of costs—tooling costs for thermoplastics being roughly 10% of those for metal construction. It is still an open question whether ABS or other such materials will have an effective lifetime of twenty years, which would be ideal. Also, ABS becomes brittle at temperatures below -25°F, which could be a problem in some regions, although not on the East or West coasts of the United States, or in Europe or Japan. However, maintenance on an ABS plastic body is remarkably simple: to take out a dent, for instance, it is only necessary to apply a little heat. Colors are permanent, requiring no painting or retouching. If the brittleness problem can be overcome or avoided, it is likely that electric cars would use plastic bodies, possibly on aluminum or steel frames. This innovation could cut the chassis weight relative to the weight of motors, controls, and batteries from 60-65% (characteristic of present-day vehicles) to somewhere in the neighborhood of 50-55%.[2] It has also been suggested—perhaps optimistically—that with the new techniques, the costs might be cut from the present figure of about $0.50 to $0.40 per pound.[3] Alastair Carter, a British industrialist planning to manufacture an electric car with an ABS thermosetting plastic body, is equally, if not more, optimistic on costs.[4] Taking the figures for chassis percentage of total weight cited earlier, for example, chassis costs would be of the order of 20 cents per pound of vehicle weight.

Costs of motors and controls, discussed in the previous chapter, still depend on some major technological uncertainties. Given further development and a

---

[2] George A. Hoffman, "Electric Motor Cars," Rand Corporation Memorandum RM-3298-FF, March 1963.

[3] George A. Hoffman, "Hybrid Power Systems for Automobiles," in *Power Systems for Electric Vehicles*, PHS Publ. No. 999-AP-37, U.S. Public Health Service, National Center for Air Pollution Control, Cincinnati, Ohio, 1967.

[4] A. Carter, Hearings, Subcommittee on Anti-Trust and Monopoly, "Economic Concentration," Part 6, September-October 1967.

resolution of these issues, $0.10 to $0.15 per pound of vehicle weight would probably span the range of likely costs.

As regards energy storage, the only commercial system capable of reasonable performance is the silver-zinc battery, which can be ruled out on grounds of short cycle life and the scarcity of silver.

Thus, if we want high-energy batteries at reasonable costs, we must look to one of a number of advanced batteries currently under development. Of these, the most immediately promising may be Ford's sodium-sulfur battery, or Gulf General Atomic's rechargeable zinc-air battery, although some of the other systems discussed in chapter 5 may be suitably developed within the same time span. It is impossible to estimate manufacturing costs accurately, but most of the basic materials involved are cheaply available in quantity (except for the noble metals now used as catalysts in the oxygen electrode of current versions of the zinc-air system). Zinc, for example, is in the same general price range in bulk as lead. Thus, in practical sizes and in quantity production, the fully developed zinc-air battery should not be much more expensive (*per pound*) than the lead-acid battery. Since the specific energy (watt-hours per pound) for the zinc-air cell is from four to six times that of the lead-acid cell, it should ultimately be cheaper in terms of dollars per kwh stored. Costs for the well-developed lead-acid battery approximate $0.50 per pound plus the basic material costs. Theoretically, since all the materials for the zinc-air or sodium-sulfur cells are equally readily available, eventual production costs for either of these new batteries might also approach this value per pound. Another way of estimating the cost was suggested by an industry expert who observed that multiplying the basic cost of reactants and electrode materials by five seems to work quite well with existing batteries in all price ranges. This method, despite obvious drawbacks, may be the best available rule of thumb for estimating future battery costs.[5] On this basis, either of the above batteries would work out to about $0.75 per pound of reactant, while the General Motors lithium-chlorine battery would presumably cost about $2.50 per pound of lithium chloride.

Actually the reactant and electrode weights may be a much less significant fraction of the total for the zinc-air and lithium-chlorine systems than for the sodium-sulfur system, owing to the presence of auxiliary components and controls. In the latter case, however, it may not be unreasonable to assume an ultimate cost of $0.75 per pound or about $5 per kwh for the whole battery. Thus a vehicle with an *overall* specific energy of 50 whr/lb would have to have at least one-third of its total weight devoted to storage cells (∿150 whr/lb) costing $500 per ton of vehicle weight. Allowing for a spectrum of possibilities ranging from under-powered vehicles to more expensive batteries, the realistic range is probably plus or minus 50% of the above estimates for an all-battery vehicle.

For trips, a battery might be rented as suggested earlier, or perhaps only the active elements or the catalysts, depending on the type and construction of the

---

[5]T. A. Ferrell, Electric Storage Battery Co., personal communication.

battery. On a rental basis, even the costly silver-zinc battery might not be altogether inconceivable. The principal manufacturer of silver-zinc batteries has argued that the cost of renting one of these batteries for use in a small car driven 12,500 miles per year would be $74 per year, and that the car could be operated for less per year than a comparable gasoline-powered car.[6] The latter conclusion is probably overly optimistic, but the rental concept, in general, deserves serious consideration. Presumably, in any such scheme, the customer would purchase the car chassis plus its electric motor or motors and controls, but rent the "energy package."

Electric motors are routinely designed for operating lifetimes of 10,000 to 20,000 hours. If a car is driven an average of 10,000 miles per year, the approximate total driving done in eight years comes to 3,200 hours, assuming a 25 mph average speed; this is probably close to the national median. This is, of course only a fraction of the useful motor life. A liquid electrolyte battery would probably have to be changed at least every two or three years, unless a cycle lifetime well in excess of 1,000 charge-discharge cycles can be achieved without sacrificing energy storage capacity. However, the new Ford sodium-sulfur battery and others of the molten-salt type might have an almost indefinite longevity. Fuel cells, also, have reported a substantial life expectancy. Both hydrogen and reformed natural gas fuel cells have been operated for 10,000 hours, although this is unusual. Thus, most of the major components of an electric car, with the exception of some types of batteries and the electronic speed control circuitry, whose reliability and longevity is basically unknown, can be expected to last for many years of normal use.

If we assume that a typical 3,000-pound car is driven 10,000 miles per year at an average of 15 miles per gallon of gasoline and a cost of $0.20 per gallon (exclusive of taxes), then the energy costs of a conventional car come to about $133 per year, or 1.3 cents per mile. For city driving, the cost might well be higher. Other costs for both conventional and electric cars include regular maintenance (oil, grease, tires, etc.), repairs, and depreciation. However, an electric car does not require oil changes, antifreeze, or tune-ups, and repair costs should be lower because the electric motor is designed for a long operating life.

The average power required to operate a vehicle was analyzed theoretically in chapter 3. If the duty cycle (or driving cycle) can be approximated reasonably well in terms of independent distribution functions of acceleration (a), slope (s) and velocity (v), with a specified fraction (F) of the time in a cruise mode, the general expression is as given in equation (3.9). The situation where energy is produced at a constant rate and stored in an intermediate reserve and where regeneration occurs during deceleration (braking) introduces further complications, however, since not only are there losses involved in storage and

---

[6]G. Dalin, "Performance and Economics of the Silver-Zinc Battery in Electric Vehicles" in *Power Systems for Electric Vehicles*, PHS Publ. No. 999–AP–37, U.S. Public Health Service, National Center for Air Pollution Control, Cincinnati, Ohio, 1967.

recovery, but these losses are, in general, functions of the rate at which these processes occur. Thus a modified expression for average power, $\bar{P}_e$, is as follows:

$$\bar{P}_e = F \underbrace{\int_0^\infty dv\, f_3(v) \int_{-\infty}^\infty ds\, f_2(s)\, P_w\, E^{-1}(P_w)}_{\text{cruise}}$$

$$- \underbrace{\frac{1}{2}(1 - F)e_R \int_0^\infty dv\, f_3(v) \int_{-\infty}^\infty ds\, f_2(s) \int_{-\infty}^0 da\, f_1(a) P_w\, E^{-1}(P_w)}_{\text{deceleration}} \quad (11.1)$$

$$+ \underbrace{\frac{1}{2}(1 - F) \int_0^\infty dv\, f_3(v) \int_{-\infty}^\infty ds\, f_2(s) \int_0^\infty da\, f_1(a)\, P_w\, E^{-1}(P_w)}_{\text{acceleration}}$$

where the negative second term (deceleration) corresponds to return of power to the intermediate reserve via regenerative braking. The regeneration efficiency is assumed to be e_R, where $0 \leq e_R \leq 1$. The function $E^{-1}(P_w)$, representing power-related inefficiencies in energy storage and release, is defined to be unity at some "nominal" discharge rate[7] and increases as a nonlinear function of P_w. For convenience we can assume the quadratic form

$$E^{-1}(P_w) = 1 + \mu \left\{ \frac{P_w}{P_5} \right\} + \eta \left\{ \frac{P_w}{P_5} \right\}^2 + \ldots \quad (11.2)$$

where P_5 is the standard 5-hour discharge rate, and μ and η are empirical constants. The average rate at which energy is withdrawn from the reserve or otherwise becomes available is given by:

$$\bar{P}_e = \bar{P}_w + \frac{\mu}{P_5} (\overline{P_w^2}) + \frac{\eta}{P_5^2} (\overline{P_w^3}) + \ldots . \quad (11.3)$$

Numerical values for the parameters μ and η can be inferred directly from published data for most types of batteries by making the plausible definition

---

[7]In the case of batteries the nominal discharge rate is generally taken to be the "5-hour" rate, i.e., the rate such that a total discharge requires 5 hours. For consistency we shall assume a standard 5-hour rate in all cases, including flywheel storage.

that the inefficiency E^{-1} is equal to the inverse of the ratio of P_w to P_5. (These values can be read directly from figure 5-8 in chapter 5.) A rough fit of the quadric form (11.2) to the above ratio for several major battery types gives the values shown in table 11-1. Note that P_5 is actually a measure of the battery capacity: battery storage capacity in kwh is equal to five times the 5-hour discharge rate measured in kilowatts. It is probably reasonable to ignore the quadratic terms for most practical purposes, since they are very small. To simplify the algebra, let us assume a strictly urban driving situation where the

Table 11-1. Parameters for the Function E^{-1} (P_w)

Battery type	μ	η
Lead–acid	+0.13	−0.003
Nickel–cadmium	+0.04	−0.001
Zinc–silver	+0.03	0.000
Sodium–sulfur	+0.05	+0.001

driver is either accelerating or decelerating at all times ($F = 0$) and assume the simple distribution functions introduced in equations (3.6), (3.7), and (3.8). In this case, the final expression for average energy consumption per mile becomes—in analogy with equation (3.11):

$$
\frac{\bar{P}_e}{v} = \left[\alpha + \frac{(1 - e_R)}{\sqrt{2\pi}} (\beta\sigma_s + \gamma\sigma_a) \right] W + \delta A v_o^2 \exp(2\sigma_v^2)
$$

$$
+ \frac{\mu}{P_s} \left\{ \left[\alpha^2 + \sqrt{2/\pi}\,(1 - e_R)(\alpha\beta\sigma_s + \alpha\gamma\sigma_a + \sqrt{2/\pi}\,\beta\gamma\sigma_a\sigma_s) \right. \right.
$$

$$
\left. + \frac{(1 + e_R^2)}{2} (\beta^2\sigma_s^2 + \gamma^2\sigma_a^2) \right] W^2 v_o \exp(\tfrac{1}{2}\sigma_v^2) \tag{11.4}
$$

$$
+ \left[2\alpha + \sqrt{2/\pi}\,(1 - e_R)(\beta\sigma_s + \gamma\sigma_a) \right] \delta A W v_o^3 \exp(\tfrac{9}{2}\sigma_v^2)
$$

$$
\left. + \delta^2 A^2 v_o^5 \exp(\tfrac{25}{2}\sigma_v^2) \right\} + \dots,
$$

where W equals mass and A equals the cross section of the vehicle. To obtain the energy consumption in kilowatt-hours per ton-mile, we multiply equation (11.4) by $[0.746 \times (2{,}000/W)]$. The numerical values of α, β, γ, δ, and A, given in chapter 3, correspond to data for a Volkswagen, while σ_s, σ_a, and σ_v were estimated for the Pittsburgh driving cycle. We return to these numerical calculations later.

Assume that an automobile weighs one ton (2,000 pounds); that rolling friction and drag coefficients are equivalent to those for a Volkswagen on gently rolling terrain (σ_s + 0.3); that operating conditions (i.e., driving cycles) correspond to $\sigma_a = 1.3$, $v_o = 16$ mph, and $\sigma_v = 0.4$, and that lead-acid batteries ($\mu = 0.13$) having a 6.5-kwh capacity ($P_o = 1.3$ kw) and a regeneration efficiency of 70% ($e_R = 0.7$) are used. Equation (11.4) then gives:

$$\frac{\bar{P}_e}{v} = 0.248 \text{ kwh/ton-mile.} \tag{11.5}$$

However, if all other factors remain constant, but we decrease μ from 0.13 to 0.06 (see table 11-1) and increase P_o to 5 kw (corresponding to a 25-kwh nominal battery capacity of intermediate power density) the energy cost drops to:

$$\frac{\bar{P}_e}{v} = 0.152 \text{ kwh/ton-mile.} \tag{11.6}$$

By introducing a still better energy storage system with performance characteristics similar to those projected for the Ford sodium-sulfur battery, ($P_o = 17.5$, $\mu = 0.03$) the energy cost drops to:

$$\frac{\bar{P}_e}{v} = 0.136 \text{ kwh/ton-mile.} \tag{11.7}$$

On nearly flat terrain, with the same basic driving cycle, the absence of regenerative braking increases the energy consumption for the case of lead-acid batteries from 0.248 kwh/ton-mile to 0.324 kwh/ton-mile. In other words, regenerative brakes with an e_R of 0.70 would add about 30% to the range of a vehicle; this is consistent with empirical observations. In hilly country the effect would be greater, of course. For the case $\sigma_s = 1.0$ without regeneration ($e_R = 0.7$), the energy consumption increases to 0.348 kwh/ton-mile; regenerative brakes (with $e_R = 0.7$) therefore result in a 35% increase in range.

Equation (11.4) is obviously quite sensitive to the median velocity v_o because of the quadratic, cubic, and quintic terms. Figure 11-1 shows the energy requirements per ton-mile as a function of v_o, with and without regenerative braking both for lead-acid batteries and for a hypothetical high-performance battery.

The energy costs given above do not include electrical and mechanical losses and inefficiencies in the vehicle itself (about 15-20%), nor do they include battery charging losses (about 15-25% depending on the rate of charging). In addition, every automobile requires a certain amount of power for accessories

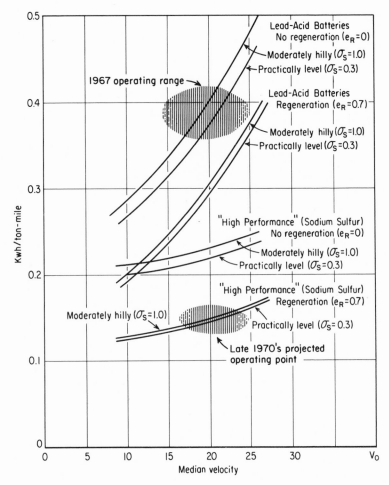

Figure 11-1. Energy requirements per ton-mile for various battery types. (From R. U. Ayres and Richard McKenna, *Technology and Urban Transportation: Environmental Quality Considerations*, Vol. II, Hudson Institute, Croton-on-Hudson, New York, HI-949/II-RR, 1968.)

such as lights, radio, heater, and air conditioner. The first two cause relatively little drain, and there is no reason why heat for the passengers cannot be derived largely from the motors, which will require cooling anyhow. Air conditioners operating from waste heat and requiring no auxiliary power can theoretically be designed, but it is probably necessary to allow extra power for this purpose.

If the vehicle is assumed to be completely battery powered, the "fuel" operating cost, based on an expected late 1970s figure of merit of ~0.15 kwh/ton-mile (see figure 11-1) and $0.12 per kwh as the Federal Power Commission's projected cost of electric power from the distribution lines, comes to $0.0018 per ton-mile or $18 per ton of car per year for 10,000 miles of driving.[8] If power is available at a cheaper rate during off-peak hours, the actual costs would be still lower.

There is an additional factor to be considered in battery cost analysis. In general, battery lifetimes are considerably shorter than those of the vehicles themselves; thus they need several replacements during an average vehicle life. It is convenient, therefore, to treat battery first costs as operating costs by amortizing them over the life of the battery. The life of a secondary battery can be measured either in length of time (as is typically done for batteries under constant charge, like the standard automotive SLI battery) or by the maximum number of charge-discharge cycles that can be accommodated. The latter method, which might be more appropriate for the case of the battery-powered vehicle, affords an insight into the relative percentage of stored battery capacity that can actually be withdrawn.

The total energy available during the life of the battery is a function of the number of cycles and the depth of each discharge. The number of charge-discharge cycles in most batteries typically decreases with rate of discharge from some maximum at a low rate. While it is in principle possible to discharge a battery completely, in practice this is never done; 100% discharges are not only inconvenient, they are usually harmful to the cells. The actual depth of discharge influences the number of cycles—the shallower the discharge, the longer the cycle life. (Figure 11-2 shows the projected relationship of the two parameters for 1972 lead-acid cells.) It is clear that the total energy of a battery is the energy capacity of the battery multiplied by a factor X, which is the product of the percentage depth of discharge and the expected cycle life at that depth. This factor for lead-acid batteries is shown in figure 11-3. Thus, the optimum discharge point for a lead-acid battery would be about 50%; at that level, the total available energy would be approximately 300 times the battery capacity (in kwh). This would be further decreased by a factor that is a function of the percentage of time spent at high rates of discharge.

It can be seen that the additional cost per ton-mile that must be added to the electrical recharge cost of a battery-driven vehicle is the total cost of the battery amortized over the above quantity (i.e., the capacity times the cycle life times the depth of a discharge). For the lead-acid case, then, the additional cost per kwh of capacity, if it is assumed that the specific weight is 10 whr/lb and the average cost is $0.50 per pound, is:

$$\$(0.15 + 0.12) \times 0.15 \text{ kwh/ton-mile}$$

or $0.05. In turn, for 10,000 miles of driving, this becomes $500 per ton of car, per year.

---

[8]*National Power Survey*, A Report by the Federal Power Commission, Vol. 1 (1964), pp. 277-85.

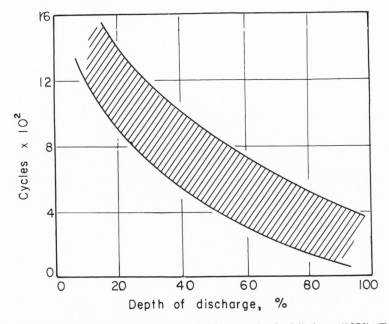

Figure 11-2. Projected lead-acid battery cycle life versus depth of discharge (1972). (From E. T. Canty et al., *New Systems Implementation Study*, A Report Prepared for the U.S. Department of Housing and Urban Development, February 1968, Vol. II, pp. 6-53. Courtesy of General Motors Research Laboratories. The data are a GM estimate of the state of the art as of 1967.)

Table 11-2 shows the *total* energy available per pound in various other batteries for typical discharge depths and expected cycle lives.

It is worth noting that more than one-third of the conventional car's annual actual fuel bill goes for road and highway taxes—about $90 for a car driven 10,000 miles a year at an average of 15 miles per gallon of gasoline. If ten million vehicles in the United States were replaced by battery-powered cars, there would be an annual tax loss of about $900 million. Obviously, to the extent that these taxes are required for road and highway maintenance, or for general revenue, some alternative form of tax would have to be devised. The problem is, how should it best be collected? With the existence of a nationwide network of charging stations that only charged vehicles, road tax collection would presumably be fairly straightforward, but this would not apply to home-charged second cars. This point raises some interesting questions of public policy. Assuming the feasibility of a hydrocarbon fuel cell–battery hybrid vehicle, the question might turn out to be academic, of course, since the existing hydrocarbon (gasoline) distribution system would still provide the basic energy input.

In view of the economic constraints, a possible future pattern in the development of electric cars seems to be discernible. Small limited performance vehicles

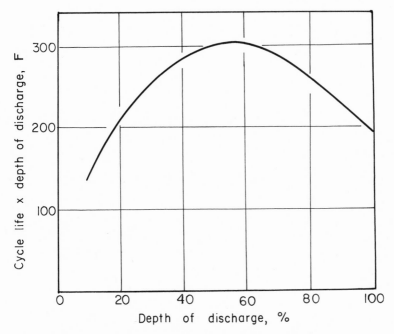

Figure 11-3. Withdrawable energy factor for lead-acid battery.

Table 11-2. Total Battery Energy Available per Pound

Battery	Energy density (whr/lb)	Discharge rate (hr)	Depth of discharge (%)	No. of cycles	Total energy available (kwh/lb)
Nickel–cadmium	14	5	30	2,000	8.4
Silver–cadmium	24	5	10	500	1.2
Silver–zinc	50	5	10	250	1.25
Zinc–air	50	10	50	20	0.5
Lithium–silver chloride	30	5	10	200	0.6

Source: Department of Defense Research on Unconventional Vehicular Propulsion, U.S. Army Engineer R&D Laboratories, Fort Belvoir, February 18, 1967.

powered solely by batteries—initially, perhaps, by lead-acid batteries—would presumably become available first; such vehicles can be produced now, in fact, at prices almost competitive with small gasoline cars. Within a period of several years, small cars, powered by higher-energy batteries, would be a possibility.

With the full development of one (or more) of these various high-energy batteries, the performance of battery-powered cars could become competitive with at least the smaller internal-combustion cars—Volkswagen, Renault, Fiat, etc. Accordingly, a range of typical electric cars and their possible performance is presented in Table 11-3. There will be many individual variations, but this illustrates a plausible "scenario," based on the authors' judgment of the tech-

Table 11-3. Profile of Possible Electric Car Development

Car	Approx. weight	Possible energy source	Motor(s)	Controls	Max. speed	Range	Feasible approx. dates	Remarks
1st generation commucar or shopping car	1,500 lb	6-7 kwh PB-acid battery	7-8 hp, DC	Switching variable resist.	~35 mph	30-35 miles	Late 1960s	This is typical of several small cars already in existence
2nd generation electric subcompact	2,000 lb	High-energy battery (zinc-air) ~20 kwh	20-25 hp, AC	SCR, inverter, regenerative braking	55-60 mph (governed)	70-100 miles av. conditions	Early 1970s	Performance comparable to today's subcompacts; still needs recharge time
3rd generation electric compact	2,500-3,000 lb	20-kwh high-energy battery + 20-kw fuel cell (reformer?)	60-70 hp, AC or DC homopolar	SCR, inverter, regenerative braking	70+ mph (governed)	200-300 miles (cruise)	Mid-1970s	Acceleration comparable to today's compacts; expensive, but possibility of rental of energy package

nology likely to be available at various points in the future. Table 11-4 compares each with an equivalent conventional ICE-powered car in various operational aspects.

The feasible dates, of course, depend also on other things, such as the resistance of the automotive and petroleum industries and their success in "cleaning up" the emissions from conventional internal-combustion engines. However, there can be little doubt that electric-powered cars can ultimately be made to perform at least as well as the internal-combustion engine car for most applications.

Regardless of the timetable for the future, a considerable number of electric vehicles using existing components and off-the-shelf hardware have been designed and built, although often just as test-beds for new concepts; indeed, many have seen a considerable amount of road testing. A number of these efforts are described in the remainder of this chapter.

ELECTRIC VEHICLE DEVELOPMENT EFFORTS

The following survey is by no means a complete catalog of all the electric vehicle efforts to date. For one thing, the field is continuously expanding, with new developments being announced often. In addition, the various manufacturers of commercial special-purpose electric vehicles such as golf-carts, fork trucks, industrial carriers, and so on, have been omitted. A large number of these devices are made and sold each year by such companies as Laher Spring Electric Car Corporation, the Revolvolator Company, and the Pargo-Columbia Car Corporation, among others—and the market is growing.

Alden Self-Transit Systems Corporation

This organization is the developer of the StaRRcar (an acronym for Self-Transit Rail and Road Car), an electrically powered vehicle designed both to run on a road under battery power and to ride on a specially designed track (or guideway) from which it also would draw distributed electricity. The vehicle weighs 1,700 pounds, has a 3-passenger capacity, and is propelled by four 2.5-hp DC motors. On the road, powered by lead-acid batteries, it has a maximum speed of 30-40 mph; on its special guideway, it is said to be capable of traveling at 60 mph.

Alden also has designed a 12-passenger bus, the StaRRbus. This vehicle, which has a curb weight of 3,000 pounds, is also capable of dual road—guideway propulsion modes.[9]

American Motors–Gulton Industries

AMC and Gulton have pooled resources to develop a small electric car using two battery systems—one for cruising and one for acceleration. Their Amitron

---

[9]Data sheets, Alden Self-Transit Systems Corporation.

Table 11-4. Comparisons of Three Types of Electric Cars with Conventional Counterparts

Characteristic	I. Very Small		II. Subcompact		III. Compact	
	Electric "Commucar" or shopping cart, Urbanina	Gasoline-powered equivalent Trident, Messerschmitt	Electric (battery powered)	Conventional (VW, Renault)	Electric (hybrid)	Conventional
Initial cost	*If rented	*If purchased	*If rented	*If purchased	*If rented	*If purchased
Operating cost	*Lower on repairs and fuel	—	*Lower on repairs and fuel	—	*Lower on repairs and fuel	—
Feeling of "leashed power"	. . . . . . Not applicable		. . . . . Not applicable		—	?
Dynamic brakes	*Possible	No	*Yes	No	*Yes	No
Split axle (independent drive)	*Possible	No	*Possible	No	Possible	Possible

220

Acceleration 0-40						
Mechanical reliability lifetime	*Very good	—	*Good	—	*Better	—
Freedom from need for service	*Very good	—	*Good	—	*Better	—
Refueling	Easy but slow	Fast	—	*Easier, faster	No difference	—
Quietness and odor free	*Yes	No	*Yes	No	*Yes	—
Instant start	*Yes	—	*Yes	—	*Yes	—
Smoothness	*Better	—	*Yes	—	*Yes	—
Operational simplicity	*Yes	—	*Yes	—	*Yes	—
Range	*Longer	*Longer		*Longer	No difference	No difference

*Indicates superiority.

vehicle will cruise exclusively on two Gulton lithium–nickel fluoride batteries (75 pounds each). When extra power is needed for acceleration or passing, a nickel-cadmium system, weighing 100 pounds and capable of high discharge rates, will take over. With regenerative braking, a range of 150 miles can be achieved.

The Amitron, weighing approximately 1,100 pounds overall, will seat three passengers. Its maximum speed will be about 50 mph.[10]

Army Engineer Research and Development Laboratories

The lab has converted two trucks to fuel cell electric power. The Mark I, with a curb weight of 5,400 pounds, is propelled by a DC series motor through the truck's original rear differential, using four 5-kw fuel cell modules. Maximum speed here is 45 mph. The Mark II has one AC motor in each of the four wheels and uses twelve 5-kw fuel cell modules. In this case, performance is fully equivalent to that achieved with its original 94-hp engine—70 mph and a 68% grade capability.[11]

Emphasis has been on the development of fuel cell powered systems using hydrocarbons, although it is not known whether any such systems actually have been installed in test vehicles.

Chrysler Corporation

Chrysler has electrified a small car built by one of its foreign subsidiaries. This car is a French Simca, which uses 1,400 pounds of lead-acid batteries and two electric motors and has a range of 40 miles with limited acceleration.

Chrysler has simulated an electric car by installing a small computer in a gasoline-driven Plymouth Barracuda. The computer measures the rate of energy expenditure and informs the driver via a red light when the hypothetical lead-acid batteries would have run out of power. Though billed as "electric car research" this project is only marginally relevant to advancing the art.[12]

ESB-Battronic

ESB, Inc. (formerly the Electric Storage Battery Company), in conjunction with the Battronic Truck Corporation (of which it is a part owner) has several electric vehicle ventures. Battronic manufactures electric delivery trucks and has recently introduced an electric minibus. The trucks, with a loaded gross weight of 9,500 pounds each, can transport some 2,000 pounds of men and

---

[10] J. H. B. George, L. J. Stratton, and R. G. Acton, "Prospects for Electric Vehicles—a Study of Low-Pollution-Potential Vehicles—Electric," A. D. Little, Inc., Report for the Department of Health, Education and Welfare, May 15, 1968.

[11] "Department of Defense Research on Unconventional Vehicular Propulsion," U.S. Army Engineer R&D Laboratories, February 1967.

[12] *IDEAS*, IR&T Newsletter, Vol. 1, No. 6 (March 1969).

equipment at city traffic speeds; the 84-volt lead-acid batteries used to provide energy have capacities that range from 28 kwh to 35 kwh and allow the vehicles a range of 40 miles and a top speed of 25 mph. The Battronic minibus, which can carry 12 passengers, is powered by essentially the same 84-volt industrial lead-acid batteries that power the trucks. Both vehicles feature solid state power controls.

ESB also operates a converted Renault similar to the Henney Kilowatt. Powered by 72-volt lead-acid batteries, the car can accelerate—according to published reports—from 0 to 29 mph in three seconds. It has a top speed of 40 mph and a range of 25–35 miles.[13]

Electric Fuel Propulsion, Inc.

Electric Fuel Propulsion, a small firm, offers its MARS series of electric cars for sale in the $5,000 range. These cars, like so many current electric cars, are converted Renaults. MARS II weighs approximately 4,000 pounds, including 1,900 pounds of lead-acid batteries, which contain a cobalt additive that apparently permits a higher energy storage capability and allows a longer life at high charging and discharging rates. The car will accelerate from 0 to 40 mph in about 12 seconds with a maximum speed of about 60 mph and a range of 70–120 miles. The propulsive unit is a 15-hp DC motor. MARS II provides an onboard charger and regenerative braking.

A number of these vehicles have been sold, principally to utility companies. In addition, Electric Fuel Propulsion has successfully operated the MARS II in several long-distance promotional trips—notably from Detroit to Washington, and from Detroit to Phoenix.[14]

The company is reportedly designing a completely new car. The new vehicle will be a four-passenger sedan, approximately the size of a Ford Falcon. Its curb weight will be 3,400 pounds, and it will feature a flat lead-acid cobalt battery, a DC motor, regenerative braking, a solid state chopper control, and onboard charging. Performance will be considerably better than that of the MARS II; the new car will have a cruising speed of 65 mph, a top speed of 85 mph, and a maximum range of 150–175 miles before recharging. It is expected to sell for under $5,000.[15]

Ford Motor Company

The Ford organization has had two widely divergent vehicle projects—the Comuta and the Lead Wedge, neither of which utilizes Ford's new sodium-sulfur battery developments.

---

[13] *Commercial Car Journal*, August 1968; and George et al., "Prospects for Electric Vehicles."

[14] Robert Aronson, "The MARS II Electric Car," SAE Paper No. 680429, Detroit, May 1968.

[15] Robert Aronson, personal communication.

The Comuta is a small urban car developed by Ford's subsidiary in the United Kingdom. Conventional in design, weighing 1,200 pounds, and powered by lead-acid batteries and two 5-hp DC motors, it reaches a top speed of around 30 mph and has a range of only about 40 miles.[16]

The Lead Wedge, which was built as a promotional high-speed vehicle by the Autolite-Ford Parts Division, derived its name from the twenty standard lead-acid car batteries beneath the wedge-shaped body. At the Bonneville Salt Flats in Utah, with the lead batteries producing a peak output of 90 kw, the car reached a momentary top speed of 138.862 mph—which has been billed as the record for an electric car. However, apart from advertising purposes, the Lead Wedge had no practical utility.[17]

General Electric

General Electric, as might be expected, has a considerable interest in the successful advancement of the electric automobile. GE's working prototype vehicle is a subcompact called the Delta. This is a town car, a small 3-door wagon, that is 130 inches long and weighs 2,300 pounds. Propelled by a 10.9-hp DC series motor, the Delta features a hybrid battery system consisting of improved lead-acid batteries (18 whr/lb) for main cruising power and nickel-cadmium batteries for accelerative peaks. The Delta has a maximum speed of about 55 mph, and a range of 40–50 miles for city-cycle driving (i.e., four stops per mile, 30-mph cruise). A 30-amp solid state onboard charger is included. The SCR chopper controls also keep the booster battery unit recharged.[18]

General Motors Corporation

The giant of the automotive industry has built three experimental vehicles in recent years: the Electrovair, the Electrovan, and the Stir-Lec I (a hybrid which is discussed in the next chapter). The Electrovair is a redesigned GM Corvair using 680 pounds of silver-zinc batteries and a single AC induction motor capable of producing 100 hp for only 130 pounds weight. Power is controlled by a solid state SCR (silicon-controlled rectifier) inverter. Performance in the Electrovair has been good (0–60 in 16 seconds, a top speed of 80 mph), but the range is only 40–80 miles. The principal drawbacks to the vehicle (other than cost, which is to be expected in an experimental device) are the limited number of charge-recharge cycles (100) and the high weight of the inverter (235 pounds).

The Electrovan, also an experimental vehicle, is a converted GM van. Overall weight is 7,100 pounds. Motive power in this case also is a high-power (125 hp) AC induction motor, but the van's energy source is a hydrogen-oxygen fuel cell

---

[16]George et al., "Prospects for Electric Vehicles."

[17]Ford-Autolite News Release.

[18]B. R. Laumeister, "The G.E. Electric Vehicle," SAE Paper No. 680430, Detroit, May 1968.

capable of producing a continuous output of 32 kw, with a peak overload capability to 160 kw. The Electrovan has a maximum speed of 70 mph and a range of 100–150 miles.[19]

Gould–National Batteries, Inc.

Gould–National, in conjunction with North Star Electric, a Minnesota utility company, has modified several Henney Kilowatts. Improved automotive starter batteries, yielding 20 whr/lb at a low discharge rate, have been installed along with a GE solid state control system. These modified Gould Kilowatts now have a range of some 65 miles instead of the original 35–40 miles.[20]

Linear-Alpha, Inc.

This company offers for sale two electrified vehicles—a converted Ford Falcon at approximately $12,000, and a modified International Harvester Metro Van for $15,000. The electrified Falcon sedan, using a 25-hp AC induction motor, SCR controls, and 360 pounds of lithium–nickel fluoride batteries, reaches a top speed of 60 mph and a range of 75 miles at a 30 mph average speed. The van, with a 40-hp AC motor and 450 pounds of the same batteries, also reaches 60 mph and has a range of 63 miles at a 13-mph average. Both vehicles utilize regenerative braking.[21]

National Union Electric Corporation

One of the earliest of the new wave of electric vehicles to appear was the Henney Kilowatt, which the company developed and built in small numbers in the early 1960s. A converted Renault carrying approximately 800 pounds of lead-acid batteries, the car has an energy capacity of about 8 kwh. It uses a single 7.1-hp series DC motor, with a series-parallel switching of resistances for control. Its unloaded body weight is 2,135 pounds. With a maximum speed of approximately 40 mph and a range of 40 miles, the Henney Kilowatt has an energy consumption of 0.3–0.55 kwh/ton-mile for normal driving. Some 50 Henney Kilowatts were built. Most of these cars are in active use in various places throughout the country. Utility companies bought a number of them, and several have been used by other companies as test-beds for more advanced electric conversion.[22]

Rowan Controller Company

This company is the developer of a prototype car called the Rowan Electric, of which four models have been built. The car's body was built by Ghia S.P.A.

---

[19] General Motors Press Release; and George et al., "Prospects for Electric Vehicles."

[20] George et al., "Prospects for Electric Vehicles."

[21] Electric Vehicle Information Brochures, Linear Alpha, Inc., Evanston, Ill.

[22] George et al., "Prospects for Electric Vehicles."

of Turin, Italy, and the chassis was designed by de Tomaso Automobili of Modena, Italy, both Rowan subsidiaries. The drive was built by Metrodynamics Corporation (another Rowan affiliate), the built-in battery charger by American Monarch, and other control components by Rowan. General Electric supplied the motors.

The Rowan Electric has a curb weight of 1,300 pounds. It is powered by lead-acid batteries of the type used in golf carts, has a top speed of about 40 mph and a reported maximum range of over 100 miles. The vehicle also features regenerative braking and an inexpensive solid state controller for its two DC compound motors.[23]

Stelber Industries

Stelber Industries, of Elmhurst, New York, has developed an electric bicycle propelled by a unique "electric wheel" under license from its inventor, Gar Wood, Jr. Power for the Stelber bicycle is drawn from two 6-volt golfcart batteries. It is claimed that the vehicle can carry a 200-pound rider at 15 mph for more than 3 hours before recharging is needed. Production models of the bicycle include a charger circuit to permit recharging from standard 110-volt household outlets. Stelber envisions the same electric drive being ideally adaptable to golf-carts, wheelchairs, delivery trucks, etc.

The company apparently also expects to make a three-wheeled version of the bike, with a large shopping basket in the rear. Prices are expected to range from $130 for the bicycle to $200 for the tricycle.

The ingenious electric wheel around which the idea centers actually consists of four to six inexpensive 1/3-hp DC motors arranged to share the drive load on a large central planetary gear. A one-way clutch transmits the torque to the front wheel.

Stelber also hopes to market a small single-seat electric car designed for short-range local driving (shopping, etc.), which they expect will be priced at about $500.[24]

West Penn Power Company

This utility company, part of the Allegheny Power System, Inc., has for some years used a small electric demonstrator called the Allectric car. The car was originally a one-of-a-kind Stuart Electric, but proved unsuccessful and was modified into the Allectric II.

A two-passenger vehicle, the car weighs about 2,300 pounds and is powered by a 7.1-hp DC motor and 900 pounds of lead-acid batteries. It features a solid state control system and an onboard battery charger.

---

[23] Rowan Controllers Press Release, March 1968.

[24] S. Berkowitz, Stelber Industries, personal communication.

Although the Allectric II has not yet been completely tested, it is anticipated that it will have a maximum speed of about 50 mph and a typical range of 50 miles.[25]

Westinghouse

The Westinghouse entry into the electric car field, announced in 1967, was the Markette, a two-passenger car with a curb weight of 1,730 pounds. Westinghouse, through its Electric Marketeer subsidiary (Redland, California) has manufactured golfcarts and personal service carriers since 1964. The Markette was a derivative of these vehicles. It is relatively limited in performance, although it was intended largely as an auxiliary vehicle for special uses—as a second car, or shopping car, for example. Powered by 800 pounds of lead-acid batteries and two 4.5-hp DC series motors, it could reach a top speed of 25 mph in 12 seconds. Its maximum range, at a rate of one stop per mile, was about 50 miles.

However, the Markette recently was withdrawn (temporarily, at least) from production. The reason given was that it did not comply with various new federal safety requirements. The company has said very little about the project since.[26]

Gar Wood

Gar Wood, the former speedboat racing champion, founder of Gar Wood Industries (Detroit), and one of the most prolific inventors in history, together with his son, Gar Wood, Jr., was developing an electric car (the Super-electric Model A) in the late 1960s at his secluded estate on Fisher Island, south of Miami Beach, Florida. An experimental car described several years ago used two 2-hp motors and 520 pounds of lead-acid batteries and had a maximum speed of about 52 mph. Largely just a chassis when first reported, the prototype car has been used principally as a mobile test-bed for Wood's patented control unit, the details of which have not been revealed, except that it offers five forward and five reverse speeds.[27]

More recently, a licensing arrangement was announced between Gar Wood, Jr., and Stelber Industries on an "electric wheel" for powering small vehicles. It is not clear whether this arrangement includes the control unit noted above, or whether Wood plans further development independently of Stelber.

Yardney Electric

Yardney, primarily a manufacturer and developer of batteries, has taken a Henney Kilowatt and substituted high-energy silver-zinc batteries for the lead-

[25]W. E. Sturm, "An Allectric Future—Now," Electric Auto Symposium, San Jose, California, 1967; and George et al., "Prospects for Electric Vehicles."

[26]Westinghouse Electric Corp., Press Release, April 4, 1967; and George et al., "Prospects for Electric Vehicles."

[27]"Gar Wood Testing Electric Auto," *Florida Trends*, December 1966.

acid cells in the "basic" car. As a result, the weight and performance characteristics of the car were notably improved. The Yardney version (the Yardney Silvercel) weighs about 1,600 pounds and uses 240 pounds of silver-zinc batteries. It has an energy capacity of 12 kwh. Able to accelerate from 0 to 30 mph in 5 seconds, it reaches a maximum speed of 55 mph, with a range of about 80 miles. (The car figured prominently in publicity surrounding the 1967 electric car hearings.)

Yardney is allegedly installing improved silver-zinc batteries in this vehicle. Using 400 pounds of the new batteries, Yardney has claimed a maximum speed of 60 mph and a range of 150 miles.

The company has also been considering the possible use of several of its newer third-generation batteries in electric autos. For example, combining its Zynoxel battery, a zinc air system, with their Nyzin (nickel-zinc) battery in a hybrid system would give a typical electric car a 200-mile cruising range and extra power when needed for peak accelerations. The Nyzin battery has already been used to power a Stelber electric bicycle, while the Zynoxel system has been successfully tested in an electric motor scooter.[28]

Miscellaneous Electric Vehicles

Perhaps one of the earliest of the modern electric vehicles was the tractor, developed by Allis Chalmers in 1959, in which natural gas or propane powered a fuel cell which supplied 15 kw to a 20-hp motor. The tractor was able to exert a 3,000-pound pull on a plow.[29]

A fuel cell also supplied power to propel a motorbike built by Union Carbide Corporation, although this system used a hydrazine-air combination. The motorbike could reach 25 mph and cover 200 miles on a gallon of hydrazine.[30]

Foreign Developments

Activity in electric car development abroad is quite diverse and extensive, notably in England and Japan, although the rationale for developing such vehicles is often different from that of most groups in the United States. Growing concern over air pollution undoubtedly plays an important role, but an acute need for smaller vehicles (to which electric propulsion is ideally adapted) and the generally higher costs of petroleum fuels have carried greater weight abroad.

In the United Kingdom, in particular, there is widespread and flourishing commercial use of electrically propelled delivery vans, lift trucks, postal vans, etc. At the end of 1967, for example, there were approximately 45,000 electric

---

[28] Yardney Electric Corp. Fact Sheet; and "Yardney Tests New Batteries for Electric Cars," *Automotive News*, November 4, 1968, p. 40.

[29] "Fuel Cell Debut," *Machine Design*, Vol. 31, No. 22 (October 29, 1959).

[30] Union Carbide News Release, November 29, 1966.

trucks in use in the United Kingdom, some 85% of them being used for home milk delivery. Among the major British manufacturers of such special-purpose vehicles are Austin Crompton Parkinson, W & E Vehicles, Smith's Electric Vehicles Ltd., Stanley Engineering, and Lansing Bagnall Ltd. Power sources for such delivery trucks are typically lead-acid batteries, which give the trucks a range of 20–30 miles at maximum speeds of about 15 mph.

A considerable number of small general-use electric road-cars, similar in level of performance to the Henney Kilowatt also have been developed abroad.

FUEL CELLS

There have been a few demonstration vehicles based on fuel cells alone (without batteries). The most notable was the General Motors "Electrovan" (mentioned earlier) powered by a 90-kw Union Carbide-built oxygen hydrogen fuel cell. Owing to the limited power outputs available from fuel cells, however, and their very poor performance under part load (not overload), it is probably unlikely that a fuel cell alone will ever be used in vehicular application.

Fuel cell costs at the moment are quite high. Estimates quoted in the Appendix to this book suggest that future prices might range from $100 per kw to less than $25 per kw for hydrocarbon reforming cells by the late 1970s or in the 1980s.[31] At $100 per kw and 40 watts per pound, the cost of a 10-kw fuel cell would be $1,000, or $4 per pound; but at $25 per kw and 50 watts per pound, the figures would drop to $250 and $0.80 per pound. A late 1970s fuel cell–battery hybrid system would presumably cost about the same per pound as the battery alone, but would be capable of superior overall performance, especially in regard to range.

---

[31] These estimates were based on the assumption that there would be continuing large-scale development effort; in practice, fuel cell efforts have slackened notably.

Chapter Twelve

Hybrid Power Systems

No single power source is ever likely to be *optimal* with respect to all the desired characteristics of an automotive vehicle. To gain in one area one must typically give ground in another. To put the situation another way, the choice of a power system for a given application is largely a question of assessing trade-offs between performance and cost or between one figure of merit and another.

The problem of conflicting objectives is most acute in the area of power supply: almost regardless of technology, it seems to be true that high specific energy (energy storage capacity) and high efficiency are incompatible with rapid response and high power output. For this reason, despite generally unfavorable economies of scale, combinations of two or more different but complementary energy conversion and storage schemes are receiving increased attention.

In general the primary energy conversion device output will be set roughly equal to the average power consumption of the vehicle over its duty cycle. (Typical driving cycles for automobiles were discussed in chapter 3.) A constant load operation permits simpler design and optional once-for-all adjustments of compression, fuel-air ratio, electrolyte circulation rate, and other salient parameters. Parasitic losses are minimized with respect to total power for any engine operating at full rated load. A constant load operation also permits more efficient fuel combustion with much less elaborate control mechanisms and less highly refined, lower octane fuel. In the case of automobiles, it also permits operating with an engine or fuel cell having less than a fifth the horsepower that would be needed if the primary system had to handle peak loads. While these savings do not reduce the weight or cost by anywhere near a factor of five, they are significant.

In ICE or ECE electric forms of hybrid systems an alternator-motor combination plays the same role as the clutch and transmission in a conventional vehicle. While alternators and motors are not cheap—especially when sophisticated electronic controls are included—it must be remembered that the mechanism they replace is also both heavy and complex. In the case of a fuel cell hybrid the alternator can be omitted entirely, resulting in a significant further saving in weight and cost.

The peak-power storage device (flywheel, or electric or thermal battery) must have a high specific power but it need not have a tremendous capacity. It must be able to meet the peak demands imposed by the duty cycle, at the minimum cost and weight.

To optimize a hybrid vehicle design we need some criterion for determining the intermediate reserve storage capacity that will be required, assuming the primary power supply is equal to the average power consumption over the duty cycle. This will be based on the statistical frequency and the magnitude of the demands on it.

The net energy withdrawn from intermediate storage in time T—assuming it was "full" to start with—can be defined as (T), where $T = N$ (for convenience we can set T equal to one hour), whence:

$$\epsilon(T) = \epsilon(N) = \tau \sum_{n=1}^{N} P_e(n), \tag{12.1}$$

where $P_e(n)$ is the effective rate of power demand in the nth time increment. Then we can ask how long it will take, on the average, before a random sequence of events in the driving cycle will cause the reserve supply (whether flywheel or battery) to be depleted to the point ($\sim 90\%$) where it limits the vehicle's performance. Clearly, the larger the reserve capacity, the later such an occurrence would be.

Under reasonable statistical assumptions, it can be shown that the variance of ξ will be given by:

$$\text{Var } \xi(N) = \sigma_\epsilon^2$$

$$= N\tau^2 \text{ Var } \left\{ [\bar{P}_e - P_e] E^{-1} (P_e) \right\}$$

$$= N\tau^2 [\bar{P}_e - P_e]^2 [1 + \frac{\mu}{P_s} (P_e)]^2 \tag{12.2}$$

$$= N\tau^2 \left\{ [1 + \frac{\mu}{P_s} (\bar{P}_e)]^2 (\bar{P}_e^2) + \frac{3\mu^2}{P_s^2} (\bar{P}_e^2)^2 \right\},$$

provided it can be assumed that the reserve can be utilized subject to a nonlinear condition such as was specified by equation (11.2). An analogous expression—representing declining returns—can be expected to hold for any storage system. Evidently, if P_e is negative (corresponding to power being regenerated), the cumulative amount of energy stored in the intermediate reserve is less than the total amount supplied, owing to internal losses. A limit on the capacity of the intermediate storage system is difficult to take into account in the formalism, but since the reserve would normally "float" somewhere below its maximum, the error is probably not significant to first order. The reserve storage capacity enters equation (12.2) through the parameter P_5(kw), which is numerically equal to one-fifth of the "nominal" storage capacity in kwh. Note that

$$(\overline{P}_e{}^2) = \overline{[P_w{}^2 (1 + \mu \frac{P_w}{P_5} + \ldots)^2]} \simeq (\overline{P}_w{}^2) + 2\frac{\mu}{P_5}(\overline{P}_w{}^3), \qquad (12.3)$$

where $(\overline{P}_e)$ was given by equation (11.3) and (P_w) was defined by equation (3.8).

To first order in powers of $[\frac{\mu}{P_5}]$ we have:

$$\sigma_\epsilon{}^2 \simeq N\tau^2 \left\{ (\overline{P}_w{}^2) + 2\frac{\mu}{P_5}(\overline{P}_w)(\overline{P}_w{}^2) + 2\frac{\mu}{P_5}(\overline{P}_w{}^3) \right\}. \qquad (12.4)$$

If the reserve capacity is very large, of course,

$$(\overline{P}_e) \approx (\overline{P}_w) \qquad (12.5)$$

and

$$\sigma_\epsilon{}^2 \simeq N\tau^2 (\overline{P}_w{}^2). \qquad (12.6)$$

Since $T = N\tau$ (or N hours), it can be seen that the discharge level, which represents, say, the 84th percentile of possible measurements (taken at random times)[1] increases as the square root of elapsed time. From this it follows that the amount of reserve energy needed to ensure an 84% ("one sigma") probability of *not* "overdrawing" in N hours of driving must be equal to:

$$E_{1\sigma} \simeq \sqrt{N[(\overline{P}_w{}^2) + 2\frac{\mu}{P_5}(\overline{P}_w)(\overline{P}_w{}^2) + 2\frac{\mu}{P_5}(\overline{P}_w{}^3)]}, \qquad (12.7)$$

while the reserve supply needed to ensure a 98% ("two sigma") probability of not overdrawing in N hours of driving must be

$$E_{2\sigma} \simeq 2\sqrt{N[(\overline{P}_w{}^2) + 2\frac{\mu}{P_5}(\overline{P}_w{}^2) + 2\frac{\mu}{P_5}(\overline{P}_w{}^3)]}. \qquad (12.8)$$

---

[1]That is to say, the chances of depleting the reserve beyond this level are only 1 in 7, given the assumed duty cycle.

Roughly a 98% probability of not depleting the reserve to the point where vehicular performance would be affected in an hour of average driving[2] would require a reserve capacity (in kwh) equal to twice the root-mean-square (r.m.s.) power demand over the given duty cycle. If the latter were 10 kw, then the reserve should be roughly 20 kwh. In the above case, P_w is equal (by definition) to the steady power provided by the primary converter, whether it be an ICE or a fuel cell. Actually, of course, the power provided by the primary converter could be arbitrarily set larger or smaller than this level, and it would be possible to determine, by a procedure similar to the above, what a rational choice of reserve capacity would be to provide a given margin.

The actual levels of energy capacity reached during the Pittsburgh driving cycle were simulated, in simplified form, by computer for a number of typical cases. In these calculations, two basic variations were predicated—one with no battery regeneration, and one with regeneration at an efficiency of 80%. Various combinations of primary converter or generator power and maximum energy-storage capacity were assumed. For the case of zero regeneration, 10-hp, 15-hp, and 20-hp generators were each matched with battery capacities of 18,000, 36,000, and 54,000 hp-sec (36,000 hp-sec = 10 hp-hr = 7.5 kwh). Power ratings of 10, 15, 20, 25, 30, and 40 hp were matched in the 80% regeneration case with capacities of 18,000, 36,000, 54,000, and 72,000 hp-sec. The net results of the computer simulation are illustrated in figures 12-1 and 12-2, which show the cumulative percentages of time for each value of generator power during which the battery in each case was below the energy level indicated in the ordinates of the figures.

The significance of the results is perhaps better seen in figure 12-3, which plots the ratio of minimum energy level reached during the cycle to maximum energy for each case, against the values of generator power. From this it would seem that 15 hp represents a probable minimum level of generator power except perhaps for very large capacities; the values of E_{min}/E_{max} drop off precipitously below it.

Furthermore, it can be seen that in the lowest values (e.g., 10 hp and 36,000 hp-sec), a small increase in generator power raises the value of E_{min}/E_{max} considerably further than does the installation of a regenerative system. In a real case, the choice of component sizes and types would be dictated by such factors as costs, reliability, etc.

However, it should be remembered that the simulation was considerably simplified. Perhaps most important, the loss of capacity with increased discharge rate (which is quite significant for most likely battery systems) was not taken into account. Furthermore, no consideration was given to actual battery discharge capabilities; on a number of occasions during the simulated cycle the

---

[2]Such an event would not be tantamount to "running out of gas"; it would merely imply that the driver would have to reduce his demands on the vehicle by driving more slowly or accelerating less sharply until the reserve was reestablished.

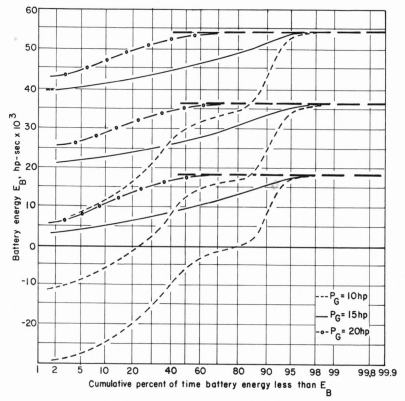

Figure 12-1. Distribution of battery energy levels in the Pittsburgh cycle for various maximum energies and generator power levels assuming no regeneration.

actual power required for short times was extremely high, necessitating very rapid battery discharge rates, which would tend to further limit the choice of both storage system and drive motor.

Although the flywheel and battery have been incorporated into the foregoing analysis without distinction, there is an important operational difference to bear in mind: an electric storage battery will hold its charge for many days without significant loss, whereas the lifetime of a thermal battery or a flywheel is measured in hours, owing to the thermal conduction in the first case and frictional losses in the other. In a vehicle that is not used around the clock, thermal or flywheel storage would have one overwhelming disadvantage: the necessity of bringing the temperature or the rotational speed up before the start of the trip or of driving for quite a while in a very conservative manner to allow accumulation of a reserve supply of energy. However, in a fleet vehicle such as a taxi or a city bus, thermal or flywheel storage might well be the more economical approach, since both devices are comparatively simple and cheap and would

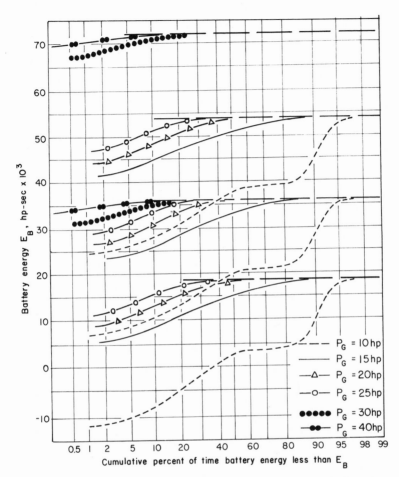

Figure 12-2. Distribution of battery energy levels in the Pittsburgh cycle for various maximum energies and generator power levels assuming 80% regeneration.

probably last indefinitely, whereas most types of electric storage batteries currently known or envisaged would "age" rather quickly and require regular replacement. Qualitatively, the economic argument for a hybrid system utilizing an engine (ICE or ECE) as a constant source of power, together with an alternator for propulsion, and with batteries or flywheel for energy storage, is based on the part-load inefficiency of the engine and the consequent assumption that fuel is saved when the engine is run at constant speed.[3] A case of sorts can be made for such savings, due to the fact that at speeds of 10 mph to 30 mph actual

---

[3] George A. Hoffman, "Hybrid Power Systems for Automobiles," in *Power Systems for Electric Vehicles*, PHS Publ. No. 999-AP-37, U.S. Public Health Service, National Center for Air Pollution Control, Cincinnati, Ohio, 1967.

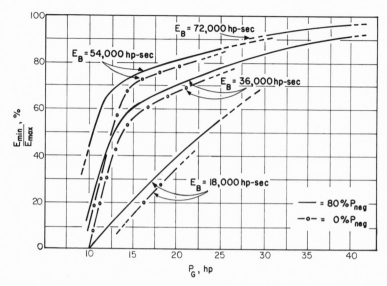

Figure 12–3. Discharge fraction (E_{min}/E_{max}) for various maximum energies versus generator power, during the Pittsburgh cycle.

road-load fuel consumption of an ICE averages about twice the full-throttle-load fuel consumption at optimum speed (mainly due to transmission losses and the power needs of parasitic auxiliaries, which become a larger percentage of the total power at low speeds).[4] Thus even fairly sizable battery (or flywheel) charging and discharging losses will not bring the overall fuel consumption of a reasonably well-designed ICE hybrid up to the ICE-only system. This argument does not apply, however, to the ECE whose internal and parasitic losses are much smaller.

Examples of hybrid systems that have been, or might be, considered are listed in table 12–1.

The indicated fuel savings of perhaps 20% or 30% overall must be paid for in terms of added cost and weight. The small ICE will be considerably more expensive per pound of weight than its larger cousin. Peak power demands will either not be met at all or will be met by an electric motor. Even if the motor were one of the most compact high-performance electric motor designs that exist today ($\sim$4 lb/hp), it would weigh as much or more than an ICE capable of producing the same shaft power, and it would cost considerably more. Alternators and batteries or flywheel must be added on top of this.[5]

---

[4] E. Pritchard, Testimony at Joint Hearings of the Senate Committee on Commerce and the Subcommittee on Air and Water Pollution, 90 Cong. 2 sess., May 28, 1968.

[5] It is true that existing ICEs are accompanied by quite small generators, voltage regulators, starter motors and batteries; however, it seems clear that gains in scaling down the engine would be vastly overbalanced by the costs of scaling up the other four items.

Table 12-1. Hybrid Combinations

Source of continuing primary power	Source of peak power
ICE or gas turbine, direct drive	Flywheel
ICE or gas turbine–alternator (+ motor)	Electric storage battery
Closed cycle turbine–alternator (+ motor)	Electric storage battery
Closed cycle turbine–alternator (+ motor)	Thermal battery
Closed cycle, turbine or reciprocating direct drive	Thermal battery
Fuel cell (+ motor)	Electric storage battery

The conclusion seems inescapable that in terms of both performance and initial cost the hybrid-electric system (in small sizes, at least) cannot hope to compete with an all-ICE or all-ECE system in performance and initial cost. In terms of residuals production, the hybrid-electric is superior to the ICE. While operating on batteries alone, as it would in city driving, the hybrid emits no exhaust products at all, but on the average an ICE-hybrid would not be as clean as an ECE by itself.

For these reasons an ICE electric hybrid cannot be envisioned as a practical system except in special circumstances. It might serve as a "supplemental power package" that could be rented by owners of all-battery electric cars for long trips, or as a power package for a dual-mode vehicle capable of utilizing a powered guideway. In this application it might play a useful role in overcoming the single major objection to all-battery powered electric automobiles—their limited range. And whereas a practical battery and fuel cell hybrid is a decade or more in the future, the ICE-hybrid could be built much sooner.

In fact, the University of Pennsylvania received a contract from the Department of Housing and Urban Development in early 1967 to design an ICE-hybrid system for an urban minicar, with assistance from the General Motors research laboratories. Development was continued with assistance from the Department of Transportation, although DOT support was withdrawn in late 1969, allegedly because it was felt that the minicar project might interfere with the development of mass transportation in the area.

Minicars, Inc., of Goleta, California, was organized to pursue the development with the University of Pennsylvania. The minicar, which has reached the prototype stage, is basically a three-passenger vehicle utilizing a small ICE with a peak of 40 hp, and a DC motor also capable of 40-hp peak output. Capable of accelerating from 0 to 40 mph in 10 seconds, it has a cruise speed of 50 mph and a maximum speed of 70 mph. On its ICE, the car has a cruising range of approximately 200 miles, with a 2–5 mile range on its lead-acid storage battery only. A considerable number of safety and pollutant-reduction features have been incorporated into its design, as well as a provision for conversion to a suitable small steam engine or other non-polluting prime mover. Table 12-2 outlines the pertinent parametric data on the minicar system.[6]

Minicar has been designed as a transit system rather than as a system of individual private vehicles. The cars, completely interchangeable, would be leased to users on a monthly basis, each user receiving a charge card. Instead of having to park, the user would simply leave his minicar at the minicar lot nearest his destination (a network of such are envisioned) and pick up another when he

Table 12-2. The Minicar Hybrid Propulsion System (ICE coupled electric motor/storage battery system)

Engine	
Cylinders	6
Displacement	164 cu in
Cooling	forced air
Speed	
Cruise	50 mph
Maximum	70 mph
Cruising range	
Internal combustion engine	200 miles
Storage	2-5 miles
Acceleration	0 to 40 mph in 10 seconds
Peak horsepower	
Internal combustion engine only	40 hp
Storage battery only	40 hp (25 for prototype)
Combined	80 hp
Weights	
Engine	375 lb
Motor-generator	100 lb
Controls	75 lb
Trans-axle	242 lb
Batteries	440 lb
Total system	1,232 lb
Battery life	>1,000 hr
Fuel mileage	24 mpg
Emissions	
Hydrocarbons	0.4 gram per mile
	(2 grams for current prototype estimate)
Carbon monoxide	0.4 gram per mile
	(16 grams for current prototype estimate)
Oxides of nitrogen	0.4 gram per mile
	(3 grams for current prototype estimate)

Source: Information furnished to the U.S. Department of Transportation by Minicars, Inc., Goleta, California.

[6] Robert U. Ayres and Roy Renner, *Automotive Emission Control: Alternatives to the Internal Combustion Engine*, Paper presented to the Fifth Technical Meeting, West Coast Section, Air Pollution Control Association, San Francisco, October 8-9, 1970.

desired to go elsewhere. Thus the system is designed to eliminate the parking problem, reduce urban crowding and pollution, and at the same time provide the freedom and convenience of the private automobile.[7]

Designed originally for the city of Philadelphia (the system was expected to be operational there by the mid-1970s if the project had been continued), the minicar system would be equally adaptable to other urban activity centers.

The outlook for the ICE-electric hybrids would be slightly better for an advanced rotary ICE, such as the Wankel or Mallory, if production costs could be brought down to the present range (about $0.50 per lb). Even $0.75 per lb for an engine capable of delivering 1 hp/lb would reduce the engine weight and costs in a hybrid by nearly a factor of 3, but the total package would be affected much less dramatically.

An experimental hybrid electric car using a Stirling cycle (external combustion) engine was built by General Motors in 1968. The "Stir-lec" is a converted 1968 Opel Kadett with a small 8-hp Stirling engine as its battery-charging unit.

The Stir-lec system utilizes a kerosene fuel and pressurized hydrogen as the working fluid. Continuous combustion of the fuel assures very low pollutant concentrations—0.01% CO, 2 ppm of hydrocarbons, and approximately one-half of the nitrogen oxide concentrations produced by a typical internal combustion engine. The heat generated in the burner is led via tubes to the engine's sealed cylinder and piston unit, which turns an alternator to recharge the vehicle's bank of lead-acid batteries.

The vehicle is propelled by a 20-hp, 3-phase induction motor. Fourteen 12-volt batteries supply about 5 kwh. The Stirling engine used in the present model—a single-cylinder 8-hp design—provides sufficient energy to trickle-charge the batteries while the car is being driven at 30 mph on a level road. In the experimental Stir-lec, which has a 5-gallon fuel tank, this provides a range of 150–200 miles.

The Stir-lec has a top speed of approximately 55 mph. Travel at this speed, however, would limit its range to 35–40 miles, since the batteries would then be discharging at a higher rate than the Stirling engine's charging rate. One advantage of the hybrid is that when the batteries are fully discharged the driver need only wait by the side of the road until the Stirling engine has recharged them sufficiently to proceed. The performance of this vehicle is roughly comparable to that of small gasoline-driven automobiles. It can accelerate to 30 mph in about 10 seconds; a standard Opel Kadett requires about 6 seconds.[8]

For large vehicles, such as buses, trucks, or earth-movers, the ICE-electric hybrid looks more promising, especially where a turboalternator can be utilized.

---

[7] "The Minicar Transit System," Minicars, Inc., Report, 1969; and *Ideas*, IR&T Newsletter, September 1969.

[8] *Motor Trend*, September 1968, p. 10; and *Ideas*, IR&T Newsletter, March 1969.

The reason is that both turbines and electric motors are more economical in larger sizes and at higher speeds. Also, for commercial (or military) vehicles the initial cost is less important than payload, reliability, and operating cost.

In West Germany, Daimler-Benz has announced the development of a diesel-electric city bus. In the city, this vehicle is driven by a 150-hp electric motor. Its 3.5 tons of lead-acid batteries provide sufficient energy for a range (on battery alone) of 37 miles and a speed of 41 mph. However, the on-board 65 hp auxiliary diesel engine extends this range indefinitely, since it can both propel the bus—e.g., outside the city—and recharge the battery at the same time. The vehicle also possesses a 25% efficient regenerative braking system which can function even when the diesel is charging the battery. Under ordinary driving conditions, the bus can negotiate 11% grades, and 16% grades for short periods of time.[9]

Lear Motors (Reno, Nevada) has proposed a turbo-electric hybrid system for use in buses consisting of a closed-cycle vapor turbine driving an alternator, which in turn provides power to an AC electric motor. The latter would be directly coupled to the differential, thus eliminating the conventional transmission.[10]

One of the most expensive, yet unreliable components of a large truck or bus is the mechanical transmission. Hence there is a strong incentive to save space and also replace the mechanical transmission by utilizing an electric drive.[11] In the case of city buses, there is an additional potential advantage to be gained from an electric transmission: the bus floor could be lowered at least 18 inches, thus eliminating the awkward, unsafe double-step, which is awkward or hazardous for many elderly passengers. The use of electric transmissions with ICE or turbine engines is not strictly a hybrid vehicle concept, however.

Over the past few years a great deal of work also has been done by U.S. Army Engineer R&D Laboratories to adapt hybrid electrical propulsion to vehicles. A number of trucks, tanks, and other vehicles have been used as test-beds for a diversity of electrical power systems.

An M–113 tank weighing 22,900 lb was equipped with a hybrid system consisting of two AC induction motors and a 250-kw alternator driven by an internal combustion engine. The vehicle was capable of 40 mph and could climb a 60% grade.

Similarly, a large earthmoving tractor was equipped with a turbine-driven 200-kw AC generator and a 50-hp AC motor in each wheel. This machine also had a top speed of 40 mph and was capable of negotiating a 60% grade.

---

[9] "Mercedes Ready with Electric Bus," *Automotive News*, September 29, 1969.

[10] "Phoenix May Rise in Reno: Lear Turns to Turbo-Electric," *Product Engineering*, March 16, 1970.

[11] W. Slabiak, L. J. Lawson, R. P. Borlund, and C. G. Puchy, "Electric Transmission with Individual Wheel Drive Proves Practical for Military Trucks," *SAE Journal*, October 1966, pp. 36–40; G. Bouladon, "Le Dumper Articule Berliet TX–40," *Revue Automobile*, No. 7 (1965).

A 12,500-lb truck was used as a test-bed in which individual AC motors were installed in each of six wheels. An ICE-driven high-frequency alternator and solid-state control provided power to the wheel motors. This truck had a 50-mph maximum speed and a 60% grade capability. A similar truck equipped with DC homopolar motors in each wheel, and a DC generator in place of the alternator, yielded approximately the same performance results.[12]

There is a fairly convincing case for the use of thermal storage in conjunction with an ICE. In this case, as mentioned in chapter 5, such a storage system (comprised, e.g., of a molten salt heat exchanger) could be inserted into the flowpath of the working fluid between the vapor generator and the engine expander. With any sudden increase in output load (and the attendant increase in flow-rate through the vapor generator), there is a drop in the net enthalpy per pound that is available for work. The loss is, of course, made up by the vapor generator, but only after a lag of several seconds. It is this gap that the thermal storage system would fill. When the demand again dropped, the energy from the vapor generator would go partly to "recharging" the thermal storage system. The computer results plotted in figures 12-1, 12-2, and 12-3 are as applicable to the thermal "battery" as to the electric one.

Such "thermal buffering" can be readily implemented in external combustion systems by completely surrounding the steam generating tubes with the thermal storage substance. A variation of this approach forms the basis for a thermal-storage-ECE system advanced by Thermo Electron Corporation of Waltham, Massachusetts (see chapter 8). In this case, the engine uses Thiophene (CP-34), which is a thermally degradable organic substance, as a working fluid. To eliminate the "hot spot" problem and protect the working fluid, TECo has surrounded the working fluid flow tube with a concentric tube containing a proprietary intermediate heat-transfer fluid. The latter effectively absorbs the high heat flux and in turn emits heat to the working fluid at a controlled rate, in addition to providing reserve capacity for peak loads.[13]

So far we have not discussed the fuel cell–battery hybrid system. This is much simpler in principle, since it involves no generator. It should also be significantly more efficient than any other system presently envisaged. A reasonable choice, at least for purposes of discussion, might be a fuel cell capable of providing the 25-kw steady power required by a 3,000-lb vehicle cruising at 55 mph. If a specific power of 40 watts/lb can be achieved in a practical fuel cell (roughly what is available from an advanced hydrazine cell at present), the unit would weigh 625 lb. To satisfy peak demands, however, additional power is needed. Thus to be able to accelerate the 3,000-lb vehicle in question at the rate of 6 mph/sec at a speed of 45 mph, and at the rate of 2 mph/sec at 75 mph

---

[12]"Department of Defense Research on Unconventional Vehicular Propulsion," U.S. Army Engineer Research and Development Laboratories, February 1967.

[13]D. T. Morgan, E. F. Doyle, and S. Kitrilakis, "Organic Rankine Cycle System with Reciprocating Engine," Intersociety Energy Conversion Engineering Conference, September 1969.

would require about 75 kw *in addition* to the 25-kw cruise power specified above, or 100 kw in all.[14]

A 750-lb storage battery with a specific power output of 100 watts/lb would adequately meet this requirement. In terms of the criterion stated subsequent to equation (12.8), assuming a root mean square power demand equal to 25 kw, the battery should have a capacity of roughly 50 kwh or 67 whr/lb to ensure a 98% probability of not being discharged more than 80% in a 2-hour period of "normal" driving.

From the above numbers it can be seen that an electric car with performance comparable to that provided by an internal or external combustion engine is far beyond present capabilities and likely to remain so for some time. However, there is every reason to believe that some day—perhaps by 1980—fuel cells with specific power capability of 40 watts/lb and rechargeable batteries capable of 100 watts/lb and 67 whr/lb will be available with reasonable lifetimes and at reasonable cost.

If it is assumed that somewhat lower levels of performance would be satisfactory for many applications, the fuel cell–battery hybrid becomes more plausible. For instance, in a small 2,000-lb city car, a 10-kw fuel cell with a power density of 25 watts/lb, would weigh 400 lb, leaving about 300 lb for the battery and perhaps 150 lb for the motor and controls. A battery with an energy-density of 50 whr/lb would provide a reserve storage capacity of 15 kwh. The battery, of course, would discharge only when the instantaneous demand exceeded the constant 10 kw supplied by the fuel cell. In a typical *urban* driving cycle the fuel cell would be charging the battery at a variable (but low) rate during the time spent in idling or decelerating—which is typically almost half the total driving time in heavy traffic. Peak-power demands of, say, 70 kw would imply intermittent withdrawal at a 15-minute discharge rate, but these would be very rare.

The above example was chosen almost arbitrarily, with no attempt to optimize the fuel cell–battery combination. Such an optimization requires consideration of the operating conditions (driving cycle), the vehicle's weight, and the detailed characteristics of the battery and fuel cell. Preliminary studies by the U.S. Army Electronic Components Laboratory, assuming a 100-whr/lb battery and a 60-watt/lb fuel cell with a 50% overload capability, indicate that a wide variety of combinations can be expected, depending on application (see table 12–3).

The kw/kwh ratio indicates the relative importance of fuel cell and battery in these applications. On closer scrutiny, it can be seen that the more variable the duty cycle (i.e., the more different kinds of driving conditions encountered) the greater the reliance on the battery and the smaller the fuel cell (charger) need be. Thus, although the assumption of 60 watts/lb for fuel cells, on which table 12–3

---

[14]Both figures ignore internal losses in the motor and control system; if these amounted to 15% the total power required would be increased by this amount.

Table 12-3. Alternative Hybrid Configuration, 2½-Ton Army Truck

Duty cycle	Ratio of fuel cell power to battery capacity (kw/kwh)	Total weight of power plant (lb)
Conventional ICE (regardless of cycle)	Not applicable	1,550
Heavy-duty convoy truck (70% of time at 24 kw; 20% at 108 kw)	0.25	1,700
Milk truck (50% idle; 50% at 20 kw)	0.5	464
Commercial highway truck (10% idle; 90% at 24 kw)	2.0	700

Source: Adapted from data in E. A. Gillis, U.S. Army Engineer R&D Laboratories, "Power Systems for Electric Vehicles," in *Power Systems for Electric Vehicles,* PHS Publ. No. 999-AP-37, U.S. Public Health Service, National Center for Air Pollution Control, Cincinnati, Ohio, 1967.

was based, is quite unrealistic at present, the effect on the argument is insignificant if we are talking about relatively low ratios of constant power to reserve energy supply.

Given advanced batteries and fuel cells engineered for optimum joint performance, important operating economies appear feasible. For instance, waste heat from the battery can be used to maintain the high temperatures needed for a thermal ammonia or hydrocarbon-steam hydrogen reforming cell, for instance. As noted earlier, parasitic power drain for auxiliaries (whose sizes are determined by the peak power capability) becomes proportionately more important for fuel cells operating much of the time under partial load than for cells operating at constant output. Most important, the fuel cell in a hybrid system is never required to operate under an extreme overload where voltages drop and internal losses become very heavy. In figure 12-4 theoretical efficiency is plotted against power for a hypothetical hybrid fuel cell–battery system.[15] The efficiency drops sharply at first with increasing load, but reaches a minimum when the voltage falls to the point where the battery cuts in and takes over. No further drop occurs until the drain is such as to cause serious polarization of the battery.

The overall efficiencies of a hybrid system involving a high-temperature hydrocarbon-reforming fuel cell and a molten-salt battery (such as General Motors' lithium-chlorine cell) have been calculated by Frysinger for a number of U.S. Army vehicles under specified military duty cycles (see table 12-4). The increased overall operating efficiency of the hybrid system compared with a conventional ICE power plant, corresponding to a decrease in calculated specific fuel consumption, ranged from a factor of 3 for the smaller vehicles to a factor of 2 for the large trucks.

---

[15]Galen R. Frysinger, "Power Sources for Electric Cars," U.S. Army Electronics Command, R&D Technical Report ECOM-2929, January 1968.

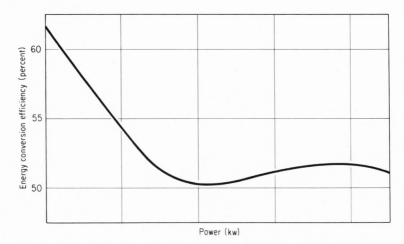

Figure 12-4. Efficiency versus power for a battery-fuel cell hybrid system. (From Galen Frysinger, *Power Sources for Electric Cars*, U.S. Army Electronics Command, R&D Technical Report ECOM-2929, January 1968.)

Table 12-4. Overall Efficiencies of a Hybrid System

Army vehicle	Conventional ICE power plant		Hybrid fuel cell-battery power plant[a]	
	Duty cycle efficiency (%)	Effective energy density (whr/lb)	Duty cycle efficiency (%)	Effective energy density (whr/lb)
1/4 ton	11	97	36	500
3/4 ton	12	129	36	332
2½ ton	18	167	38	630
5 ton	19	229	38	550

Source: Galen R. Frysinger, "Fuel Cell-Battery Power Sources for Electric Cars," in *Power Systems for Electric Vehicles*, PHS Publ. No. 999-AP-37, U.S. Public Health Service, National Center for Air Pollution Control, Cincinnati, Ohio, 1967.

[a]A high-temperature hydrocarbon fuel cell and a molten-salt battery.

Chapter Thirteen

Technological Review

In this chapter, the detailed technical discussions of Part II are summarized as a preliminary to comparisons of performance and cost. Although the tables and illustrations contained in the ten technical chapters (chapters 3 through 12) are not presented here again, essentially the same material is covered in the same order but in condensed form. If the reader has chosen to bypass those chapters, he can find the salient points covered here.

ENERGY REQUIREMENTS FOR AUTOMOTIVE PURPOSES

In order to assess vehicular power requirements, a representative driving cycle must be developed. Such a cycle is a typical sequence of driving actions (acceleration, cruising at various speeds, deceleration, idling, etc.). Obviously, such a cycle will vary from one place to another. The first one to be defined was the California cycle, which was used to develop federal emissions standards.

One approach to driving cycles involves representing the sequence of actions in terms of trapezoidal speed-time profiles, with a smooth rise to cruise speed followed by a steady cruise and an eventual deceleration to zero. For various accelerations, cruise speeds, numbers of stops per mile, and idling time, an average speed can then be found. For large numbers of stops per mile, the average speed can be shown to approach a limit that depends strictly on the number of stops and the idling time.

Driving cycles can also—and perhaps more conveniently—be represented by using continuous frequency distributions. The distribution of gradients and accelerations has been plotted for the Pittsburgh driving cycle, and the values of the logarithm of velocity deduced from the Pittsburgh data. Plotting the values of $(ln\ v)^2$ leads to results that indicate a normal distribution for this function.

Driving patterns are also functions of the power capabilities of the vehicles themselves. Basically, the kinetic energy involved in accelerating to any given speed divided by the time required to do this is the power required. For example, to accelerate a 3,200 lb car to 60 mph in 10 seconds requires an average power of approximately 70 hp (neglecting losses). The actual tractive force is delivered by the torque (the force times the radius of the wheel), which is related to the power in hp by:

$$P = 0.684NT,$$

where N is rotational speed in rpm and torque T is in ft-lb.

In practice, power is not only consumed in accelerating, but also in climbing gradients and in overcoming frictional losses, rolling resistance, and air drag. Internal frictional losses, which may account for 40% of the rated brake horse-power, arise from mechanical inefficiencies (e.g., in transmissions), and from the power needed to operate the fan, generator, oil pump, etc. Rolling resistance, due mainly to the constant flexing of the tires, increases with speed, while air resistance increases with the square of the speed.

Since the torque available is a function of rotational speed (which can be readily translated into a linear car speed), it can be seen that there is a maximum speed at which no further torque is available for propulsion. The power actually available to the driver for this purpose is equal to the maximum power minus that necessary to maintain the speed of the car at that time.

Despite the large drag forces at high speeds, the highest torque is needed at standstill, hence the need for a transmission and starter motor in the internal combustion engine, which generates no torque when its shaft is not rotating. This, in fact, is one of the characteristic weaknesses of existing propulsion systems. Rankine-cycle (steam) engines and electric motors, on the other hand, deliver maximum torque at zero rotational speed.

While a graph of acceleration versus speed reveals a brief acceleration rise to about 0.4g in low gear for a typical case, an approximate average accelerating capability can be set at 0.2g (or 6.4 ft/sec²). Earlier expressions for power can be combined to yield a power equation in terms of velocity, acceleration, weight, and other parameters.

Because the actual road-load demand thus determined increases with the cube of the speed, it takes much more power to produce a given acceleration at 60 mph, for example, than it does at zero speed. The engine power available must be larger than the actual acceleration demand, of course, to allow for the 40% internal losses and for the 5-10% auxiliary equipment requirements.

Average power used and energy consumption per mile over a driving cycle can be calculated by integrating the power equation over the distribution functions corresponding to the frequency with which the various speeds, slopes, and accelerations occur. Energy consumption per mile by a car of weight W is in particular a useful measure of efficiency.

In an actual computer evaluation of a typical Pittsburgh driving cycle, for a peak power of 185 hp, it was found that the actual instantaneous power was less than 50 hp for 90% of the time, and negative for 40% of the time. For 2% of the time, on the other hand, it was above 100 hp. It is these brief power surges that any propulsion system must be able to supply.

Energy consumption per unit of distance can also be derived in terms of engine performance by averaging fuel consumption over the driving cycle. Actual fuel consumption, of course, varies with the power being delivered and the instantaneous speed of the vehicle. In general, both energy and fuel consumption per ton-mile vary inversely with average speed.

ENERGY CONVERSION

The term "energy conversion" is used in the present discussion to mean only the conversion of chemical energy into mechanical or electrical energy in a self-contained or transportable mechanism. Although a large number of fuels (energy sources) fit this prescription, the mechanisms themselves fall into only a few basic categories. For example, they may be either combustion engines (internal or external), or direct conversion (noncombustion) engines. Perhaps more simply, they may either require conversion of the fuel energy to heat to be transferred to a working fluid which runs through a full cycle and returns to its initial state (Q-engines, e.g., the steam engine), or they may utilize a continuous process or "open" cycle (E-engines, e.g., internal combustion engines, gas turbines).

The paramount example of the E-engine is the familiar reciprocating spark-ignition engine (generally called *the* internal combustion engine), which operates on the so-called Otto cycle. In the standard version, the engine takes 4 strokes to draw in air and fuel, compress the mixture, ignite it, and expand and exhaust it. A 2-stroke form is more efficient in terms of weight and volume, but less so in thermodynamic terms as it is particularly prone to discharge unburned or partially burned fuel. The 2-stroke engine is generally relegated to such relatively minor applications as lawn mowers and outboard motors.

Under optimum conditions (disregarding accessories, transmission losses, etc.), Otto engines today operate at 20–30% efficiencies, consuming less than 0.5 lb of fuels per hp-hr. However, under nonoptimum conditions, efficiency is markedly degraded, and efficiencies of 10% or so are more typical in actual urban driving cycles.

Another familiar E-engine is the diesel engine. Operationally similar to the Otto engine, it relies on ignition from very high compression rather than from a spark. The fuel is injected only after the air in the cylinder is already compressed and heated. This system eliminates the spark plug and the carburetor, but it requires a carefully timed fuel injection system and very heavy-duty construction is essential. Diesel engines are more efficient and longer lived—but more expensive—than spark-ignition engines.

A third basic E-engine is the gas turbine, in which the expanding gases act against turbine blades to do work. This device utilizes the Brayton cycle, which is also applicable to closed cycle or Q-engines. Numerous variations on the turbine theme have been devised to improve poor performance at low speed; these include compound or split-shaft designs, and the application of waste exhaust heat to the input gas (regeneration). The thermal efficiencies are generally lower for gas turbines than for reciprocating ICEs of comparable horsepower.

Q-engines, the archetype of which is the steam engine, rely on the compression and expansion of a working fluid in a closed cycle to convert heat energy into useful work. Maximum efficiencies are thus limited by the temperatures of the working fluid before and after expansion. Such machines may do work by altering pressure, volume, enthalpy, entropy, and temperature in various sequences, either in a reciprocating or a rotating element. The Rankine, Stirling, and other cycles on which Q-engines operate all represent attempts to approximate the ideal Carnot cycle, in which isothermal and adiabatic compression is followed by isothermal and then adiabatic expansion.

Like the Carnot cycle, the Stirling and Ericsson cycles are both theoretically reversible and therefore can be considered ideal. Unlike the Carnot cycle (which has never been embodied in an engine), modified versions of the Stirling and Ericsson cycles have been used in experimental engines, using air, helium, or hydrogen as the working fluid.

The Rankine cycle is adapted to condensing fluids and includes the following operations: adiabatic compression, heating, further heating at constant temperature and pressure, superheating at constant volume, adiabatic expansion during which work is done, and condensation and heat rejection. For some fluids, notably steam, "wetting" occurs—i.e., condensation takes place spontaneously. For other fluids, principally organic compounds, the vapor becomes more superheated with expansion and "drying" occurs. It can be shown that for given maximum and minimum temperatures a fluid intermediate between wetting and drying offers maximum efficiency.

Under proper conditions, the energy released by a chemical reaction can appear directly as electricity instead of being manifested as heat. The separation of reactant molecules into oppositely charged ions in an appropriate solution (electrolyte) will produce an electric potential which can drive an external electric circuit. This, in fact, is the essence of electrochemistry. The energies from such reactions are theoretically quite impressive, in terms of energy per lb of reactant. In practice, when inefficiencies and losses and the structural weight requisites of the reactant cell are taken into account, the effective specific energy density is decreased manyfold.

Batteries and fuel cells—the two classes of electrochemical energy converters—are E-engines with efficiencies not limited by temperature and the Carnot cycle. Maximum efficiency, however, normally occurs at the slowest possible reaction rate, whereas the demands of vehicular applications make it desirable to extract the energy at a very high rate. Since this rate tends to be rather limited by polarization at the electrolyte-electrode interface, the reaction couples in the cell should be chosen so as to have the highest possible potential difference, all else being equal.

Once all the reactions have proceeded to completion, the fully charged condition in an electrochemical cell can be restored either by replacing the reactants or by recharging the cell electrically. Ideally, the electrolyte should be a good conductor and should not change during reaction; the reaction products should be safe and nontoxic; and unless the electrodes actually participate in the reaction they should be inert, which usually requires a catalyst.

A large number of reactants (fuel and oxidizer combinations), electrode materials, and electrolytes fulfill these requirements to some degree at least. In the last analysis, choice of practical cells will probably be determined by such considerations as cost and commercial availability of the components.

Still other modes of energy conversion exist in principle. Thermionic and thermoelectric generators, for example, convert heat directly into electricity, the former by boiling electrons off hot metal surfaces and the latter by using the voltage difference produced by a temperature difference in dissimilar metals. Another electrothermal technique, undeveloped as yet but theoretically interesting, utilizes a charged aerosol in a moving gas flow. In principle, such systems could be linked to electric drives, perhaps in parallel with a storage battery to handle peak loads. However, the efficiencies of these techniques are still very low (less than 10%, typically) and they are unlikely to be serious contenders in the near future.

In conclusion, the principal energy converters of practical interest for vehicular propulsion, besides the familiar gasoline engine, are the diesel engine, the gas turbine, the Rankine-cycle engine (e.g., steam), the Stirling cycle engine, and the various electrochemical systems (e.g., fuel cells and batteries). The latter, however, must be considered energy-storage schemes, since they still require an electric motor to convert their outputs to tractive force.

ENERGY STORAGE

In general, energy storage here means any of three classes of systems that require some energy-consuming means of bringing the system back to its initial energy state other than by refuelling: electrochemical, thermal, and mechanical.

The familiar lead-acid battery of automotive use is a well-developed, reliable, and fairly inexpensive energy storage system. Its specific energy density of 10–13 whr/lb limits the range of even a small vehicle using it as an energy source to 40–50 miles at about 30 mph. However, it is the most important commercial battery, and it has been used in a number of rather specialized electric vehicles.

The basic lead-acid cell consists of lead and lead dioxide plates immersed in a sulfuric acid and water solution. Three major types are made: the starting-lighting-ignition (SLI) batteries used in conventional automobiles; traction types used in golf carts and forklift trucks; and stationary types. More recently, the addition of cobaltous sulfate and the use of tripolar construction has apparently resulted in improvements in both energy and power densities.

The second main storage battery is the nickel-iron alkaline cell, which exhibits roughly the same energy density as the lead-acid cell, although its voltage drops sharply at high discharge rates. This, combined with its greater cost and poor performance at low temperatures, makes it an unlikely choice for automotive propulsion systems.

The nickel-cadmium battery is perhaps the most versatile commercial battery, and only the high cost and scarcity of cadmium keep it from dominating the market. Its energy of 15-20 whr/lb is not much higher than that of the lead-acid and nickel-iron cells, but it can offer high power at very fast discharge rates, and it can be recharged in a short time and recycled many thousands of times. Recent bipolar designs actually allow discharge at rates comparable to those of capacitors.

The silver-zinc cell has the highest energy density commercially available (40-55 whr/lb), and it can be made to offer power densities up to 150 watts per pound. The high cost of silver, coupled with the cell's limited cycle life (about 100 charge-discharge cycles) militate against it being widely used for automotive power. However, the silver-zinc cell has been used to power experimental electric cars, and recent developments suggest that cycle life may be improved by a factor of ten or more.

The silver-cadmium cell, which uses even more costly materials, and which has a lower energy density, is also an unlikely automotive candidate.

A nickel-zinc cell has recently been investigated by a number of companies. This cell substitutes zinc for the cadmium of the nickel-cadmium battery. In doing so, however, it reduces cell lifetime to 100-200 charge-discharge cycles.

The large number of anode-cathode couples theoretically offering high energy densities fall into three general groups: metal air cells, lithium-organic electrolyte cells, and molten electrolyte cells.

In the metal-air category (an air electrode paired with a metal), many combinations are receiving attention: lithium, magnesium, aluminum, sodium, and, in particular, the zinc-air cell. Many of the former zinc-air cell problems involved in electrical recharge (e.g., dendrite formation) have largely been solved, and zinc-air secondary cells with energy densities of roughly 50 whr/lb have been achieved. It is estimated that 70-80 whr/lb are possible.

The calcium-air cell has also been explored to some extent, notably by Yardney Electric. A nonaqueous electrolyte must be used because calcium tends to combine spontaneously with water, which complicates the picture; but some positive results have been obtained.

Similarly, development of other metal-air cells is in an early stage generally. Interesting results have been obtained in a number of cases—sodium-air and iron-air, for example. The high theoretical energy densities (2,566 whr/lb for lithium-air) make it probable that efforts will continue.

A considerable effort has also been expended in combining various alkali metals in organic electrolytes with different cathode materials. The use of lithium with such materials as silver difluoride, copper fluoride, and nickel fluoride in such electrolytes as propylene carbonate has been extensively researched, and energy densities up to 100 whr/lb have been achieved at low discharge rates.

The inability of such cells to withstand high discharge rates can be circumvented to some extent by using a molten salt component or components. Thus, the General Motors lithium-chlorine cell, with a molten lithium anode, a molten lithium-chloride electrolyte, and a gaseous chlorine cathode, has been projected to be capable of 100–200 whr/lb at 615°C, with a peak power density of 60 watts per pound. Ford's well-known sodium-sulfur battery is another example of the genre. With demonstrated energy density of 80–90 whr/lb, the sodium-sulfur cell produces up to 100 watts per pound with little voltage drop.

Bimetallic molten cells on which work has been done include lithium-bismuth, lithium-tellurium, lithium-selenium, and lithium-sulfur. With all components molten (operating temperatures range from 360°C to 470°C), the electrolyte is diluted with a ceramic powdered filler into a paste. All indicate high energy densities and high power densities.

It can be seen that none of the batteries in existence, or on the horizon, is particularly suited for vehicular applications. Some batteries yield high energies, and some yield high powers; some do both, but these are usually quite undeveloped and/or expensive. The relative values of various characteristics for some typical batteries are shown in summary form and very general terms in table 5–3, chapter 5.

Energy may also be stored in the physical state of a material and released by a change in that state. Thus, if thermal energy were stored in a molten salt, for example, it could be released via a heat engine, most likely in a hybrid propulsion system wherein the molten-salt container would serve as the analog of a storage battery in an electric hybrid. The heat engine in the system serves to "recharge" the thermal storage battery when it is not discharging (i.e., during nonpeak requirements). Such a system would be smaller than an all-external combustion system for an equivalent cycle. A large number of salts exist (e.g., $LiNO_3$, or $NaNO_3$) that would suffice for the above function.

Finally, there is the possibility of storing energy mechanically. There are several ways to do this in principle (in deformable solids, in compressed gases, etc.), but in practice, the only method that is likely to be convenient for vehicular uses is the storage of energy in a high-speed flywheel. Here, kinetic energy is built up in the wheel and slowly transferred to the vehicle via a generator and motor. Energy densities are not large (about 10 whr/lb for a flywheel withstanding 100,000 psi on its rim), but they are comparable with those of present storage batteries. Several such flywheel systems have been used, notably the Swiss Oerlikon gyrobus. With modern developments in materials technology, the flywheel concept is getting renewed attention. Investigators under D. W. Rabenhorst at the Applied Physics Laboratory of The John Hopkins University claim 40 whr/lb or so, using radial bars made of filaments, with higher values indicated for possible future materials.

NEW DEVELOPMENTS IN INTERNAL COMBUSTION ENGINES

Efforts to improve ICE technology have been largely directed toward two areas: improving the ratio of power to weight, size, and cost; and increasing efficiency while reducing emissions. The attempts to achieve the first goal have mainly consisted of the development of various rotary engines, which usually require some form of speed changer but not a transmission. The most celebrated of these is the Wankel engine, invented by Felix Wankel in 1953. This patent, owned by NSU Motor Werke in Germany and licensed to many firms, including Curtiss-Wright and General Motors in the United States, consists of a solid rotor with a rounded triangular cross section of constant diameter, rotating eccentrically in a two-lobed chamber so that the tips of the rotor always meet the chamber walls. There are thus always three combustion chambers. Early problems with lubrication and cooling have largely been solved, and Wankel engines with good power-to-weight ratios have been manufactured. Efforts to produce a clean exhaust have also met with considerable success. The quenching effect of the large wall areas results in cooler running conditions, and the engine geometry permits use of a thermal reactor next to the combustion chamber, thus allowing further combustion of any unburned hydrocarbons. NSU and a Japanese licensee, Toyo Kogyo, both plan to produce Wankel-powered cars, and Mercedes-Benz and Rolls Royce are experimenting with Wankels, the latter company in a diesel form.

There are a number of other rotary engine designs, among them the Kauertz, the Virmel, and the Tschudi, and several variations of the epitrochoidal concept. A novel design by Kal-Pac Engineering of Vancouver uses crossed pistons to produce rotary motion. An adaptation of the vane-type air engine by P. R. Mallory and Company features a number of sliding vanes in a rotor rotating eccentrically in an elliptical combustion chamber; with continuous combustion, this engine is akin to a gas turbine in some respects.

A novel adaptation of the standard reciprocating internal combustion engine is the constant pressure reciprocating engine concept developed by Mechanical Technology, Inc. In this engine, one of each pair of pistons in a standard ICE is used as a compressor. Combustion of the fuel and compressed air takes place in a combustion chamber between the pistons, the gases then expanding through the opposite piston. The developers claim improved fuel economy and virtual elimination of unwanted exhaust pollutants. In addition, the engine can be readily built with standard ICE production techniques.

In general, the two approaches to eliminating pollutant emissions are to prevent them from forming, or to remove them from the flow once formed. The basic techniques for the first method are designed to ensure more complete combustion by more precise control of fuel flow under all conditions, by reducing cool wall surfaces, or by increasing the dwell time of the gases—or all three.

Fuel injection, which is one means of ensuring more complete combustion, is now beginning to make itself felt in the automotive market, notably in the

Volkswagen 1600. By more nearly matching the fuel input to the engine's demands, fuel injection reduces carbon monoxide output. Improved combustion can also be effected by varying the fuel-air ratio. This can be achieved by the stratified charge technique, wherein the injection of fuel backward into a high-velocity swirl of air concentrates the fuel near the central spark plug. Combustion spreads outward and, with proper design, never reaches the cool outer walls, and thus is never "quenched." This technique not only improves combustion and does away with the problem of unburned hydrocarbons, it prevents knocking, allows lower-grade fuels to be used, and improves fuel economy. The principal work in this area has been done by Texaco, although others have been active. A variation of the stratified charge principle, developed by the Azure Blue Corporation, provides a pre-combustion chamber in which a rich mixture is diluted by the lean mixture from the main chamber during compression, so that ignition occurs at a near-optimum ratio of 13.7 : 1. The developers claim significant emissions reductions.

Of the sources of pollutant emission, crankcase blow-by is the easiest to deal with. Gases are simply returned to the engine air intake and recycled. This scheme forms the basis for more than 60 systems now certified for use in California.

Pollutants can also be removed after they have left the combustion chamber by providing extra air for more complete oxidation. The usual methods are either by injecting air into an exhaust manifold reactor or by more precise control of the carburetion. Emission values well below the early (1968) federal standards of 275 ppm HC, and 1.5% CO can thus be obtained. Use of a fuel-rich mixture with an exhaust manifold will reduce NO_x in the exhaust, as well as HC and CO. Recirculation of the exhaust gas through the engine will also reduce the NO_x level.

Further downstream, the muffler or tailpipe of an automobile can be replaced by a catalytic or direct-flame afterburner. Although these devices have some problems (the gases are cooler at this point, requiring more heat; catalysts tend to become poisoned by lead in gasoline, etc.), growing interest in reducing nitrogen oxide pollutants (which can, in principle, be done with a catalytic converter) has accelerated development. Various companies, for example, claim to have at least lessened the catalyst–lead poisoning problem.

Evaporative losses from the carburetor and gas tank have been cut very significantly by a number of techniques, which include the use of a charcoal absorption canister, crankcase storage, and pressurization of the fuel tank. Fuel injection, of course, also eliminates carburetor hot-soak losses by eliminating the carburetor. Evaporative losses must be controlled in some way on all 1971 model autos sold in the United States.

While the 1968 and 1970 air pollution standards have been approximately met by the industry, the ultimate decrease in emissions can never be more than about 50% of the total (although 80% is removed from each car) because of the increasing number of cars, degradation of controls with age, and large numbers of uncontrolled cars. In any case, the low point will not occur until 1980. The goal of the new legislation introduced in 1968 in California was to reduce emissions levels to what they were in 1940 (the last year before "smog" became a serious problem in Los Angeles).

The Inter-Industry Emission Control Program was mounted jointly in 1967 by Ford in conjunction with several oil companies and foreign car manufacturers. This group developed various devices (programmed by-pass computer, catalytic converters, modified fuels) and embodied them in modified test vehicles.

To achieve substantial improvement above the 1970 California standards, it would seem necessary to eliminate carburetors completely, and to provide afterburners, gas-tank vapor-loss preventatives, and more sensitive fuel inlet control techniques. Estimates of the cost per car of meeting current and proposed future standards range from $40–$60 or so for 1970, to as much as several hundred for 1973.

Claims made by the automotive industry that present control devices will be effective in lowering pollutant levels in the future seem unduly optimistic. Efforts to meet the more rigid pollution standards are likely to make the internal combustion engine increasingly more complex and expensive, and undoubtedly less reliable. Furthermore, little has been said about the necessity of inspecting and maintaining these devices. It has been estimated that it would cost $15,000 per lane to install inspection equipment in existing inspection facilities—and considerably more if the facilities themselves must be built.

In principle, any combustible fuel could be used in a car engine, and many have been considered, including ammonia, hydrazine, liquid propane gas, natural gas, and hydrogen. All would seem to have some advantages and disadvantages, notable among the latter being problems of logistics and supply. Only in the cases of propane (LPG) and liquefied natural gas (LNG) has any substantial experimental work been done. These fuels do indeed result in much cleaner combustion. However, under present conditions, they are only appropriate for fleet operations. In addition, it is not clear that reserves of natural gas or natural gas liquids would be adequate if any sizable fraction of the auto market were converted from gasoline.

To summarize, the diverse approaches being taken to improve the performance and emission levels of the internal combustion engine include novel engine designs, improvement of the combustion process in conventional engines by such means as fuel injection and stratified charge techniques, removal of pollutants from the exhaust stream via afterburners and manifold reactors, and use of alternate fuels. Considerable further improvement along these lines is undoubtedly possible. It is difficult to know whether this approach will be sufficient to stay abreast of the problem, as more and more cars appear and the performance of existing engines deteriorates with age. Significant improvement is also possible in other types of power plant that have much lower pollutant emissions than the ICE to start with.

THE GAS TURBINE

A considerable amount of work has been done by Chrysler, Ford, GM, American Motors, Caterpillar Tractor, International Harvester, the Rover Company, and others in trying to adapt the gas turbine to automotive use. These efforts date back to World War II, and even earlier. Chrysler, for example, started work in 1939, and was developing its sixth-generation turbine by 1968. In general, the most immediate area of application for automotive gas turbines would seem to be for long-haul trucks and buses and off-highway earth-moving machinery, and it is toward this market that most of the manufacturers currently engaged in turbine development appear to be aiming.

While the gas turbine has several obvious advantages as an automotive power plant, it is not without serious drawbacks. Because of the low torque of the single-shaft turbine at low speed, it becomes necessary to use a split-shaft turbine, wherein the exhaust from the compressor turbine drives a second turbine, which provides power. Thus, the compressor is free of the load, and torque is maximum at standstill. This solves one problem but creates another: average efficiency is drastically reduced, because the engine often runs at top speed when power is not needed. This in turn can be solved to a considerable extent by the differential turbine configuration, which interconnects the input compressor with the output turbine via a differential gear. This variation provides up to 65% reduction in fuel consumption with decreasing load while still keeping maximum torque on the output shaft.

Insertion of a blow-off valve in the flow between the captive and free turbines in such an arrangement provides a negative torque which can be used for vehicular braking.

In general, the gas turbine (in a differential configuration) is an attractive vehicular engine. It is relatively simple and efficient, and exhibits power/weight ratios ranging from 0.6 lb/hp up to 3-4 lb/hp depending on size. It can also be used with a wide range of fuels, and, because of the large air flow the exhausts are comparatively free from smoke and carbon monoxide.

It is not yet certain, however, that practical gas turbines can be built to operate efficiently and cleanly in the small-size ranges required in typical automotive applications (e.g., 100 hp and under), although several developers have done this experimentally. Cost is also a somewhat cloudy issue. It seems probable that the gas turbine, even in mass production lots, will cost slightly more per hp than does a comparable ICE.

The need for turbine exhaust pollution control is not yet fully determined. Turbine exhaust is undoubtedly lower in carbon monoxide and hydrocarbon content than a reciprocating gasoline engine exhaust even when an adjustment is made for the turbine's greater mass flow, but evidence regarding nitrogen oxide emissions is far from clear. Some tests indicate levels as high as those from piston engines. However, it is widely assumed that NO_x output could also be reduced relative to the reciprocating ICE.

RANKINE-CYCLE EXTERNAL COMBUSTION ENGINES

Steam was an early competitor among power sources for automobiles, but by the 1920s the mass-produced gasoline engine had virtually driven it out of business. The classic early steam car, the Doble, was still produced in small numbers for some years after the Stanleys and the Whites had disappeared. Numerous steam vehicles were developed and built in Europe during the 1920s and 1930s—the Sentinel truck in England, for example, and the Henschel bus in Germany. After World War II, the experimental McCulloch steam car and, later, the Yuba steam tractor appeared. Perhaps the most well-developed and publicized private steam project is that of the Williams Brothers, who have built a number of engines, at least two of which are still in operation. Many other efforts have been expended in recent years on Rankine-cycle engines and vehicles. Among them are the developments of the Thermo Electron Corporation, Pritchard in Australia, General Steam Corporation, Lear Motors, and Kinetics, Inc. Also, in 1969, a Doble-type steam plant built by the Besler Corporation (which acquired the Doble assets in the 1930s) was applied to a converted Chevrolet for General Motors.

The desired goal of high thermodynamic efficiency in Rankine-cycle engines can be achieved by ensuring that the temperature at which vaporization occurs is as high as possible while the condensation temperature is as low as possible, and by preventing heat losses, e.g., by condensation of water during the expansion process. A common measure of steam engine efficiency is the water rate, or amount of steam required per hp-hr output; this may range from as low as 6 lb/hp-hr up to 30 lb/hp-hr, the rate decreasing as thermal efficiency goes up.

Various techniques have evolved over the years to improve steam engine performance. The uniflow engine principle, in which the expanding steam is always exhausted through a cool port at the end of the piston opposite the steam inlet valve was invented by Strumpf in 1908. This reduces unnecessary heat losses through condensation by ensuring that the inlet valve always remains hot. Compound expansion cycles may also be used to increase the average temperature of the expanding fluid and reduce condensation by providing two or more expansion stages, separated by reheat or recompression stages.

In actual operation, peak power surges may demand increased torque. This can be implemented by changing the engine's "cutoff" (the point at which steam inlet valves are closed). Early cutoff means high efficiency, late cutoff means high torque because of the higher mean effective pressure. Thus, a variable cutoff device makes an engine more adaptable to actual demand conditions. A number of specialized valving techniques have been developed to take advantage of variable cutoff. Many designs permit the engine to be reversed instantaneously, simply by the valve relationships. Early steam engines used capacity-type (teakettle) boilers which required much preheating. The monotube or "once-through" type of steam generator was first introduced by Doble in 1921 to reduce start-up time by reducing the amount of water in circulation. Whereas capacity-type boilers required up to 30 minutes to warm up, Doble cars needed only 45 seconds even for a cold start.

It is, of course, axiomatic that the steam generator must produce sufficient steam for the demands of the engine. In practice, this means relying on stored heat (in the form of steam) to handle peak power surges. In the future, a molten salt and a suitable heat exchanger may be used to supply the peak heat energy. The Rankine-cycle engine has a high overload capacity by virtue of its heat storage capability.

The critical components in the fluid circulation system of any steam system are the vapor generator and the condenser, which are the high-temperature and low-temperature heat exchangers. Design objectives are to maximize heat transfer efficiency while minimizing mass flow (and therefore heat-exchanger mass and costly materials). Use of modern materials and heat transfer technology can reduce condenser size and weight by a large factor over the state of the art in Abner Doble's day.

A large number of substances other than steam have been considered as working fluids for Rankine-cycle engines, each having some advantages and disadvantages, depending on system conditions such as maximum and minimum temperatures. The most advanced program involving the use of organic fluids is at Thermo Electron Corporation, where considerable work has been done on the use of Thiophene (CP34), a compound of carbon, hydrogen, and sulfur. A number of other organic fluids such as Freon 11, benzene, hexafluoride, MIPB (mono-isopropyl-biphenyl), and so on have also been suggested. Thermal stability at high temperatures, problems of toxicity or flammability, possible need for a desuperheat stage in the cycle, possible high volume flows leading to large pumping requirements, and numerous other problems need to be resolved.

In general, however, a Rankine vapor-cycle external combustion engine appears to be a potentially most attractive vehicular power plant. It has very high torque at zero speed, meaning that no clutch or transmission is needed (except, perhaps, to operate auxiliary equipment). This aspect is particularly significant for buses and other large commercial vehicles, which normally require large and heavy transmissions with standard internal combustion engines. In addition, the Rankine-cycle engine eliminates the need for such components as the starter motor, the carburetor, the engine-block cooling system, and so on. The instantaneous reversal capability implies the possibility of using the engine to brake the vehicle (while recovering part of the energy as heat). This too would be of great value for trucks and buses. As with any external combustion engine, an inexpensive hydrocarbon fuel (such as kerosene or diesel oil) can be burned, yet pollutant emissions—including the oxides of nitrogen—are intrinsically very low.

NONCONDENSING EXTERNAL COMBUSTION ENGINES

The Stirling cycle was the first of the so-called "air engines." Development began in the early 1800s, was abandoned by mid-nineteenth century, but revived and improved by the N. V. Philips Research Laboratories in the 1950s and 1960s. Essentially, a Stirling-cycle engine utilizes alternate heating and cooling of the working fluid (nowadays, compressed helium or hydrogen gas), at constant volume, using a thermal regenerator between the two compartments, in which the gas both gives up and regains heat in passing through in different directions. Practical units developed by Philips utilize a displacer piston to move the fluid back and forth, in addition to the power piston. An ingenious mechanism known as the "rhombic drive" converts the reciprocating piston and displaces motion into a rotary output for a single cylinder; thus no unbalanced crankshaft is necessary.

To date, the Stirling engine has been rather heavy (on the order of 10 lb/hp in prototypes) and mechanically complex. Its torque-speed characteristics are slightly more favorable than those of the diesel engine, with which it would probably compete initially, but not favorable enough for the transmission to be eliminated. Stirling engine emission characteristics are extremely good: even nitrogen oxide production is considerably lower than that for a diesel engine of comparable size.

A modified Stirling engine patented recently by Vannevar Bush utilizes a pair of double-ended hot and cold pistons, the hot pair being enclosed in a heater and the cold pair in a refrigerator. Use of bypass coils between each regenerator and its hot and cold pistons allows improved performance.

The noncondensing external combustion principle can also be applied to the Brayton (gas turbine) cycle, for example, in a rotary expander of the vane type utilizing air as the working fluid. In one configuration similar to the previously discussed vane-type rotary internal combustion engine developed by the P. R. Mallory Company, vanes in an eccentrically mounted rotor divide the crescent-shaped space between the rotor and the housing into a series of chambers. Air is introduced at the large volume section and compressed by rotation, the air leaving the compressor near the minimum volume, and entering a suitable burner, where heat is added at constant pressure. This higher temperature air is then introduced into the expander at minimum volume. Subsequent expansion does work on the rotor, after which the air is exhausted and a fresh charge is introduced.

Preliminary analyses indicate potentially favorable power/weight ratios (perhaps approximating 2.5 lb/hp) for such a device, although, like the ICE, it would probably need a starter motor and a transmission in an automotive application. Its excellent characteristics at higher speeds, however, suggest its possible value in a hybrid-electric vehicle, wherein the heat engine could run constantly under optimum conditions.

ELECTRIC PROPULSION

At present, the choice of an electric drive unit for a vehicle is limited to the series DC motor, and the AC squirrel-cage induction motor. The latter is simpler and more reliable, but the need for a frequency control means heavy and expensive solid-state controls. In the long run, however, there are other choices also.

Essentially, all electric motors rely on the interaction of an electric circuit and a magnetic circuit. Either one may be the stationary or the moving element. Thus, a number of basic configurations are possible. In the DC homopolar motor, an axial magnetic field may be provided either by a magnet or a winding. In principle, an AC homopolar motor should also be possible, although none seems to have been developed.

In conventional DC motors, the rotor is electrically active, and wound in such a way as to orient the magnetic fields longitudinally in an opposite manner to the stator fields (for example, N to S). If shunt wound, the motor has a limiting speed; if series wound, it has none.

The basic AC squirrel-cage motor has a passive rotor in which torque is induced by the action of currents induced by the external magnetic field of the stator current. Other variations utilize electrically active rotors.

While the torque-speed characteristics of the series-wound DC motor are almost ideal for vehicular uses (torque is maximum at zero speed), the torque characteristics of either AC or DC types can be custom-designed to suit virtually any requirement. AC induction motors and DC shunt motors both have built-in top speeds, while the DC series motor tends to run away under no-load conditions unless it has a governor.

We cannot seriously evaluate weight and cost in electric drives without also considering the necessary control systems. The DC homopolar motor, for instance, yields excellent hp/lb ratios, and, although the problems of switching very large currents at the required low voltages appear almost insurmountable, homopolar motors are again receiving serious attention. Control problems connected with a hypothetical AC homopolar device would be similar, although current and voltage control would be simpler than in the DC case. For the AC multipolar inductive motor, the control problem lies in the need to vary the frequency of the current, a problem analogous to the classical mechanical problem of designing a continuously variable-ratio transmission.

A power/weight ratio of about 3.5 lb/hp for AC inductive motors, including all necessary controls, cooling equipment, etc., is currently attainable. DC series motors, on the other hand, are limited by the mechanical limitations of the commutator, which does not work well at high rpm. Thus, much interest has been focused on "brushless" or solid-state control techniques for these machines.

The rated output of an electric motor used for vehicular propulsion can usually be smaller than that of other engines because many electric motors have notable overload capacities—as high as 250% for short periods of acceleration in some cases. However, overload capability is likely to decrease as motor designs become more efficient, using less metal and relying more heavily on solid-state electronics for switching.

In DC motors, voltage variation (and hence speed control) has typically been achieved by varying the resistance in the motor circuit, either by a rheostat or by switching battery cells into various series-parallel combinations. These techniques, which have been used on virtually every "commercial" vehicle to date, have many disadvantages, including large power losses, and mechanical discontinuities in switching. Electronic switching, on the other hand, is free of these problems. For DC motors, the average voltage is varied by "chopping" the current into pulses of variable length, by means of a solid-state diode (switch). In AC induction motors, this "chopper" output controls the input to a three-phase inverter, resulting in a variable-frequency output from the inverter that controls the speed and power of the motor.

Regenerative braking, in which the motor is driven as a generator by deceleration, can also be controlled by solid-state circuitry. For shunt motors, the field strength simply needs to be increased. Series motors may need either a field reversal or an additional shunt field coil. (For the motor to function as a generator, current must be able to flow out of the terminals.) AC induction machines, which automatically generate power when driven above synchronous speeds, need only a shaft-speed sensing device.

It is desirable, of course, that motor weight be minimized. This can be effected in a number of ways, notably by increasing the operating speed (power increases with speed), by reducing the required amounts of copper or other metals, and by design ingenuity (e.g., in reducing the length of flux paths). The rated torque, and hence power, can be derived by using known values of stall torque vs. motor weight. Typical weight/power ratios are 8 lb/hp for standard DC types and 2 lb/hp to 4 lb/hp for AC motors (depending on speed), although ratios approximating unity are possible with some newer AC types.

While it is difficult to assign a single rule for calculating costs, an average of $3/lb for conventional types seems typical. Leading authorities believe that enormous improvements in performance (e.g., hp/lb) will be achieved without increasing the cost per pound significantly.

In summary, electric motors have certain characteristics such as high starting torque that make them ideal for vehicular propulsion. The cost of the necessary motor controls makes electric motors somewhat more expensive per horsepower than most other tractive engines, but for the relatively primitive types now available the higher cost is somewhat offset by their high overload capabilities, which permit a lower horsepower rating for a given acceleration, and by traditionally great reliability. However, it remains to be seen whether these desirable features of relatively simple, heavy-duty industrial motors will be obtainable from the more sophisticated, light-weight, high-performance motors needed for vehicular power plants.

ELECTRIC VEHICLES

The all-electric vehicle is considered here to be a battery-powered vehicle, since a fuel cell is unlikely to be a practical power source within the next decade.

Perhaps the main drawback to battery-electric vehicles is the need to recharge them (although the recharge capability can be of value at times). At the very high charging rates that would be desirable on the road (e.g., 10 or 15 minutes), very high charging currents would be required, thus necessitating massive specialized electrical equipment. One solution would be a network of charging stations on highways, but such a grid is not likely to spring up overnight, even with large numbers of electric cars on the road. There are other alternatives, however. Batteries could be recharged slowly at home for local use, but exchanged for fresh fully charged ones for use on long trips—in effect, rented for the trip. Or trailer-mounted ICE- or ECE-powered battery chargers might be rented or sold.

Electric vehicles would probably have plastic bodies on aluminum or steel frames, which could cut the ratio of chassis weight to motor-control-battery weight to 50–55% as against 60–65% in present vehicles. Estimates of chassis costs (perhaps optimistic) range from 20 cents to 40 cents per pound. It is impossible at this time to estimate the ultimate manufacturing costs of the high-energy batteries still undergoing laboratory development. However, costs for conventional lead–acid batteries range from 25 cents to 50 cents per pound, plus the basic material costs. It has been estimated by one authority that battery costs in all price ranges come to five times the cost of reactants and electrode materials. The use of low-cost anode materials, such as zinc or sodium, would presumably result in battery costs of the order of about 75 cents per pound of reactant. Batteries for an all-electric car can be presumed to account for 35% to 40% of vehicle weight.

It has also been suggested that the user might buy the car and its motor and controls, but rent the active battery material, which would allow the use of even the energetic but expensive silver–zinc battery. Silver rental costs on a small car driven 12,500 miles per year would be of the order of $75 per year, assuming 1967 costs and interest rates.

It has been estimated that typical modern cars, with an average life of eight years, are driven an average of 3,200 hours during that time. This is considerably less than the average lifetimes of most electric car components except some battery types, which in any case probably would not have to be replaced any more often than present starter-batteries. Average vehicle life for electric cars could well be longer than present conventional vehicles.

A model has been developed to estimate the withdrawal of stored energy for driving under various conditions. For a vehicle weighing one ton, having rolling friction and drag coefficients equal to those of a Volkswagen, and using conventional lead–acid batteries with a 6.5 kwh capacity and a 70% regeneration efficiency, energy withdrawals would be roughly 0.25 kwh/ton-mile on gently rolling terrain. Without regenerative braking the energy withdrawal increases to 0.324 kwh/ton-mile. In contrast, use of a hypothetical storage system with characteristics similar to those claimed for Ford's sodium-sulfur battery lowers the withdrawal rate to 0.136 kwh/ton-mile. These figures do not, of course, include losses in the vehicle, charging losses, or power required for accessories.

Based on a projection of 0.15 kwh/ton-mile (by the late 1970s) and antici-pated power costs of $0.012/kwh, actual energy costs (for driving) would be approximately $0.0018/ton-mile, or $18/ton of car for 10,000 miles of driving (a typical year). However, several battery replacements would be required in the average life of a vehicle. If the first costs of each battery are amortized over the expected life of the battery, costs would be significantly higher. For the lead-acid battery, they would come to roughly $500/ton per year for 10,000 miles of driving.

Since one-third of today's car fuel bills goes to taxes, it is clear that the replacement of any reasonable percentage with all-electric cars would create the need for some new form of legislation or tax-collecting mechanism.

Almost certainly, small low-performance electric cars will appear first; in fact, many have already done so. These might be followed by small cars powered by higher-energy batteries. As these batteries become more fully developed, higher performance electrics may begin to compete successfully with vehicles powered by engines.

HYBRID VEHICLES

The development of hybrids is an attempt to capture the advantages and avoid the disadvantages of individual power sources by combining two or more conversion and storage methods that complement one another. In such systems, the primary energy conversion mechanism (ICE, ECE, turbine, or fuel cell) would operate at close to the *average* power level of the vehicle; the secondary system (electrical storage battery, thermal storage battery, flywheel, etc.) would act as a power storage device to meet peak demands. A primary system designed to operate at nearly constant load can generally be both simpler and more efficient than one that has to meet rapid and unpredictable variations in load.

It can be demonstrated that, to ensure a 98% probability of not depleting the reserve in one hour of driving, an energy capacity equal to twice the root-mean-square power demand is needed in a hybrid, that is, an energy capacity of 20 kwh if the r.m.s. power is 10 kw. In a computer simulation of the Pittsburgh driving cycle, various combinations of power and capacity (e.g., 10 hp through 40 hp and 3.75 kwh through 15 kwh) were tested for the two cases of 80% power regeneration and zero regeneration. From the results, it would appear that about 15 hp represents a probable minimum primary power unit output in a typical urban-suburban situation unless very large energy capacities are provided. Actual choices of systems, however, would tend to be dictated by other practicalities not included in the simulation—e.g., the decrease in capacity with increase in discharge rate in most real batteries, etc. For automotive use, the electric battery may be the most practical power source because it tends to hold its charge for a considerable length of time while the flywheel and thermal batteries both tend to "run down" in a few hours, whether being used or not.

A typical hybrid configuration using an ICE as a constant power source with a battery for energy storage and an electric drive operates at a considerable fuel savings over a conventional ICE drive, owing to the part-load inefficiencies of the ICE, although the smaller ICE required in the hybrid will tend to be costlier on a per pound basis than the larger conventional one. It is probable that hybrid electric cannot compete with either the absolute performance or the initial cost of an all-ICE or all-ECE drive, at least in small sizes, although pollutant-emission and fuel consumption would be lower for the ICE-hybrid than for the all-ICE.

Considerable work has been done on hybrid systems. The "Minicar," developed as a system of interchangeable small cars, which would be leased to users on a monthly basis for commuting but which would double as "U-drive" taxis during the day, was developed by the University of Pennsylvania and Minicars, Inc. It utilizes an ICE and a DC motor with a small lead-acid battery, and has the capacity for conversion to a small steam engine (or other ECE) as the prime mover.

Another experimental hybrid which has received some publicity is General Motors' "Stir-lec." This system utilizes a converted Opel Kadett as a vehicle, with an 8-hp Stirling engine as a battery-charger, and a 20-hp AC induction motor as drive unit. Its performance is to be comparable to that of a small underpowered conventional gasoline car.

On a larger scale, the ICE-electric hybrid is more promising because of improved economies both in electric motors and prime movers (e.g., turbines) in larger sizes. In West Germany, Daimler-Benz has been developing a diesel-electric city bus which is to feature a 150-hp electric motor and a 65-hp diesel, while Lear Motors in the United States has proposed a turbo-electric hybrid, also for use in buses.

The U.S. Army Engineer R&D Laboratories have applied hybrid electrical propulsion to a variety of trucks and tanks, using ICEs and turbines as the prime movers.

A thermal battery, in the form of a molten salt heat exchanger is also a possibility for use with an ECE. The thermal storage system in this case serves the same function as an electric battery in the other cases, supplying surge peak energy to the working fluid upon demand. One such system has been proposed by Thermo Electron Corporation, using Thiophene as the working fluid.

Perhaps the simplest conceptually of all possible hybrids would be the fuel-cell battery hybrid, which would require no electric generator. To provide power for normal driving to a 3,000-lb car would require a reliable fuel cell output of 40 watts/lb, and a rechargeable battery of 100 watts/lb and 67 whr/lb. This is considerably beyond present capabilities, although it may be possible by 1980 or so. At lower performance levels, however, the fuel cell battery combination is more plausible. A 2,000-lb car might be powered for urban use by a 400-lb, 10-kw fuel cell and a 300-lb, 50 whr/lb battery.

The important aspect of using a fuel cell in a hybrid configuration is that the fuel cell operates constantly at optimum load, never reaching high overloads where voltage drops and losses become severe. When the voltage falls below a certain point as the load increases, the battery takes over.

When tested over typical military-duty cycles, hybrid systems using high-temperature fuel cells and molten-salt batteries (e.g., the lithium-chlorine cell) have had efficiencies of 36% to 38%, compared with 11% to 19% for conventional ICE powerplants.

Chapter Fourteen

Comparisons of Performance and Cost

PERFORMANCE

An unfortunate characteristic of the internal combustion engine in its present form is its complexity and the aggravating maintenance problems that result. This problem arises partly from the basic characteristics of the reciprocating piston engine, which requires a large number of moving parts (some of which move in an unbalanced manner) plus the gearshift and clutch. It also arises in part from the elaborate controls necessary to time the ignition and monitor the fuel-air mixture in response to the changing demand for power output. Equipment needed to reduce noise and unwanted gaseous emissions will add to the complexity and reduce performance.

The conventional gasoline engine performs best at a constant speed of 3,000 to 4,000 rpm. As the speed drops, torque drops sharply while specific fuel consumption rises sharply, as noted in chapter 4. At idling speed—roughly 500 rpm—the engine gulps fuel, but produces only enough power to overcome internal friction. Diesel engines and gas turbines exhibit qualitatively similar torque-speed characteristics.

From a technical standpoint, the ideal engine should be self-starting and should produce maximum torque at zero speed, with torque dropping to zero at the maximum speed at which the engine will turn. The engine should require only a minimum of power-consuming auxiliaries (fan, water pump, transmission, muffler, etc.), and should achieve very nearly the same specific fuel consumption or conversion efficiency under conditions of low speed and heavy load as it does at high speed and no load. Finally, it should produce smooth, vibrationless power in the form of rotary motion, preferably transmitted directly to the wheels (to avoid the "universal" gears).

A change in automotive vehicle propulsion technology would be accompanied by changes in cost (initial cost and maintenance), operation complexity, and performance. "Performance" covers a multitude of subtleties, not merely power available on demand—i.e., responsiveness, acceleration, and speed. Since one can only rate power plants with respect to any given figure of merit if all other factors are equal, it is extraordinarily difficult to compare the various systems because they differ in very fundamental ways. The following attempt to do so is far from definitive. However, it must be judged in the light of the existing technical literature, which is notably deficient in meaningful comparisons between different kinds of power plants, at least for automotive applications.

The instantaneous specific fuel consumption (in lb/bhp-hr) for any engine is a function both of the power being delivered momentarily and the instantaneous speed of the engine. Figures 4-6, 4-12, and 4-16 displayed the ranges of specific fuel consumption as functions of percent of maximum engine power for Otto-cycle, diesel-cycle, and Brayton-cycle (gas turbine) engines, respectively. These curves all resemble a hyperbolic function involving two free parameters χ and ω that will be assumed to have different constant values based on the characteristic behavior of each engine. Thus, specific fuel consumption vs. engine power can be described roughly by the formula:

$$C_f = \frac{\chi P_m + \omega}{P_b}, \tag{14.1}$$

where P_m is the bare maximum hp rating of the engine and P_b is the brake hp. From an examination of numerous references, representative fuel consumption values have been selected for various engines, from which values of the parameters were determined, as shown in table 14-1.

It can be seen from table 14-1 that the specific fuel consumption at constant load increases sharply with decreases in engine power, except for the hybrid system, which consists of an ICE that is assumed to operate at some constant, near-optimum speed (with regard to fuel consumption and emissions output). Under light loads, the difference between the traction power required by the

Table 14-1. Specific Fuel Consumption

			Fraction of peak power			
	Parameters		.20		.50	
Engine type	χ	ω	SFC[a]	EFF[b]	SFC	EFF
Otto	.05	.35	.60	22	.45	30
Diesel	.05	.30	.55	25	.40	33
Gas turbine	.08	.40	.80	17	.56	23
Hybrid	.02	.36	.46	29	.40	33
Rankine	.03	.50	.65	21	.56	23

Source: M. L. Walker, Jr., "A Methodology for Estimating Fuel Consumption for Various Engine Types in Stop-Go Driving," International Research and Technology Corporation Paper No. IRT-R-20, Washington, D.C., March 24, 1970.

[a]SFC = Specific fuel consumption (lb/bhp-hr).

[b]EFF = Efficiency = $\dfrac{2545}{C_f H}$, where H is the heating value of the fuel (Btu/lb).

vehicle and that available from the engine is used to charge a secondary battery; peak power is, in turn, provided by the battery on demand. Actual propulsion is provided by an electric motor.

The variation in fuel consumption and energy consumption with average driving speed that is implied by table 14-1 is displayed graphically in figure 14-1. In general, both energy and fuel consumption per ton-mile vary inversely

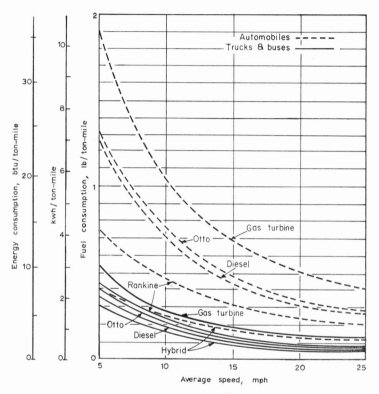

Figure 14-1. Energy and fuel consumption for various engines. (From M. L. Walker, Jr., *A Methodology for Estimating Fuel Consumption for Various Engine Types in Stop-Go Driving*, International Research and Technology Corp., IRT-R-20, March 1970.)

with average speed. The calculated curves are based on somewhat arbitrary assumptions with regard to engine characteristics and are not necessarily in agreement with the values compiled in this book and exhibited in the text. However, the curves shown in the figure have the correct qualitative behavior.

Table 14-2 compares other performance statistics for a number of power plant types, normalized to 100 hp at the rear wheels. In table 14-3, various power plants are compared in terms of their emissions characteristics.

As regards air pollution, the main cause of the problem with internal combustion engines is incomplete fuel combustion, which results from the intermittent and explosive ignition. The lead additives used to produce the slow-burning, high-octane fuel needed by Otto-cycle engines also create difficulties. Most pollution-control techniques involve improved tuning, recycling blow-by and evaporative losses, and adding a second stage of combustion, either in the manifold or (catalytically induced) in the exhaust system. However, continuous combustion, as opposed to intermittent, in the presence of excess oxygen can be achieved in certain types of engines. This approach is used in gas turbines (Brayton cycle) as well as external combustion engines based on the Rankine (condensing vapor) cycle or the Stirling cycle. With proper design of such engines it is probably possible to hold emissions of carbon monoxide and unburned hydrocarbons down to levels that would virtually eliminate them as problems. Tetraethyl and tetramethyl lead additives could be eliminated from the fuel. Only oxides of nitrogen would still be present in significant quantities, although at a fraction of the present level (depending on detailed burner design, of course).

Continuous combustion also reduces the noise output of an engine—or reduces the cost of acoustic "muffling." Excessive noise is mainly associated with buses and heavy-duty trucks with large gear-ratios. Thus, noise production would be reduced not only by developing inherently quieter engines but also by improving their torque-speed characteristics. However, this leads back to the matter of "complexity," which arises from adapting an engine by means of add-on devices to a purpose for which it is basically inappropriate.

The impact of new forms of propulsion on the overall level of environmental pollution depends, of course, on the level of emissions from other sources. In this connection, one important point must not be overlooked: it is sometimes assumed that electric cars are completely "non-polluting." Whether this is literally true depends on whether new electric power production capability comes from fossil fuel plants or from nuclear plants. Although it is possible—even probable—that the latter will account for the great bulk of new capacity installed after 1980, there are still lingering unresolved doubts as to reactor cost and reliability, fuel availability, radioactive waste disposal and reactor safety. Moreover, it must be pointed out that nuclear plants are currently somewhat less efficient than fossil fuel plants, i.e., they generate more waste heat per kilowatt hour (or per unit of energy generated). Also, nuclear fuel reprocessing plants that must accompany the rapid growth of nuclear power also have a unique residuals problem in Krypton-85, a long-lived radioactive isotope accounting for about 5% of all fission products, which is currently allowed to escape into the atmosphere. Full conversion to nuclear fuel (for new capacity) may still be a decade away.

COST COMPARISONS

Virtually all of the power plants discussed in this book can provide satisfactory (if not equal) performance for at least some automotive application, and efforts to reduce emissions from the standard spark-ignition internal combustion

Table 14-2. Comparison of Alternative Prime Movers Providing a Maximum of 100 hp at Wheels

Engine	Transmission	Overload capacity (hp)[1]	Parasitic losses[2,3]	Rated hp to supply 100 hp maximum at wheels[4]	Max. brake thermal efficiency (at normal full load)	Max. energy conversion efficiency (including parasitic losses)[5]	Total weight of power train (lb)[6]
Otto-cycle, reciprocating (4-stroke)[7,8]	3-speed auto., or manual and clutch	0	~33% (transmission, cooling system, muffler, air intake, pollution controls)	150	27–30%	19%	750
Diesel, reciprocating[7–9]	Same as Otto	0	33% (same as Otto)	150	30–33	21	900
With supercharger		30	~35% (same plus supercharger)	115	28–30	19	750
Differential regenerative gas turbine[7–10]	Speed reducer differential	25	~25% (regenerator, transmission, combustor)	110	20–26	17	400
Steam Rankine (partial venting)[11–13]	Differential	25	~28% (steam generator, condenser, fan, feed-water pump, transmission)	110	18–24	14	715
Thiophene (CP-34) Rankine	2-speed, clutch, differential	Not specified	28% (same as above)	135	16.7	12.5 (13.7 excl. transmission)	1,100
Future organic Rankine	As above		As above	~130	25	19	800

(See notes at end of table)

Table 14-2. (Continued):

Engine	Transmission	(f) Overload capacity (hp)[1]	(g) Parasitic losses[2,3]	(R) Rated hp to supply 100 hp maximum at wheels[4]	Max. brake thermal efficiency (at normal full load)	Max. energy conversion efficiency (including parasitic losses)[5]	Total weight of power train (lb)[6]
Electric[13,15-18] 13 hp (present) 40 hp (future)	Traction motors, speed reducer	n.a.	n.a. (battery charge and discharge, electromechanical energy conversion)	n.a.	30–39% (central elec. power generation)	1.0 kwh/mi 0.5 kwh/mi	1,500 800
Hybrid[13,15-18] 80 hp max. (present)	Speed increaser, electric generator, traction motors, speed reducer	200	~42% (same as electric, plus Otto ICE losses, ~35% electromechanical drive)	40 (balance supplied by batteries)	30	17%	1,275 (present)
Stirling (closed cycle)[13,19]	2-speed and clutch	50	~40% (combustion blower, heat exchanger, cooling water, fan and pump, transmission)	110	32–38	22	1,100

[1]Overload capacity normally refers to inherent flash or short-term horsepower output available as a consequence of reserve energy stored in the heat exchanger of an external combustion engine, or batteries of a hybrid system. Although overload capacity is not inherent in internal combustion engines, it can be achieved by supercharging, provided the engine has been designed structurally to accommodate increased pressure resulting from fuel ignition at higher than normal pressures.

speed characteristics consistent with the traction demands of vehicles. In addition, present battery-electric systems cannot realistically provide 100 hp at the wheels in a practical-sized vehicle.

[5] The ranking of the prime movers considered from the standpoint of efficiency may alter, depending on the nature of the driving cycle. For example, gas-turbine and Otto-cycle efficiencies tend to drop rather sharply with decreasing load; the Rankine cycle does not. In an urban driving cycle, the prime mover is operating at part load rather than optimum conditions most of the time. Efficiencies for an electric vehicle are not computed directly.

[6] Includes 2–5 lb/hp for transmission and other engine accessories where appropriate.

[7] Robert Kirk and David Dawson, "Low-Pollution Engines: Government Perspectives on Unconventional Engines for Vehicles," ASME Paper No. 69–WA/APC–5, July 1969; and L. C. Lichty, *Combustion Engine Processes* (New York: McGraw-Hill, 1967).

[8] E. F. Obert, *Internal Combustion Engines* (International Textbook Co., 1968).

[9] T. Baumeister and L. S. Marks, *Standard Handbook for Mechanical Engineers* (New York: McGraw-Hill, 1966).

[10] Frank L. Swartz, "Vehicular Gas Turbines," *Gas Turbine Engineering Handbook* (Stamford, Conn.: Gas Turbine Publications, 1966).

[11] "Automobile Steam Engine and Other External Combustion Engines," Joint Hearings before the Senate Committee on Commerce and the Subcommittee on Air and Water Pollution of the Committee on Public Works, 90 Cong., 2 sess., May 1968.

[12] J. L. Dooley and A. F. Bell, "Description of a Modern Automotive Steam Power Plant," SAE Paper S–338, January 1962.

[13] "Study of Unconventional Thermal, Mechanical, and Nuclear Low-Pollution Power Sources for Urban Vehicles," Research Report to NAPCA by Battelle Memorial Institute, Columbus Laboratories, March 1968.

[14] D. T. Morgan and R. J. Raymond, "Conceptual Design, Rankine-Cycle Power System with Organic Working Fluid and Reciprocating Engine for Passenger Vehicles," Report to NAPCA by Thermo Electron Corporation, Waltham, Massachusetts, June 1970.

[15] *Power Systems for Electric Vehicles*, U.S. Public Health Service Publication No. 999–AP–37, 1967.

[16] "Prospects for Electric Vehicles—a Study of Low-Pollution Potential Vehicles," Research Report to NAPCA by Arthur D. Little, Inc., 1968.

[17] NAPCA, Arlington, Virginia, May 1968.

[18] R. U. Ayres, R. P. McKenna, and M. L. Walker, "Electric Propulsion for Transit Vehicles," International Research and Technology Corporation Paper No. IRT–P–20, presented to American Transit Conference, May 4, 1970.

[19] R. A. J. D. VanWitteveen, "The Philips Stirling Engine, Present and Future," Philips Research Laboratories, Eindhoven, Netherlands, August 31, 1966.

271

engine may succeed in making it comparable to the ECE in this respect. In a situation such as this where there are only narrow differences in the performance and side effects of various engines, cost becomes a particularly important factor.

Unfortunately, cost data derived from careful well-documented engineering studies based on first principles are virtually nonexistent. Most of the published cost estimates are not suitable for comparing the costs of different systems. Some are based on standard (unverifiable) sources, some offer too wide a range of costs, and some appear to be designed to prove a particular point. The approach that seems most useful for our purposes is based on the assumption that the cost of a manufactured object is proportional to its weight so long as similar items are compared.

If the ICE with its transmission has a retail "cost" of $1.00 per pound, it can be expected that power systems of similar construction would cost a like amount per unit weight. This view must be modified to allow for a certain fraction of the weight of competing systems being composed of nonstandard materials. Costs for that fraction may be higher or lower than $1.00 per pound, depending on the cost of the materials and on the fabricating techniques. For example, refractory materials that are resistant to high temperatures may be a factor of 1.5 to 10 times higher in price than basic aluminum, iron, and steel and commensurately more costly to fabricate. Heavy-duty lead-acid batteries, on the other hand, can be purchased in finished form for about $0.50 per pound, or half the standard cost. One problem, of course, is that current high prices for

Table 14-3. Comparative Levels of Automobile Exhaust Emissions

(kg per thousand vehicle miles)

Power system	HC	CO	NO$_x$	SO$_x$	Particulates	References and notes
ICE, no control	11.5	85	4 to 6	0.27	0.36	Ref. 1[a]
ICE, 1970 Calif.	2.2	23	6[b]	No control	No control	Assembly Bill 357, 1968
1975 U.S.	0.5	11	0.9	No control	0.1	Projected
1980 U.S.	0.25	4.7	0.4	No control	0.03	Projected
Diesel	3.5	5	4 to 12			Refs. 1, 2
Gas turbine	0.2 to 0.9	2 to 8	1.0 to 1.6	0.3	c	Refs. 1, 2
Rankine cycle (steam or organic working fluid)	0.013 to 0.4	0.35 to 2.0	0.25 to 0.6	0.2 to 0.3	c	Refs. 3, 4, 5

Table 14-3. (Continued):

Power system	HC	CO	NO_x	SO_x	Particulates	References and notes
Stirling	0.006 to 0.1	0.3 to 1.0	1.0 to 2.6	0.2	c	Refs. 2, 4
Battery electric (present)	Negl.	Negl.	1.75	0.002	0.07	Natural gas fuel at plant
Battery electric (future)	Negl.	Negl.	0.83	0.001	0.03	Natural gas fuel at plant
Battery electric (present)	Negl.	Negl.	2	10	2.5	Coal-fired plant (2.5% sulfur)
Hybrid (ICE-electric)	0.4	4	0.4	–	–	Design goals, not measured[d]

[a]Values can be considerably higher for uncontrolled engines.

[b]Emission of NO_x was found to increase when controls were applied to CO and HC.

[c]Published data not available. Smokeless combustion should be readily attainable.

[d]Information supplied to the U.S. Department of Transportation by Minicars, Inc., Goleta, California.

References:

1. U. S. Department of Commerce, *The Automobile and Air Pollution: A Program for Progress*, Part II. Subpanel Reports to the Panel on Electrically Powered Vehicles (Washington, D.C.: U. S. Government Printing Office, 1967).

2. A. F. Underwood, "Alternative Automotive Propulsion Systems: Requiem for the Piston Engine?" *Machine Design*, August 6, 1970, pp. 20–34.

3. D. T. Morgan and R. J. Raymond, "Conceptual Design: Rankine-Cycle Power System with Organic Working Fluid and Reciprocating Engine for Passenger Vehicles." Report No. TE4121-133-70, prepared for the U. S. Department of Health, Education and Welfare by the Thermo Electron Corporation, Waltham, Mass., June 1970.

4. Technical Advisory Committee to the California Air Resources Board, "Control of Vehicle Emissions after 1974," November 19, 1969.

5. "Automobile Steam Engine and Other External Combustion Engines," Report 90–82, Joint Hearings of the Senate Committee on Commerce and Subcommittee on Air and Water Pollution of the Committee on Public Works, 90 Cong., 2 sess., May 1968.

some materials represent an artificial scarcity. Possible future battery anode materials such as lithium may be in this category. We have, therefore, tried to visualize cost reductions that might result from the widespread use of some of these materials.

The cost of a given power system relative to the cost of an ICE is given by the formula:

$$C/C_{ICE} = (1 - f + Rf)(W/W_{ICE}) \qquad (14.2)$$

where

Table 14-4. Calculation of Power-System First Cost, Relative to ICE

| Power system | Rated power to provide 100 hp at rear wheels[a] | W, Weight (lb) | $\dfrac{W}{W_{ICE}}$ | Premium or non-standard material | | $F = (1 - f + Rf)$ | Cost $\dfrac{\text{Cost}_{ICE}}{} = F\,\dfrac{W}{W_{ICE}}$ |
				$f =$ Fraction of W	$R =$ Ratio of cost to basic cost		
ICE (spark ignition)	150	750	1.0	0	n.a.	1.0	1.0
Diesel	150	900	1.2	0.2	1.5	1.1	1.32
Gas turbine (regenerative)	110	440	0.59	0.8	4.0	3.4	2.0
Rankine (steam)	110	715	0.95	0.4	2.0	1.4	1.33
Rankine (organic)	135	1,100	1.47	0.1	1.5	1.05	1.55
Rankine (future organic)	130	800	1.07	0.15	2.0	1.15	1.23
Stirling	110	1,100	1.47	0.4	3.0	1.8	2.16
Electric:							
Less battery	–	500	0.67	0.5	4.0	2.5	1.68
Present battery	13[a]	1,000	1.33	1.0	0.5	0.5	0.67
Future battery	40	300	0.4	1.0	1.2	1.2	0.48
Hybrid (ICE-electric)	40	775	1.04	0.25	4.0	1.75	1.82
Present battery	–	500	0.67	1.0	0.5	0.5	0.34

Source: R. U. Ayres and Roy Renner, "Automotive Emission Control: Alternatives to the Internal Combustion Engine," paper given at the Fifth Technical Meeting, West Coast Section, Air Pollution Control Association, San Francisco, October 8-9, 1970.

[a]Present battery-electric systems cannot realistically provide 100 hp at the wheels in a practical-sized vehicle.

f = that fraction of power system weight representing notably nonstandard materials or fabricating techniques,

R = the ratio of nonstandard to "basic" cost applying to the fraction (f),

C = the cost of the power system in question,

C_{ICE} = the cost of the basic ICE, and

W and W_{ICE} = the respective weights.

The results of such a calculation, given in table 14-4, are carried into table 14-5.

Estimated costs for fuel (or energy) and for maintenance over a standard ten-year vehicle life are given in tables 14-6 and 14-7, respectively. Battery replacement is included with energy cost for electric vehicles. Lead-acid battery life was taken to be 30 months, and it was assumed that three replacements would be needed. The same assumption was made for future batteries, though it is obvious that a longer-lived battery would make a great difference in the total cost picture.

Table 14-5. Estimated First Cost of Power Systems

Power system	Rated power[a] (hp)	System weight[b] (lb)	Est. first cost	Incremental cost over standard ICE
ICE (spark ignition)	150	750	$ 750	$ 0
Diesel	150	900	990	240
Gas turbine (regenerative)	110	440	1,500	750
Rankine cycle (steam)[c]	110	715	1,000	250
Rankine cycle (organic fluid)[d]	135	1,100	1,160	410
Rankine cycle (organic fluid, future)[d]	130	800	925	175
Stirling cycle	110	1,100	1,990	1,240
Battery electric (present)[e]	13	1,500	1,760	1,010
Battery electric (future)[f]	40	800	1,620	870
Hybrid (ICE-electric)	40-80[g]	1,275	1,620	870

Source: Ayres and Renner (see source for table 14-4).

[a]Except for electric and hybrid, to give 100 hp at wheels.

[b]Including transmission, if any.

[c]Incomplete condensing.

[d]Complete condensing.

[e]Lead-acid battery.

[f]Possible future battery, such as sodium-sulfur.

[g]40 hp continuously from ICE, up to 80 hp for very short periods with assistance from battery.

It is noteworthy that electrical energy costs drop very significantly for future batteries (see table 14-6). A 2,500-lb vehicle using present-day lead-acid batteries can be expected to require, per vehicle-mile, close to 1.0 kwh of energy at the central generating plant, assuming typical transmission and charging losses, driving cycles, and accessory loads. Very heavy internal battery losses resulting from high discharge currents (in relation to battery capacity) account for much of this. Future high-energy batteries would result in such dramatic savings in

Table 14-6. Estimated Ten-Year Costs for Fuel or Energy

Power system	Miles per gallon of fuel	Total fuel cost[a]	Electrical energy cost[b]	Battery replace-ment cost	Est. total energy cost	Cost difference, from standard ICE
ICE (spark-ignition)	13[c]	$2,560	–	–	$2,560	$ 0
Diesel	16[d]	2,080	–	–	2,080	-480
Gas turbine	10	3,320	–	–	3,320	+760
Rankine (steam)	12[e]	2,770	–	–	2,770	+210
Rankine (organic)	10[f]	3,320	–	–	3,320	+760
Rankine (future organic)	14	2,380	–	–	2,380	-180
Stirling	15	2,210	–	–	2,210	-350
Battery electric (present)[g]	–	–	$1,440	$2,250	3,690	+1,130
Battery electric (future)[g]	–	–	720?	2,250?	2,970	+410?
Hybrid[g]	24	1,390	–[h]	750	2,140	-420

Source: Ayres and Renner (see source for table 14-4).

[a]Base case for ICE is 730 gallons of fuel per year at 35 cents per gallon. No difference in price per gallon for alternative fuels is assumed.

[b]Assume a U.S. average of 1.5 cents per kwh at customer's location.

[c]Medium-displacement V-8 engine in urban driving.

[d]U.S. Department of Commerce, *The Automobile and Air Pollution: A Program for Progress, Part II.* Subpanel Reports to the Panel On Electrically Powered Vehicles (Washington, D. C.: U.S. Government Printing Office, December 1967).

[e]Based on a ratio of thermal efficiencies of 0.18/0.15, compared to organic Rankine.

[f]D. T. Morgan and R. J. Raymond, "Conceptual Design: Rankine-Cycle Power System with Organic Working Fluid and Reciprocating Engine for Passenger Vehicles." Report No. TE4121-133-70, prepared for the U.S. Department of Health, Education and Welfare by the Thermo Electron Corporation, Waltham, Mass., June 1970.

[g]Substandard performance: 13 hp (present electric), 40 hp (future electric), 80 hp (hybrid).

[h]Some driving patterns would require supplementary battery charging.

(?) Future values exceptionally uncertain.

internal efficiency that the energy required per vehicle mile would drop to the neighborhood of 0.5 kwh. (A discussion of the variation of battery capacity with rate and depth of discharge can be found in chapter 11.)

The fuel costs shown in table 14-6 are based on the assumption that the relationship between average fuel consumption under urban conditions and full-load fuel consumption would be the same for gas turbine and Rankine-cycle and Stirling cycle engines as it is for the gasoline engine, even though this is unrealistic, as figure 14-1 very well illustrates. Regrettably, there are no data for making quantitative fuel comparisons on a firm engineering basis, but there is good reason to believe that a gas turbine will lose even more in efficiency than a reciprocating ICE under variable and part-load conditions, while ECEs will lose

Table 14-7. Estimated Ten-Year Maintenance Costs for Power Systems

Power system	Major repairs	Transmission maintenance	Minor repairs, tune-up	Working fluids, oil, filters	Est. total cost	Cost difference from standard ICE
ICE (spark ignition)	$500	$200	$800	$280	$1,780[a]	$ 0
Diesel	300[b]	250	200	320	1,070	-710
Gas turbine (regenerative)	300	150	400	100	950	-830
Rankine (steam)	600	50	400	400[c]	1,450	-330
Rankine (organic)	300	150	200	340[d]	990	-790
Rankine (future organic)	300	150	200	340[d]	990	-790
Stirling	500	150	200	300	1,150	-630
Battery electric (present)[e]	200	–	100	50	350	-1,430
Battery electric (future)[e]	200	–	100	50	350	-1,430
Hybrid[e]	600	150	900	280	1,930	+150

Source: Ayres and Renner (see source for table 14-4).

[a]This independent estimate concurs with 1964 statistics (Bureau of Public Roads) of $2,087 excluding all taxes for entire automobile for ten years.

[b]Reports from owners of Mercedes-Benz diesel-powered automobiles indicate that a majority of reasonably maintained engines exceed 100,000 miles without a major overhaul. However, the overhaul is more expensive for a diesel than for the spark-ignition ICE.

[c]Separate cylinder lubrication and more types of filters required than for the ICE.

[d]Includes four changes of working fluid at $30 per filling, plus lubricants and filters totaling $220. A long life for lubricants is foreseen because they are compatible with the organic fluids used.

[e]Battery replacements are given under "electrical energy cost" in table 14-6.

less. The fuel consumption costs in table 14-6 may therefore be underestimated by 10% to 20% for the gas turbine, and overestimated by a like amount for ECEs.

Maintenance costs are likely to be lower for most of the alternative propulsion systems than they are for the ICE primarily because they do not share the ICE's dependence on a complex automatic transmission and its need for frequent tuneups. There is little experience to support the estimates in table 14-7, but the general level seems to be borne out by the assertion from one source that maintenance on existing electric vehicles (fork trucks, delivery vans, etc.) is only one-fifth to one-half that on equivalent ICE-powered vehicles.[1]

Table 14-8. Ten-Year Net Cost Differences From Baseline ICE

Power system	Difference from baseline ICE ($)	Difference from 1975 ICE ($)
ICE (spark ignition)	0	0
Diesel	-902	n.a.
Gas turbine	+830	+390
Rankine (steam)	+180	-260
Rankine (organic)	+462	+22
Rankine (future organic)	-760	-1,200
Stirling	+508	+68
Battery electric (present)	+912	+472
Battery electric (future)	+24?	-416?
Hybrid	+774	+334

Source: Ayres and Renner (see source for table 14-4).

Note: Ten-year costs include: first cost plus 20% interest on the first cost if it is over $50; cost of fuel energy (shown in table 14-6; and maintenance costs for the power system (shown in table 14-7).
n.a. = not available.
(?) = Future values exceptionally uncertain.

The relative-cost values of all the engine systems can be seen in table 14-8 which lists the ten-year difference of each engine in dollars from a standard baseline ICE. The ten-year costs include an incremental first cost, plus a flat 20% interest on the incremental first cost if it is over $50, plus ten-year costs for fuel or energy and for power system maintenance.[2] It can be seen that for a ten-year period only the diesel and a future organic Rankine-cycle engine can offer a cost saving over a 1970-model ICE. For the 1975-model ICE we have assumed that additional catalytic exhaust emissions control equipment would be needed, and

---

[1] D. R. Adams et al., *Fuel Cells—Power for the Future*, Fuel Cell Research Associates, Willow Grove, Pennsylvania, 1960.

[2] R. U. Ayres and Roy Renner, "Automotive Emission Control: Alternatives to the Internal Combustion Engine." Paper given at the Fifth Technical Meeting, West Coast Section, Air Pollution Control Association, San Francisco, October 8-9, 1970.

that this would add $200 to the initial cost plus $40 for interest and $200 for additional maintenance and higher fuel consumption over the life of the engine. All other engines except the diesel are assumed to meet the 1975 standards (although there is some doubt that gas turbines will meet the NO_x standards). The cost of modifying diesel engines to meet 1975 standards has not been estimated owing to lack of research in this area.

PART III

Conclusion

Chapter Fifteen

Comments on
Effects and Policies

Ideally, one should conclude a study like this with a complete quantitative assessment of the social benefits and costs of alternate policies now and into the foreseeable future. Unfortunately, we are not able to do this because even making a start in this direction would require a study effort at least as large as the technologically oriented one we have undertaken. However, we do not wish to conclude without making some comments on these important matters.

It is difficult to apply standard benefit-cost techniques to public programs that call for major changes in technology in an important industry. The effects of such changes go far beyond the industry itself. Changes in the location or nature of the manufacturing process may leave workers unemployed and specialized facilities unused, for example, and such social costs are hard to identify and quantify. Nor are we any better able to estimate the benefits to aesthetics and health of reducing air pollution or noise by various amounts. There is, in short, much uncertainty about the impacts of new technology. All that is clear is that the introduction on a large scale of any substitute for the internal combustion engine would have very far-reaching effects, although these would be rather different for different possible alternate engines. It is these differences in impact that may determine the institutional policies chosen.

In any case, simply assessing the total costs and gains of changes induced by public policy for the population at large does not reveal the distribution of these effects among the population, which may be as important as the sizes of the effects.

In view of the towering difficulty of fully evaluating and comparing impacts, direct and indirect, on individuals and on a variety of institutions and organizations, a clear, quantitative discussion eludes us. We can only comment briefly on the potential impact of possible technological alternatives and conclude with a short discussion of some policy issues.

IMPACTS

The automobile pervades American life today at every level, and thus any attempt to put a somewhat different mechanism in its place must inevitably make itself felt throughout. The most obvious recipients of any costs or benefits accruing from such a substitution are the public and industry, but the ramifications would also be felt in such areas as labor unions, state and local governments, and so on. Unions, for example, have traditionally opposed major technological changes. The union representing workers in an industry that is being threatened by a technological substitution (spark plugs or tetraethyl lead, for example, might be displaced by the introduction of steam engines) probably would have little recourse except perhaps through political channels. In addition, while there are cities and states whose economies are strongly entwined with the automotive or oil industries—Detroit, for example, or Texas and Oklahoma—virtually all cities (and states) have a growing interest in controlling air pollution for the sake of their own future growth and prosperity.

The effects of an alternate engine on the public at large would, in a sense, be twofold, because there are really two different "publics"—those who use automotive vehicles and those who do not. The two groups overlap, of course: all users become nonusers as soon as they park their cars or get off the bus, and most (but not all) nonusers are users on some occasions.

From the standpoint of automotive users, we have already suggested that at least three technological alternatives—the gas turbine, Rankine-cycle (steam) engine, or hybrid—could be adapted to automotive purposes with no significant sacrifices in terms of performance. Minor variations would be noticed to be sure: the gas turbine's high-pitched whine and somewhat sluggish response, the steam engine's characteristic "whoosh" and dynamic braking capability, differences of "feel," etc. However, not all of these differences would be disadvantageous in any case. Probably capital cost, maintenance cost, reliability, and fuel cost will be the dominant factors in comparing the alternative technologies from the user standpoint. Capital cost and fuel cost have been discussed already, though the results are far from conclusive. Maintenance and reliability cannot be forecast easily for systems not yet in production (or, for that matter, even in final prototype form). At this stage one can only speculate.

From the standpoint of nonusers, or even users stuck in a traffic jam, the major concern is air pollution and noise. This is the point of view taken by this book. The technical issues have been explored fairly thoroughly in Part II, and need not be repeated here.

The major industry to be affected by a change in the powerplant technology would be the motor vehicles and equipment industry, although petroleum drilling, refining, and distribution operations would also be strongly affected by almost any important technological change. In 1963, for example, auto production alone accounted for 4.2% of GNP, while related services, tires, gas and oil, batteries, and accessory products accounted for an additional 4.1%, for a total of 8.3%.[1]

---

[1] *Economic Review*, Federal Reserve Bank of Cleveland, March 1965.

The auto industry also has very high indirect effects on the economy. In terms of direct and indirect purchases of intermediate products and services per dollar of "final demand" in the automotive sector added to the GNP, and in terms of value added per dollar of gross output, the auto industry is ranked seventh out of 84 sectors comprising the two-digit Standard Industrial Classification. Hence, it can be argued that the industry has an exceptionally pervasive impact on other industries and the economy in general, even though 29% of its inputs are purchased from within the auto sector itself (see table 15-1).

Of course, many of these relationships would remain largely unaffected by a technological change involving only the power plant and possibly the source of energy. The crude petroleum, petroleum refining, automobile repair and services, and, possibly, electric power sectors are most likely to be affected rather broadly. Apart from marginal changes in the overall demand for petroleum fuel, due to possible increases or decreases in thermal efficiency, the major impact of an external combustion engine on the petroleum industry would be to eliminate the requirement for high-octane gas antiknock fuel. Not only would additives such as tetraethyl lead be unnecessary, but the costly catalytic cracking and "reforming" stage of the refining process could be avoided. This would reduce the capital investment needed to produce motor fuel. It might also release some aromatics (benzene, toluene, xylene, etc.) currently needed for gasoline manufacture for use as petrochemical feedstocks.

A shift to battery-electric propulsion (as opposed to ICE-hybrid or fuel-cell propulsion) would imply a major shift in the sources and suppliers of basic energy in the economy. Petroleum companies would sell to electric utilities, or not at all. Utilities would have to expand their output, even beyond currently projected levels. A new type of service station would be needed. Mechanics would have to retrain as electricians. Specific industries are also highly vulnerable to any replacement of the ICE by another technology: among them are those dealing with lead additives for gasoline, lead mining, spark plugs, clutches and transmissions, mufflers, special lubricants and lubricant additives, and antifreezes.

Distribution of a Rankine-cycle external combustion engine would result in significant increases in sales of heat exchangers and radiators, stainless steel tubing, high-temperature lubricants, and organic (fluorocarbon) working fluids. The chemical industry would probably be a net gainer from such a shift.

A hybrid or electric system would *add* new markets for electrical equipment, controls, and batteries. A high-performance battery, for example, is likely to require zinc, sodium, calcium, or lithium as an anode material, lithium or zinc being perhaps the most probable. Greater quantities of copper, aluminum, and special alloys would be utilized, and the requirements for basic steel would probably drop somewhat.

In any event, regardless of such effects, both population and government are becoming increasingly aware of the present automotive system's principal

Table 15-1. Relations of Automobile Manufacturing to Other Industries

	Direct purchases as % of inputs of auto industry	Direct purchases as % of output of named industry	Purchases per dollar added to final auto demand
Manufacturing			
Motor vehicles & equipment	29.5%	29.0%	$1.43
Primary iron & steel manufacturing	8.5	10.3	0.20
Other fabricated metal products	3.5	12.5	0.06
Stampings, screw machine products & bolts	3.0	18.8	0.05
Rubber & miscellaneous plastics products	2.8	9.1	0.05
Misc. electrical machinery, equipment & supplies	1.5	21.0	0.02
Primary nonferrous metals mfg.	1.1	2.6	0.05
Metalworking machinery & equipment	1.1	7.0	0.02
Glass & glass products	1.0	10.6	0.02
Misc. fabricated textile products	0.7	6.6	0.01
Machine shop products	0.6	8.4	0.01
General industrial machinery & equipment	0.6	2.8	0.01
Radio, TV, & communication equipment	0.5	1.9	0.01
Aircraft & parts	*	0	*
Farm machinery & equipment	0.1	0	*
Nonmanufacturing			
Wholesale & retail trade	3.1	0.7	0.08
Business services	2.4	2.3	0.05
Gross imports of goods & services	2.3	4.9	0.06
Transportation & warehousing	2.0	1.2	0.07
Electric, gas, water & sanitary services	0.5	0.5	0.03
Business travel, entertainment & gifts	0.4	1.5	0.02
Auto repair & services	*	0.1	*
State & local government enterprises	*	0.1	0.01
Livestock & livestock products	0	0	*
Other agricultural products	0	0	0.01

Source: "Input-Output Relations of The Auto Industry," *Economic Review*, Federal Reserve Bank of Cleveland, March 1965.

*Negligible

impact—the pollution of the atmosphere. Some of the ways in which automotive externalities might be controlled are discussed below.

POLICIES FOR CONTROLLING AUTOMOTIVE EXTERNALITIES

The two classic remedies for externalities are "internalization" and "regulation." Internalization means drawing institutional boundaries in such a way that all interests affected by a transaction are included within the institutional framework where their interests are (presumably) taken into account in the making of policy. Regulation is, of course, the other method of protecting the interests of third parties.

In the case of transportation the only institution broad enough to include all the parties affected by externalities is the community as a whole. Thus a theoretical solution of the problem would be community ownership of all transportation facilities. Actually, highways, airports, and many mass transit systems are already community-owned. However, complete internalization would also require public ownership of all automobiles, as well as railroads, buses, trucks, and airlines. This prospect is so unlikely that it seems pointless to analyze its implications.

Under the broad umbrella of "regulation" falls the spectrum of policies discussed below. Basically they range from the "utility" approach—formal control and supervision under the aegis of a public regulatory body—to relatively subtle forms of influence by a variety of means. In accordance with the limited focus of this book, we do not discuss some rather obvious possibilities for control such as policies to substitute mass transit for private cars or imposing operational restrictions on private automobiles.

Public Utilities

The utility concept deserves thoughtful consideration. Producers of primary energy (gas and electricity) are acknowledged to be public utilities, whose private interests are coordinate or even subordinate (if need be) to the public interest as represented by the state public service commissions and the Federal Power Commission. Similarly, common carriers of communication services—telephone, telegraph, radio and TV broadcasting—have a quasi-public character, with the Federal Communications Commission acting as a kind of public representative (with veto powers) on the board of directors. The case of railroads and airlines is not much different, with the Interstate Commerce Commission and the Civil Aeronautics Board representing the public interest.

However, private automobiles, which account for nearly 90% of all passenger mileage on public roads and highways, are subject to only limited control and licensing. Admittedly there are federal, state, and local regulations concerning speed limits, signals, rules of the road, air pollution control, and so forth. But important parts of the "system" and the forces that shape it remain essentially uncontrolled. Until the 1965 federal safety legislation there was no national authority with any statutory control whatever over the activities of the automobile manufacturers. Such constraints as exist today are still very tenuous.

In the very long run, automotive transportation may evolve into a "utility" form of industry with a regulatory agency having jurisdiction over all the major problem areas. However, there seems to be no likelihood of significant progress in this direction during the next decade owing to the fragmented nature of existing authority in the field. Also recent years have seen a considerable disenchantment with public utility regulation, since the regulatory bodies seem to have been particularly ineffective in dealing with externalities generated by the industries they regulate "in the public interest."[2]

Emission Standards

The present approach to controlling emissions and associated external costs is based primarily on emission standards set by the federal government. Such standards have been set for hydrocarbons and carbon monoxide, and are to become effective for oxides of nitrogen in 1973. This approach is designed to force the automobile companies to develop and adopt engine modifications and devices to control emissions. It has the advantage of definiteness but permits little flexibility in response.

This approach, in conjunction with the industry's current strategy of emission control by means of add-on devices, implies some form of effective inspection or verification. One possibility would be to set up a network of perhaps 10,000 inspection stations across the country. However, such a system would not only be expensive ($350 million or so per year), but it would put the onus of responsibility and the costs on the user rather than the manufacturer. The latter would have little or no incentive to build efficient, reliable, or long-lived control devices, since he profits from the sale of replacement parts. The manufacturer's economic interests are not likely to coincide with the interests of the general public in minimizing the production of residuals and improving the efficacy of control unless the manufacturer is required to extend the warranty on pollution control devices to cover the life of the car.

Perhaps the most extreme legislative approach so far seen was an amendment to the Clean Air Act of 1969 offered by Congressman Leonard Farbstein of New York. This proposal, which was defeated, called for prohibiting, after January 1, 1977, the manufacture or sale of any new motor vehicle powered by an internal combustion engine that could not meet very strict emission standards. This effort paralleled a move in California where, by a vote of 26 to 5, the state senate passed a bill to ban the sale of gasoline and diesel powered internal combustion engines starting January 1, 1975; the bill failed to pass the California Assembly, but it did alert automobile and petroleum companies to the seriousness of public concern.

Subsidies and/or Tax Incentives

A system of direct public investment and/or subsidy is a fairly straight-forward approach, but one that usually gives rise to some controversy. These methods are usually opposed by those with vested interests (though not so violently as policies that would reduce the demand for their products), as well as

---

[2]Mower H. Bernstein, *Regulating Business by Independent Commissions* (Princeton: Princeton University Press, 1955).

those who feel government should not compete with private industry, and those who simply prefer not to spend public money in this way. Quite naturally, support comes from those who stand to gain. The major drawback to subsidies is the high cost of investment in "hardware" and operating systems, which results in an extremely low benefit-to-cost ratio for this use of public funds. Other ways of obtaining the same objective would generally be both less expensive to the taxpayer and preferable on other grounds.

Effluent charges or taxes are another method for causing external costs to be reflected in the decisions made by producers and consumers. This device has received considerable attention as a way of dealing with air and water pollution problems and has been implemented for water quality management in several countries. In principle, the charge would reflect the actual monetary value of the external costs imposed by an emission of a particular magnitude, and it would rely on an economic incentive to achieve an optimal level of control. In practice, the charge, which would presumably be levied at the time of manufacture and at annual inspections, would have to be determined by rule of thumb. This method is one that deserves serious consideration.[3]

Sponsorship of New Technology

An important weapon in the public arsenal is the sponsorship of new technology through research, development, demonstration, and procurement. The high-speed ground transportation program in the Department of Transportation is an example of this approach. Its advantages are that most people are (now) in favor of research, on the grounds that it contributes to technological progress and the scientific strength of the nation, as well as to solving the particular problem. The disadvantages of this policy are that research sometimes takes a very long time to yield significant results, and there is something of a built-in tendency for the research establishment to perpetuate itself by postponing action in order to carry out more and yet more studies and tests. In private industry this tendency can be counterbalanced by the desire of management to show a profit, but in a government bureaucracy the research activity can sometimes become an end in itself.

On the other hand, public sponsorship of both basic and applied research is the best guarantee of preserving economic competition in the private sector—which cannot exist without technological options—and is therefore probably a sounder investment for the federal tax dollar than propping up a technologically obsolescent, but socially desirable, transportation system by direct subsidy.

The most interesting class of policy choices is concerned with modifying the competitive environment by speeding up the process of technological penetra-

---

[3]For a discussion of effluent charges as they might be applied to automobiles, see Edwin S. Mills, "Economic Incentives in Air Pollution Control," in Harold Wolozin, ed., *The Economics of Air Pollution* (New York: W. W. Norton & Co., 1966).

tion. Normally a technological innovation remains stalled for some time in the laboratory or engineering prototype stage, where it may wait years or decades for a sponsor willing to invest the capital and take the risks necessary to put it to practical use. In a static or declining industry such as rail transit, the availability of private risk capital is very nearly zero, which tends to further stultify an already anemic rate of technological change. An imaginatively administered government program may successfully bypass this obstacle by demonstrating the practical feasibility of ideas whose expensive and uncertain teething pains no private sponsor is in a position to underwrite. The Department of Transportation (Urban Mass Transportation Administration) is currently engaged in a series of such projects that are modest departures from the strictly conventional, but that range from a high-speed rail transit system in a low-density suburban area outside Chicago (the Skokie Swift) to turbine-powered cars on the Long Island Railroad, to minibuses in downtown Washington, D.C. Many of the projects are already showing signs of success; indeed, an upgraded suburban passenger service on the Harlem Division of the Penn Central Railroad has already become permanent and apparently is to be followed by similar improvements on other divisions.

One of the most interesting of the publicly supported research and development efforts getting underway while this study was in preparation is the San Francisco Bus Project sponsored by the Department of Transportation. This demonstration project entails the operation of one or more steam-powered transit buses that will be tested and evaluated as they operate over regular Bay area bus routes. By the end of 1969 steam engine producers had been selected to produce power systems for the buses. By the end of 1970, design work had been completed and actual component fabrication had begun. In 1971 a prototype was demonstrated. This project complements a parallel effort by the state of California to test steam-propelled highway patrol cars—a project that was ordered by the California Legislature in 1969.

Apart from feasibility demonstrations the competitive climate for new technological alternatives can be improved in other ways. One approach worth mentioning is for federal or local government to underwrite some of the ancillary facilities and services that would have to accompany any radical departure from existing technology. Thus, if it were thought desirable to speed up the public acceptance of electric city cars, it would be enormously helpful to provide electric outlets at selected downtown parking locations. Still better, a number of small downtown parking garages could be leased by a city, fitted with meters and electric outlets, and reserved for electric minicars. An assured place to park at an attractive rate (based on a small vehicle) might prove a strong incentive for a city dweller to buy or rent an electric car. If public service commissions would go further and permit—or insist upon—electric utilities reducing their rates at night to encourage overnight recharging of battery-powered cars, another useful incentive would be created.[4] These in turn might increase the potential demand for electric vehicles enough to attract new entries to the market, and so on. An even more radical step in this direction might be for a city such as New York to create an electric car "utility" that would buy, maintain, lease, rent, or operate electric vehicles in the congested midtown or downtown areas. The utility might be operated by the city itself or by a private concern under contract to the city.

Although electric propulsion is not a competitive alternative given the present pattern of vehicular ownership and use, it could become competitive for some applications if there were a substantial rate of technological progress and a marginally more favorable environment. It can probably be safely assumed that the inherent dynamism of the electrical and electronics industries is such that they will quickly exploit any potential opportunities in the transportation market.

A similar assumption cannot be made for the external combustion engine propulsion alternative, however. Although a modern ECE, such as a steam engine, for example, appears to be already competitive in every important respect with the standard internal combustion engine—and superior in some ways—there is no dynamic ECE industry ready to take the opportunity that beckons. The few companies that still manufacture reciprocating steam engines (for ferry boats and the like) are small and conservative, with no apparent interest in research and growth. The public image of steam is compatible with the condition of the industry. If it is considered desirable to convert to steam, there will either have to be a radical change within one of the country's largest and most influential industries or the government will have to try to stimulate the creation of a whole new industry from scratch. If the government should decide to act—as it did with the domestic uranium mining industry, the nuclear reactor industry, the nuclear weapons effects monitoring industries (notably EG&G, Inc.) and the communication satellite industry (COMSAT)—its principal tools would be the awarding of research and development contracts, demonstration grants, and government procurement. The federal government has no direct means of supplying working capital, except through the Small Business Administration and its subsidiaries. However, a firm with a sizable government contract can often raise money from the private capital markets, and the government might do well to consider contracts as a way of encouraging private capitalization.

The fact that the government is a major buyer and operator of motor vehicles was recognized in a bill presented to the Senate in October 1969 by Senator Warren Magnuson of Washington. This bill (which was later incorporated into

---

[4] It is worth noting that this step has not been taken anywhere in the United States or the United Kingdom, despite the utilities' obvious interest in furthering electric vehicle technology.

the Amendments to the Clean Air Act of 1970) provides a ready market for motor vehicles that meet specified standards of exhaust emissions of hydrocarbon, carbon monoxide, and nitrogen oxide. It stipulates that federal agencies are to purchase any low-emission vehicle meeting these standards if it costs no more than 125% of comparable vehicles with higher exhaust emissions. A five-member board certifies vehicles that meet the rigorous standards stipulated in the bill. Such certified, low-emission vehicles are, in effect, guaranteed entry to the 400,000-vehicle federal government market.

A Perspective on the Present R&D Effort

Despite the marked increase in efforts to reduce air pollution from automobiles in the United States, R&D activity is still infinitesimal in comparison with the dimensions of these industries and activities. Estimated governmental R&D expenditures for control of air pollution were $47 million in fiscal year 1969 and $61 million in fiscal year 1970, with perhaps half being devoted to pollution from stationary sources. Although these figures are limited to R&D expenditures under the wing of the Environmental Protection Agency, this agency controls the bulk of the funds; thus it indicates well enough the size of the total governmental effort.

Expenditures by industry are not a matter of public record. But the order of magnitude in the past is indicated by such items as a 2½-year, $7 million program on emission control, carried out by six petroleum firms and one automobile manufacturer, or a 3-year, $10 million research program jointly undertaken by the Automobile Manufacturers Association and the American Petroleum Institute and funded through the Coordinating Research Council. In addition, individual companies are financing their own research.

If government and industry were to increase their expenditures for research and development on air pollution from mobile sources to, say $75–$100 million per year (probably an optimistic guess), the R&D outlay would still be less than 1% of the wholesale purchase bill for new motor vehicles, which comes to $25 billion or so per year. The R&D spending would look even smaller if compared with a total that included highway and other related expenditures. And it still looks small when compared with the $1 billion a year that is spent on model changes alone.

There is indeed a great clamor today over air pollution, on all levels. As a society, however, we still seem unable (or unwilling) to do anything about it on a suitable scale. As we have remarked, the probable cost of such an effort, while large, appears quite trivial in comparison with various frivolous expenditures which we seem to take for granted.

Appendix — Fuel Cells

Appendix—Fuel Cells

HYDROGEN-OXYGEN FUEL CELLS

This system utilizes the simple reaction whereby hydrogen and oxygen combine to produce water:

$$H_2 + \frac{1}{2}O_2 = H_2O .$$

In the standard hydrogen-oxygen cell, with KOH as the electrolyte, hydrogen atoms are absorbed on the surface of the electrolyte side of the electrode (assisted by a catalyst), and then react with hydroxyl ions (OH^-) in the electrolyte to form water ($H^+ + OH^- = HOH$), giving up one electron to the electrode and the external circuit. Similarly, oxygen is absorbed at the cathode. When the external circuit is closed, the combination of oxygen plus the surplus negative charge from the circuit plus the water in the electrolyte form hydroxyl ions ($O^{--} + H_2O = 2\,OH^-$), thus completing the cycle.

The efficiency of the reaction increases as the temperature is raised, as Bacon was the first to demonstrate in the early 1930s. At 200°C to 250°C (390°F to 470°F), the hydrogen molecule H_2 can be easily dissociated and ionized within the presence of inexpensive catalysts such as nickel. Operating at these temperatures and pressures up to 600 psi, Bacon achieved power densities of 1,100 watts per square foot of electrode surface, a figure that has still not been surpassed. However, there are some disadvantages in the Bacon cell. Preheating time is too long, for one thing, and electrodes corrode rapidly in the hot KOH electrolyte. Also, the system requires very pure hydrogen; although, for very short operational times, Bacon found that the nickel electrodes were protected from contaminant "poisoning" by the presence of up to 10% carbon monoxide in the hydrogen.[1]

---

[1]C. G. Peattie, "A Summary of Practical Fuel Cell Technology to 1963," *Proceedings, IEEE*, Vol. 51, May 1963, pp. 795–806.

The hydrogen-oxygen cell in other temperature regimes has been fairly well developed over the last decade by a number of organizations, among them General Electric, Pratt and Whitney division of United Aircraft, Ionics, Inc., Leesona Moos Laboratories, Union Carbide, and Allis Chalmers. A variation of the basic hydrogen cell, developed in particular by GE and Ionics, Inc., uses an ion-exchange membrane as the cell "electrolyte." Such ion-exchange fuel cells, using both hydrogen and oxygen and hydrogen and bromine, have perhaps been advanced more than any other fuel cell.[2] The ion-exchange technique forms the basis of a practicable 200-watt portable power pack developed by GE for the U.S. Navy Bureau of Ships and the U.S. Army Signal Corps.[3]

However, a Yugoslav group has shown that ion-exchange membranes do not always lead to the best results.[4] Significant results have been achieved using a rather coarse pore-size plastic matrix, with a compressed and powdered electrode made from treated carbon powder impregnated with small quantities of platinum. Two novel features of this cell result in substantial weight reductions: (1) The lateral feeding of the reactant gases into the electrode and the elimination of the usual gas chambers, screens, and similar gas distribution devices, and (2) the use of a lightweight electron-conducting plastic (e.g., a polyethylene-graphite mixture) as the current-collector material. Extrapolating the results from a single cell to a large fuel cell battery, a power density of 68 watts/lb has been predicted.

A variation of the basic hydrogen-oxygen fuel cell, developed by Westinghouse Research Laboratories, makes use of a solid electrolyte consisting of a composition of zirconium-calcium oxides or zirconium-yttrium oxides $(ZrO_2)_{0.85}(CaO)_{0.15}$ or $(ZrO_2)_{0.9}(Y_2O_3)_{0.1}$. These are ceramics permeable to the O^{--} ion but not to other ions or atoms.[5]

The ionic conductivity is an increasing function of temperature. Hence these cells exhibit maximum power output around $1,000°C$ ($1,800°F$), which would be something of a drawback in vehicular applications.

Hydrogen gas is a questionable fuel for a practical automotive application. While it is used routinely and relatively safely in industry, its bulk (in gaseous form), explosiveness, and extreme inflammability are drawbacks. However,

---

[2] C. Berger, "Fuel Cells Incorporating Ion Exchange Membranes—Current State of Development"; W. Glass and G. H. Boyle, "Performance of Hydrogen-Bromine Fuel Cells"; L. M. Litz and K. V. Kordesch, "Technology of Hydrogen-Oxygen Carbon Electrode Fuel Cells," in *Fuel Cell Systems*, Advances in Chemistry Series 47, American Chemical Society, 1965; and J. W. Plattner, "Allis Chalmers Capillary Matrix Fuel Systems," Intersociety Energy Conversion Engineering Conference, 1968.

[3] E. A. Oster and L. E. Chapman, "Description of Fuel Cell Power Pack Operating on Hydrogen and Ambient Air," *Fuel Cells*, Vol. 2, Reinhold, New York, 1963.

[4] D. M. Drazic, R. R. Adzic, and A. R. Despic, "New Methods of Obtaining Fuel Cell Electrodes," *Journal of the Electrochemical Society*, Vol. 116, No. 6 (1969), p. 885.

[5] D. H. Archer et al., "Westinghouse Solid Electrolyte Fuel Cell," *Fuel Cell Systems*, Advances in Chemistry Series 47, American Chemical Society, 1965.

hydrogen can be produced as needed by a number of means. Thus the possibility presents itself that some other safer carrier might be employed and hydrogen generated only as needed. This concept, in fact, underlies a number of current efforts to utilize hydrocarbons and other fuels in fuel cells.

In theory, hydrogen can be generated in the following ways:

1. by the electrolysis of water;
2. from hydrocarbons such as methane (CH_4) via steam reforming or the so-called water-gas reaction;
3. from red-hot iron or zinc and steam;
4. from methanol and steam;
5. via the thermal cracking of ammonia;
6. by heat from light metal hydrides;
7. from the action of acid and an active metal; or
8. it may be absorbed on the surface of a finely divided metal such as a porous nickel, which can be transported as a "slurry," suspended in water.

The electrolysis of water is probably the cleanest and simplest means of generating hydrogen. However, an electrolytically regenerative fuel cell is, in effect, a rechargeable or secondary battery;[6] hydrogen is generated on recharge and must still be stored and carried as hydrogen.

Of the other methods of hydrogen generation, the only ones that are really attractive from an economic standpoint are those involving the cracking or reforming of hydrocarbons, ammonia, etc., and possibly the absorption technique. Techniques using metal hydrides or active metals to produce hydrogen are attractive because they are low-temperature processes and basically very simple, but they are too expensive in terms of pounds of reactants used per kwh finally generated. (In general, it would be more efficient to generate electricity directly—via a metal-air battery, for example.)

An interesting concept, developed at the Battelle Institute (Geneva), uses a hydrogen-absorbing metal powder in the form of a "slurry" suspended in the electrolyte fluid as a means of storing hydrogen conveniently and conveying it physically to the active electrode. The specific power of the cell, including "fuel," would be about 36 watts/lb in a 20-kw size. One of the most attractive features of the slurry cell is that the metallic particles apparently help the oxidation reaction and large amounts of noble metals are not required. An electrode consisting of 90% nickel and 10% silver-cadmium is apparently satisfactory. The estimated capital cost of this system (assuming development is successful) would be only about $20 per kw for the cell plus $17 per kwh for

---

[6]E. Findl and M. Klein, "Electrolytic Regenerative Hydrogen Oxygen Fuel Cell Battery," *Proceedings, 20th Annual Power Sources Conference*, Power Sources Conference Publication Committee, Red Bank, N.J., 1966.

the hydride slurry. Rehydrogenation would cost about $0.0125 per kwh if done chemically or $0.025 per kwh if done electrolytically (external to the fuel cell).

Whatever the hydrogen generation technique, once the hydrogen is generated, the cell has all of the performance characteristics of a hydrogen cell.[7] If the equipment will break down one or another of the cheap hydrogen compounds as needed and if its weight will not, in effect, reduce performance, the more attractive features of both the hydrogen and the hydrocarbon cells can be combined without some of the present disadvantages. It is worth noting that a GE ion-exchange membrane portable power pack was developed for the military in two models, one of which carried stored hydrogen under pressure and the other a sodium-borohydride hydrogen generator. Both of these models weigh approximately the same overall. Thus, hydrogen generation need not impose any extra weight burdens on a system. Whether or not this will prove true of systems that generate hydrogen from hydrocarbons remains to be seen, however.

HYDROCARBON FUEL CELLS

Hydrocarbons such as gasoline, kerosene, propane, natural gas, ethane, and acetylene rank among the most potentially desirable fuels. They have a high energy content (up to about 6,000 whr/lb), and they are readily available, cheap, and relatively easy to handle. Thus, many efforts have been made to exploit them.[8] As early as 1896, Jacques was achieving outputs as high as 100 amp/ft^2 of electrode, in a 1.5-kw cell, over extended periods using liquid hydrocarbons as fuel, with a molten alkali electrolyte.[9]

Unfortunately, despite this promising start, a hydrocarbon fuel cell with a high output and no noxious exhaust fumes still eludes us today, although a great many hydrocarbon cells of various types have been operated with considerable success.

One of the principal difficulties has been sluggish reactions at low (ambient) temperatures, resulting in low voltages and low current densities per unit area. Performances have typically been in the range from 5-15 watts/ft^2. Considerable amounts of platinum catalyst have usually been required: 1-2 gm/ft^2 of electrode surface or $>$ 100 gm/kw. In addition, that traditional bugaboo of hydrocarbon combustion—various incompletely oxidized carbon compounds in the exhaust—has not yet been overcome.

---

[7]Except in the case of the "slurry cell" the electrochemical reactions may take place on the surface of the carrier particles, rather than on the electrode-electrolyte interfaces. Theoretically this may greatly reduce the need for expensive catalysts.

[8]In fact, the classic problem of the "coal cell"—the dream of extracting electricity directly from coal—has attracted electrochemists for over a century, though so far without signal success.

[9]H. A. Liebhafsky and E. J. Cairns, "Hydrocarbon Fuel Cells—A Survey," AIEE Pacific Energy Conversion Conference, 1962.

Because of the problems associated with low temperatures, much research has been directed toward higher temperature regimes. Unfortunately other difficulties arise. It is no longer possible to utilize aqueous electrolyes, and it becomes more difficult to satisfy both the reactivity and invariance requirements simultaneously. High-temperature cells using both molten alkali electrolytes (around 500°C) and molten carbonate electrolytes (500°C–800°C) have been explored, although both, particularly the former, have short lifetimes due to difficulties in maintaining the invariance of the electrolyte. In addition, high-temperature cells have been developed using "paste" electrolytes such as a paste of magnesium oxide held in a matrix and solid electrolytes such as the zirconium-oxide ceramic described earlier for the hydrogen cell. The oxidizer in most of these cases has been oxygen extracted from the surrounding air. These cells, like the zirconium solid electrolyte cell, require such high temperatures that it is hard to envision their use in vehicles.

Rather than oxidize hydrocarbons directly, it is possible—as mentioned previously—to treat them as sources of hydrogen, or as mixtures containing hydrogen and something else, usually carbon monoxide. This has been perhaps the most frequently used approach to the utilization of hydrocarbons in fuel cells, particularly high-temperature fuel cells. This is usually done by "steam reforming" (700°C–1,000°C) of the hydrocarbon, or by partial oxidation (1,400°C). The initial product gas usually contains a mixture of hydrogen, carbon dioxide, carbon monoxide, and other substances, which must be passed through a series of further reactors to remove the carbon monoxide. The technique as currently practiced is too complex and costly to be considered in vehicular application.

A promising hydrocarbon technique has recently been developed jointly by Bolt, Beranek and Newman, Inc., and Atlantic Richfield[10] using only gasoline and steam at about 500°C (950°F). Gasoline is fed into a carburetor-like device in which it is atomized and mixed with incoming steam. The mixture is then passed through a catalytic reforming bed enclosed in a palladium foil anode. The hydrogen thus released diffuses through the foil into a surrounding electrolyte (a 4-to-1 anhydrous mixture of KOH and NaOH), forming water and releasing electrons. A prototype fueled by natural gas and steam produced 0.6 to 0.7 volts and approximately 500 amps per ft^2, which is roughly 0.3 kw/ft^2 of anode area.

SOLUBLE FUEL CELLS

A number of potential fuels are soluble in water, and therefore in aqueous electrolytes. These include ammonia, hydrazine, and various alcohols. The fuels can simply be dissolved in an aqueous electrolyte, into which both electrodes

---

[10]W. Juda, "Hydrocarbon Fuel Cell," Electrochemical Society Meeting, Montreal, October 1968.

can then be inserted. In addition, such cells are potentially operable at ambient temperatures, which further simplifies design.

In the case of hydrazine (N_2H_4) results to date have been promising. The hydrazine is dissolved in a circulating aqueous potassium hydroxide electrolyte. The byproducts of the cell reaction (oxidation with air) are only gaseous nitrogen and water.

Two portable units, with outputs of 300 and 600 watts, have been developed by Union Carbide Corporation.[11] The 600-watt unit weighs 73 pounds overall, including the weight of a 6-hour fuel supply, and occupies 1.9 cubic feet.[12] Thus this unit offers a specific power of some 8 watts/lb and an energy density of 48 whr/lb for a full six hours of use. A 5-kw unit developed by Monsanto Research Corporation for the U.S. Army weighs 200 pounds without the fuel supply, and with fuel for ten hours operation it gives an output of about 18 watts/lb.

An even greater improvement is gained by further scale-up in size. A 20-kw unit installed in an Army truck-bed is estimated to weigh 600 pounds without its fuel.[13] Allowing 150 pounds for tanks and fuel for eight hours of operation gives a power density of 26 watts/lb. With further improvements in design, it should be possible within a few years to develop hydrazine fuel cells of moderate size (i.e., 10 kw and above) that would have specific power ratings of more than 50 watts/lb of total system weight.

At the moment, hydrazine costs too much for commercial use. At $1.15 per pound for hydrazine and 50% efficiency, it costs $0.97 to produce 1 kwh—about four orders of magnitude more than the cost per kwh of natural gas, for example.[14] Only if hydrazine prices can be brought down drastically can the hydrazine fuel cell be an important commercial contender. There seems to be no fundamental reason why hydrazine could not be produced in bulk directly from ammonia, at a comparable cost (say, $0.05 per pound). However, at present this is merely a pipe dream.

Another obvious avenue to achieving lower costs leads naturally to the consideration of ammonia (NH_3) as a fuel. Like hydrazine, ammonia is soluble in an aqueous electrolyte, and its reaction products consist only of nitrogen and water. Also it is relatively inexpensive (about $0.03 per pound). In spite of this, it has not been as extensively explored for fuel cell use as has hydrazine. Ammonia cells have been operated at $180°C$ ($356°F$) and at normal pressures with inorganic nonaqueous electrolyte matrices and with air as the oxidizer. A

[11] E. A. Gillis, "Hydrazine-Air Fuel Cell Power Sources," *Proceedings, 20th Annual Power Sources Conference*, 1966.

[12] G. G. Fee and E. Storto, "20-Ampere, 28-Volt DC Hydrazine-Air Fuel System," *Proceedings, 23rd Annual Power Sources Conference*, 1969.

[13] E. A. Gillis and L. D. Gaddy, Jr., "Fuel Cell Electric Propulsion Test Rig," AD 666763, U.S. Department of Commerce Clearinghouse, February 1968.

[14] G. C. Szego, "Economics, Logistics and Optimization of Fuel Cells," Research Paper P-208, Institute for Defense Analyses, December 1965.

power/volume density of 1.1 kw/ft^3 overall has been achieved for short periods. Furthermore, the slightly (and tolerably) elevated temperature improves electrode kinetics and facilitates by-product (water) removal, thus saving somewhat on weight and cost.[15]

A lower-temperature (120°C or 250°F) cell developed by GE has achieved a power density of 50mw/cm^2 with only 2mg/cm^2 of platinum on each electrode. Teflon-bonded platinum black electrodes provide the necessary electrocatalytic activity.[16] The indirect use of ammonia in a hydrogen "reforming" cell also has some potential. As discussed earlier, hydrogen can be produced quite readily by the thermal cracking of ammonia, which produces a 75–25 mixture (by volume) of hydrogen and nitrogen respectively; if desired, the nitrogen can be removed by liquefaction. If portable ammonia-cracking equipment is developed, this technique may represent a most convenient way of obtaining hydrogen without actually carrying it around: one cylinder of ammonia is roughly the equivalent of nine or ten cylinders of hydrogen. One small ammonia-cracking unit has been reported as having a total weight of 15 pounds and producing 60 cubic feet of hydrogen in twelve hours.[17] There are not yet enough results on the overall performance of ammonia systems to evaluate them on a specific power basis. However, on the basis of fuel cost and availability, they would seem to be worthy of consideration.

Various other electrolyte-soluble fuels have been investigated to some extent; in particular, a considerable amount of work has been performed on cells using methanol dissolved in an acid electrolyte.[18] As a further step, the Illinois Institute of Technology Research Institute has successfully operated methanol cells in which the oxidant was dissolved also, thus reducing the size and weight of the oxidizer electrode.[19] In this case, a sodium-chlorite oxidant was dissolved in an alkaline solution. The cell produced about 140 amp/ft^2 at 0.6 volts, or about 85 watts/ft^2. But such systems are not developed enough to evaluate the power/weight ratios that may be ultimately attainable.

As with ammonia, it is also possible to reform methanol with steam to produce hydrogen, which can then be reacted in a hydrogen cell. The economics

[15] R. A. Wynveen, "Preliminary Appraisal of the Ammonia Fuel Cell System," *Fuel Cells*, Vol. 2, Reinhold, New York, 1965; B. S. Baker, "The Ammonia Fuel Cell, An Economic Approach," AIEE Pacific Energy Conversion Conference, 1962.

[16] E. L. Simmons, E. J. Cairns, and D. J. Surd, "The Performance of Direct Ammonia Cells," *Journal of the Electrochemical Society*, Vol. 116, No. 5 (1969).

[17] D. Singman and A. Forziati, "Hydrogen Sources for Fuel Cells," Harry Diamond Laboratories, TR-1168, November 1963.

[18] B. L. Tarmy, "Methanol Fuel Cells," *Proceedings, 16th Annual Power Sources Conference*, 1962; G. Cirpios, "Methanol Fuel Cell Battery," *Proceedings, 20th Annual Power Sources Conference*, 1966; C. E. Heath et al., "Methanol-Air Fuel Cell," Final Report, Esso Research and Engineering Co., Technical Report ECOM-02387-F, U. S. Army Electronic Command, Fort Monmouth, N.J., February 1968.

[19] D. B. Boies and A. Dravnieks, "Methanol Fuel Cells with Dissolved Oxidants," *Fuel Cell Systems*, Advances in Chemistry Series 47, American Chemical Society, 1965.

are questionable, however, since cheaper, more energetic hydrocarbons such as propane can be used as hydrogen sources if need be.

PRESENT STATUS

It is apparent from the foregoing that there are as yet no "advanced" fuel cells: the fuel cell concept, despite its considerable history, is still in an early phase of development. The best specific power figures are about 25 watts/lb of overall system weight from hydrogen and oxygen and possibly 30 watts/lb from hydrazine and air, in 25-kw sizes; these leave much to be desired as practical power sources. In addition, for civilian commercial application, there are other considerations that are as important as performance. The reactants must be inexpensive and easily handled, for example, and the cell must be simple, reliable, and long-lived. While hydrogen-oxygen cells offer the best overall compromise, both hydrogen and oxygen are potentially hazardous to handle and transport, and cryogenic auxiliary equipment is too complicated, expensive, and difficult to maintain for vehicular applications. The hydrazine-air cell performs as well or better than the hydrogen cell at ambient temperatures and is simpler and more easily operated under ordinary conditions, but the fuel is expensive and also somewhat toxic. On the other hand, hydrocarbon cells might utilize convenient, cheap, and readily available fuels such as propane, but they require more costly catalysts, operate moderately well only at high temperatures, and tend to produce undesirable waste products such as carbon monoxide unless the reactants are recycled several times.

Even though most present-day fuel cells are largely experimental, it is possible to compare the various types in a general way. For this purpose, the basic parameters have been itemized and rated for the major categories of cells in table A-1.

Note that none has been given a "good" rating in specific power (watts/lb). The maximum values achieved barely suffice for vehicular propulsion. The table does not contain a rating for specific energy (whr/lb), which is the most important single parameter for batteries but of minor importance for fuel cells. Values for energy density of 125-150 whr/lb for fuel cells are often cited, usually in literature of a nontechnical nature. Specific energies can generally be made arbitrarily high, at a trivial increase in weight, simply by adding more fuel. This shows clearly in figure A-1. It should be recalled that applications requiring ten hours or more of continuous operation are often envisioned for fuel cells, whereas three hours is usually adequate for automotive purposes. Looking at it another way, the hardware required to make practical use of a theoretical liquid or gaseous fuel-oxidizer couple is typically a much larger percentage of the total system weight than is normally the case for batteries where "fuel" and electrodes are the same. But considering the weight of the reactants only, the theoretical energy content of hydrogen and oxygen is more than 1,600 whr/lb; using oxygen scavenged from the air, it is more than 14,000 whr/lb. For various

Table A-1. General Fuel Cell Ratings

Fuel cell	Power density, watts/lb	Life-time	Cell cost	Fuel cost	Fuel storage and handling	Exhaust product quality
Hydrogen cells	Fair	Good	Poor	Fair	Awkward	Excellent
Low-temperature hydrocarbon cells	Very poor	(?)	Very Poor	Excellent	Excellent	Poor
High-temperature hydrocarbon cells	Poor	Poor	Poor	Excellent	Excellent	Poor
Hydrocarbon reforming cells	Poor	Good	Poor	Excellent	Excellent	Fair
Hydrazine cells	Fair	Good	Fair	Poor	Good	Excellent
Hydride slurry cell	Fair	(?)	Good	Good	Excellent	Excellent

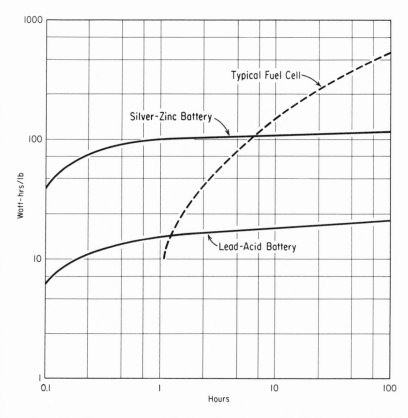

Figure A-1. Specific energy versus discharge time for several batteries and a typical fuel cell.

hydrocarbons, the theoretical energy available ranges up to about 6,000 whr/lb of reactant, using atmospheric oxygen. In other words, the value of K_r (the reactant efficiency) for fuel cells is, at best, very small. Thus, values like 125 whr/lb, while reasonable from an electrical propulsion standpoint, constitute a much smaller fraction of the energy potentially available than is the case for some high-energy batteries currently under development.

The structural weight penalties can perhaps be even better illustrated by a few examples. Tabulated in table A-2 are the characteristics of several complete fuel cell systems that are available, at least for limited uses and special purposes. Consider, for example, Union Carbide Corporation's Model E2F-1-468.[20] The cell stack alone for this unit, without fuel, weighs 42 pounds; since the output of the unit is 1,000 watts, the specific power for the stack alone is 24 watts/lb, without including the auxiliary components (the electrolyte pump, heat exchanger, gas coolers, regulators, instruments, etc.). This cell uses 0.1 pound of H_2 and 0.8 pound of O_2, or a total of 0.9 pound of reactants, for each kwh output, which amounts to a theoretical energy capacity of about 1,100 whr/lb based on the reactants alone. To enable this cell to operate for 10 hours, only 9 pounds of reactants would be required. The weight of reactants for one hour's operation amounts to roughly 2% of the stack weight, exclusive of auxiliaries, and even for ten hours this figure would only rise to 20%.

Another Union Carbide hydrogen-oxygen cell, which was developed in conjunction with General Motors and installed in a delivery-type van (ElectroVan), uses 32 individual cells to produce a design peak output of 94 kw, with short-duration peaks up to 160 kw.[21] Since this unit weighs approximately 3,650 pounds, its average specific power is 26 watts/lb, which implies an energy density of $26T$ whr/lb, where T is the desired operating time (in hours).

General Electric has developed a portable 200-watt hydrogen-oxygen fuel cell, which comes equipped with either a hydrogen supply in containers or with a chemical sodium boro-hydride hydrogen generator, either of which will supply sufficient fuel for seven hours of operation. The overall weight of this 200-watt unit is 60 pounds; thus its specific power is only 3.3 watts/lb, although this system (which utilizes an ion-exhange membrane technique) is in some respects the most advanced of all the hydrogen-oxygen fuel cells. Over its full operating time of seven hours per tank of fuel, its specific energy is 23 whr/lb, but if it operates for only one hour, this figure drops to 3.3 whr/lb.

Both Union Carbide and General Electric have presently abandoned their fuel-cell efforts until specific commercial markets can be recognized. Evidently, a fuel-cell system with a specific power in excess of, say, 75 watts/lb would offer satisfactory performance even for a short operational period such as an hour, while a very slight increase in total system weight would permit ten hours (or

---

[20] Union Carbide Corporation, Technical Data Sheets 66-3-2133 and 66-4-2133.

[21] General Motors News Release, October 28, 1966.

more) of operation. Over a four-hour period such a system would have an effective energy density of 300 whr/lb, which would make it competitive with other prime movers for long-distance transportation purposes.

Fuel costs, which were rated qualitatively in table A-1, are quantified in table A-3. The weight and cost advantages of scavenging oxygen from the air are evident. It is also clear that cost and availability favor the use of natural gas, or some petroleum-derived hydrocarbon like gasoline.

Table A-4 lists the approximate costs per kwh for power generated by hydrogen produced from various typical sources, assuming 100% conversion and utilization of all reactants and the use of scavenged oxygen from the air as oxidizer, and based solely on the commercial costs per pound of the materials involved, except in the case of electrolysis, where the necessary input of electrical energy is the cost.[22] The general characteristics of most of these hydrogen generation techniques have been discussed earlier.

Reducing these methods to practical hardware requires equipment of varying complexity and weight penalty, but they all suffer from disadvantages of one kind or another. For instance, the use of any secondary system to produce hydrogen also reduces overall system efficiency considerably. If the basic cell operates at about 60% efficiency, for example, and a hydrogen generation unit is introduced operating at 50% efficiency, the overall system efficiency drops to only about 30%.

Reduced efficiency would not really matter too much—except as it affects total system weight—if a system could be developed which permitted the reforming of a cheap hydrocarbon like natural gas to hydrogen without any appreciable carbon monoxide content in its exhaust and which yielded a high watts/lb rating (e.g., 50 to 75) for the entire system, including the reforming unit. It is difficult at this stage to derive reliable figures comparing the weights of hydrogen generation systems versus storage systems. Overall weights depend both on the hydrogen generation technique used and on the size of the system. It is clear, however, that weight ratios become more favorable with increase in system size.

A typical hydrogen tank containing one pound of hydrogen—the equivalent of about 0.65 kwh assuming 50% efficiency—weighs perhaps 80 pounds. Including the tank weight thus reduces the specific energy to about 8 whr/lb. For automotive propulsion, the appropriate system size is larger. In this size range, given high pressures and light materials, it can probably be assumed that 50-100 watt-hrs equivalent of hydrogen can be stored per pound of tank weight.

At 50 whr/lb, a 10-kwh equivalent of hydrogen could be stored and carried in tanks weighing 200 pounds or less. For hydrocarbon reforming systems, it is estimated by Frysinger that some 40-45 pounds must be assigned to the

---

[22] Szego, "Economics, Logistics and Optimization of Fuel Cells."

Table A-2. Some Complete Fuel Cell Systems

Company	Output (kw)	Weight (excl. fuel) (lb)	Volume (cu ft)	Fuel and oxidant	Electrode and electrolyte	Operating conditions	Application
Union Carbide H2R-1-278	0.3	33 (inc. fuel)	0.87	Hydrazine, air	Carbon plated on nickel, circulating KOH	Ambient	Commercial
E2F-1-468	1.0-2.5	42 + aux.	3.36	Hydrogen, oxygen	do.	120°F-150°F	Commercial
E2F-4-864	3.74-9.4		28	Hydrogen, oxygen	do.	do.	
	94	3,650		Hydrogen, oxygen	do.	do.	GM Electrovan
Allis-Chalmers	0.5			Methanol, oxygen	Porous nickel electrodes, platinized anode, silver cathode, asbestos matrix, KOH	Ambient	Demonstration
	3			Hydrazine, air	do.	Ambient	Demonstration
	4.5	1,190	34	JP-150 (reformer)	do.	Reformer 1,450°F (cell 160°-180°F)	U.S. Army
	2-5 (overload)	169 (complete)	<5.3	Hydrogen, oxygen	do.	190°F	Space
	15		20 + aux.	Propane, oxygen	do.	Ambient temp.	Mounted on tractor
Monsanto Research	0.06	14.5	0.35	Hydrazine, air	Circulating KOH	200°F	
	5	200		do.	do.	200°F	

Manufacturer				Fuel/oxidant	Electrolyte	Temperature/pressure	Application
General Electric	1			Hydrogen, oxygen	do.	~500°F, 55 psi	NASA (LEM)
	2			do.	do.	~500°F, 55 psi	Apollo
	0.03	7	0.3	Lithium hydride, air	Ion-exchange membrane sulphonic acid	35°F–110°F	Military field use
	0.06	10	0.7	Active metal/hydrogen, air	do.	do.	Military field use
	0.2	60	–	Hydrogen, air	do.	do.	
	1.0	70 + aux.		Hydrogen, oxygen	do.	do.	NASA (Gemini type)
	1.5	140	8	JP–4 reformer, air	do.	do.	Battery charger
Energy Conversion Ltd., U.K.	6	1,300	2,813	Methanol, air	–	–	Electric truck
Chloride Group, U.K.	–	–	–	Hydrazine, air	–	–	Demonstration
Shell, U.K.	–	–	–	Methanol, air	Acid electrolyte	–	Demonstration
ASEA, Sweden	50	–	–	Ammonia (reformer)	–	–	Submarine
Varta, A.G., Germany	2	–	–	Hydrogen, oxygen	–	–	Forklift truck

Table A-3. Costs per Pound and per kwh Produced of Some Common Fuel Cell Reactants

Reactant	Cost per pound	Cost (fuel only) per kwh(100% conversion efficiency)
Hydrogen (g) (liquefied bulk)	$0.110	$0.006
Oxygen (liquefied bulk)	0.01–0.02	–
Ammonia (g) (gaseous)	0.03	0.011
Natural gas	0.005	0.0007
Gasoline	0.024	0.0044
Hydrazine	1.15	0.49

Source: G. C. Szego, "Economics, Logistics and Optimization of Fuel Cells," Institute for Defense Analyses, December 1965.

Note: Costs are for very large quantities.

Table A-4. Cost per kwh of Energy Produced by H_2 From Various Sources, Based on the Costs of the Required Fuels Used

Source of Hydrogen	(Based on cost per lb of fuel only; O_2 scavenged from the air) Cost per kwh generated by H_2
Reforming of natural gas	$0.001–0.002
Thermal cracking of ammonia	0.016
Electrolysis of water (elec. energy costs)	0.023
Metal hydride storage (excluding initial cost of metal "carrier")	
Regenerated chemically	0.012
Regenerated electrically	0.025
Active metal–acid reaction (including basic cost of metal)	0.35

reformer for each kw of hydrogen.[23] Thus, a reforming unit for a 5 kw system would weigh about 200 pounds. If the system is to operate for a short time—say, two hours—this would be 10 kwh for 200 pounds, neglecting the weight of the hydrocarbon fuel, which would be small but not negligible. The overall weight of either system described above would reduce the overall specific power to about 10 to 12 watts/lb.

A tremendous amount of effort over the last decade or so has gone into improving the technical characteristics of various fuel cell components (for example, watts/ft² of electrode, lifetimes of electrodes, catalyst loading, invariance of electrolytes, etc.), which have unquestionably tended to improve system performance in many respects but have not yet had too much outward effect in terms of specific power (watts/lb) of complete systems. Some selected performance trends are illustrated graphically in figures A-2 and A-3.

---

[23] G. Frysinger, personal communication.

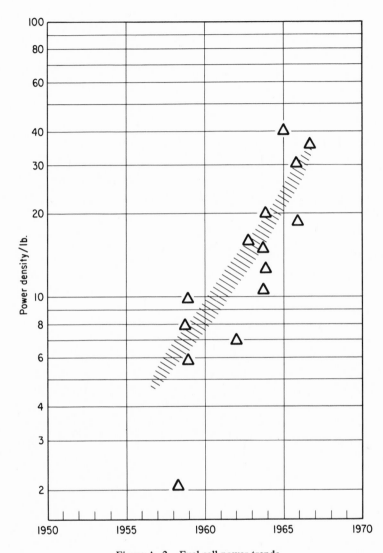

Figure A-2. Fuel cell power trends.

Accurate figures on specific power have only been available in recent years—over the last decade or so, in fact—and they are scattered over too wide a range to form a reliable basis for forecasting the future, although the trend shown in figure A-2 is suggestive.

There has already been a great deal of progress in reducing the amounts, and thus the costs, of the catalysts required to effect a given reaction. Fuel cells that cost $6,000 per kw in 1959 cost only about $100 per kw by 1966, according to

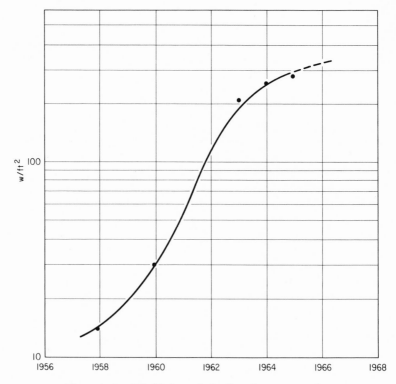

Figure A-3. Allis-Chalmers fuel cell performance trends.

the U.S. Army Engineer R&D Laboratories.[24] There has been a singular lack of improvement in current and power densities per square foot. Those achieved by the Bacon-type hydro cell in the 1930s, for example, are at least as high as anything currently available, and the power densities of 100 watts/ft² recorded by Jacques in 1896 for a 1.5-kw cell using hydrocarbons and a molten alkali electrolyte have not been improved upon for hydrocarbons. However, there has been a marked improvement in power per square foot of electrode, particularly during the last decade. The experience of the Allis Chalmers group in this regard has been fairly typical (see figure A-3). And, while the materials and details of construction will vary considerably depending on the particular cell, increases in power density per unit surface area of electrode will inevitably be manifested as improvements in specific power for the whole system. Notwithstanding the optimism of many researchers in the mid-1960s about eventual cell performance and initial capital costs, it is apparent that fuel cells have not yet proved to be a practical power source for vehicular propulsion.

---

[24] "Department of Defense Research on Unconventional Vehicular Propulsion," U.S. Army Engineer Research and Development Laboratories, February 1967.

The experience of Battelle Memorial Institute is noteworthy in this regard.[25] With as many as 44 sponsors, Battelle's research program was aimed at finding a substitute catalyst for platinum and a satisfactory method of oxidizing inexpensive fuels like propane and natural gas. An economically attractive solution was not found for either problem, and Battelle discontinued research in this area as of June 1969.

Allis Chalmers, which has been the major contractor for NASA's fuel-cell effort, has perhaps the most realistic appraisal of the future. They feel that the ultimate role of the fuel cell is yet to be defined and that intensive work for perhaps five more years is needed before a realistic appraisal of its commercial potential is likely to unfold.[26]

However, United Aircraft (Pratt and Whitney Division) is betting that the fuel cell can competitively convert natural gas directly to electrical energy in a stationary facility. In this application, the size or weight and the initial capital costs of the fuel cell would not be as inhibiting as they would for a vehicular application, and, because fuel cells convert fuel directly to electricity, a generator would not have to be part of a future fuel-cell power-production facility.

[25] *Chemical and Engineering News*, July 21, 1969, p. 42.
[26] Ibid.

Index

ABS (acrilonitrile-butadiene-styrene), 208
AC induction (squirrel cage) motors, 190–205, 240–41, 258–60, 263
Acceleration, 16, 31, 245–46; in energy conversion equations, 36–39, 42–49, 210–12
ACID test cycle (New Jersey), 132
Acton, R. G., 205n, 222n
Adams, D. R., 204n, 278n
Adams, J., 101n
Adzic, R. R., 296n
AEI Ltd., 201
Afterburners (mufflers), 6, 127, 131, 253–54
Air: as electrodes, 93–98, 250; as working fluid, 248
Air cells: iron, 105; metal, 93–98, 250; zinc, 93–96, 209, 217, 228, 250
Air conditioning, in electric cars, 214
Air engines *See* External combustion engines; noncondensing; Stirling-cycle engines
Air-fuel ratio, 12–13, 55–57, 123–25, 127–28, 253
Air injection systems, 126–27, 130
Air pollution, from automotive emissions, 6–11, 15–31, 206, 267–68, 272–73, 284; effluent charges as a method for controlling, 289; research re, 122–36, 292. *See also* Emissions, automotive; Residuals
Air Programs, Office of, 16, 25, 30, 129
Air resistance, and automotive performance, 41, 246
Air-standard cycles, 52–53, 55–61, 64. *See also* Otto-cycle engines
Aircraft, 3, 7; engines for, 120, 138–39, 149, 202
Alcohols, as fuel, 134–35

Alden Self-Transit Systems Corporation, 208, 219
Alkali-metal anodes, 98–100
Allectric car, 226–27
Allied Chemical, 171–72, 174
Allis Chalmers, 228, 296, 306, 310–11
Alternator-motor combination, 230, 235–37; turboalternator, 239–40
Altitude, effect on emissions, 20
Aluminum, 51, 93, 98, 102, 170, 202, 208, 250, 261, 285; oxide, 168
American Monarch, 226
American Motors Corporation, 116, 139, 219–22, 255
American Oil Co., 130
American Petroleum Institute, 292
Amitron electric car, 219–22
Ammonia, as fuel, 51, 133–34, 243, 254, 299–301, 308
Archer, D. H., 296n
Argon, as working fluid, 177
Argonne National Laboratory, 104
Arizona, 21
Armengaud, René, 138
Army, U.S., 153, 201; Electronics Laboratory, 99, 242–43; Engineer R&D Laboratories, 222, 240, 263, 310
Aromatics, 21–22, 134, 285
Aronson, Robert, 88n, 223n
"Athodyd" (propulsive duct) system, 139
Atlantic Richfield Oil Co., 130, 299
Austin Crompton Parkinson (U.K.), 229
Autoignition, 124
Automobile Manufacturers Association, 292
Automobiles, 3–4, 6, 15–21, 28; electric, 87, 92–93, 105, 206–29, 260–61; performance characteristics of, 40–50, 265–79; state and federal standards for, 129–33,

313

253, 279, 287–88, 292. *See also* Automotive transportation; Batteries; Emissions, automotive; Residuals

Automotive transportation: application of gas turbines to, 137–47, 255, 284; application of noncondensing gas cycle ECEs to, 180–89, 257–58, 284; application of Rankine-cycle engines to, 148–79, 256–57; assessement of effects and policies on, 283–92; electric propulsion systems for, 190–205, 258–59 (*see also* Electric vehicles); energy conversion requirements for, 35–50, 245–47; external effects of ICEs, 12–31; hybrid power systems for, 230–44, 262–64, 284; ideal engine for, 265; new developments in ICEs, 113–36, 252–54; performance comparisons, 265–68; public interest in, 3–11, 283–86. *See also* Automobiles; Batteries; Buses; Emissions, automotive; Trucks

Ayres, Robert U., 11*n*, 37*n*, 179*n*, 238*n*, 278*n*

Azure Blue Corporation, 125, 253

B.M.W. (Germany), 136
Bacon cell, 295, 310
Baker, B. S., 301*n*
Baker, Milton C., 134*n*
Barber, John, 137
Bates, J. J., 197*n*
Battelle Memorial Institute, 85, 297, 311
Batteries, for automotive transportation, 86–108, 248–49; charging of, 86, 206–8, 213, 242, 249, 261, 290–91; costs, 93, 105–8, 250, 272–79; discharge rate of, 211–12, 215, 233–34; in electric systems, 198, 206–18, 260–63; in experimental electric vehicles, 219–29; Ford's sodium-sulfur, 103–4, 209–10, 213–14, 223, 251, 261; in hybrid systems, 84, 113, 231, 233–37, 242–44, 262–64, 285; ideal, 105; lead-acid (*see* Lead-acid cell); lithium-chlorine, 101, 207, 209, 217, 243, 251, 264; lithium-nickel, 222, 225; thermal, 234, 237, 241, 251, 262–63. *See also* Fuel cells

Battronic Truck Corporation, 222–23
Bauman, H. F., 100*n*
Bell, A. F., 151*n*
Benefit-cost analysis, difficulties of, 283–84
Benzene, 171–73, 257, 285
Berger, C., 296*n*
Berkowitz, S., 226*n*
Bernstein, Mower H., 288*n*
Berry, D. S., 20*n*
Besler Corporation, 148–49, 256

Bicycle, electric, 226, 228
Binary vapor cycle, in steam engines, 160–61, 177
Biphenyl, 176–77
Bjerklie, J., 75*n*, 121*n*, 177*n*, 178–79
Blemel, K. G., 22*n*
Blomme, G. W., 20*n*
Blow-by, crankcase, 12, 15, 126, 253
Boies, D. B., 301*n*
Bolt, Beranek and Newman, Inc., 299
Borgeson, G., 148*n*
Borlund, R. P., 240*n*
Bouladon, G., 240*n*
Bowerman, E. R., 90*n*
Boyle, G. H., 296*n*
Braking: "dynamic," 48*n*; with electric engines, 201, 210–13, 240, 260; with gas turbines, 255; with steam engines, 257. *See also* Deceleration
Brayton, George, 62
Brayton-cycle engines, 52, 62–63, 121–22, 177–78, 186–88, 247, 258, 266, 268. *See also* Gas turbines
Brobeck (W. A.) Associates, 156
Brown, Boveri and Co., 138–39
Brubacker, M. L., 35*n*
"Brushless" motors, 197, 201, 259
Buchi, Alfred, 139
Bugliarello, G., 25*n*
Buses, 3–4, 17, 20, 40; electric, 219, 222–23; flywheel system for, 111, 234, 251; gas turbines for, 141, 147, 255; hybrid systems for, 239–40, 263; steam-powered, 149–50, 156, 256–57, 290; Stirling-cycle engines for, 183
Bush, Vannevar, 185–86, 258
Butane, as fuel, 134–35

C.A.V. Ltd., 201
Cadmium, 90–93, 168, 206, 217, 222, 224, 250
Cairns, E. J., 81, 84*n*, 104*n*, 105*n*, 298*n*, 301*n*
Calcium, 93, 96–97, 250, 285
California, 21, 253; Air Resources Board, 128; emission requirements in, 12*n*, 31, 126, 129–32, 154, 254, 288; testing of steam-propelled patrol cars in, 290. *See also* Los Angeles
California driving cycle, 35, 154, 245
Campbell, John, 207
Carbon absorption canister, 128
Carbon dioxide, 16
Carbon monoxide, in automotive emissions: from diesel engines, 62; federal standards re, 288; from gas turbines, 146–47, 255;

from ICEs, 12, 15–20, 26, 28–31; reduction of, 122–23, 126–30, 134–35, 253, 268, 272–73; from Stirling engines, 184, 239; from Williams engines, 154

Carburetor: dual-ratio, 125–26; evaporation from, 13–15, 128–29, 253

Carnot: cycle, 69–71, 75, 84–85, 248; efficiency, 74; limit, 78

Carter, Alastair, 208

"Cascading" of fuel cells, 85

Catalysts, need for cheap and effective, 83–84

Catalytic converters, 127–28, 130

Cater, R. D., 150n

Caterpillar Tractor Co., 139, 255

Chalmers, Bruce, 110n

Chapman, L. E., 296n

Charging, of electrical batteries, 86, 206–8, 213, 242, 249, 261, 290–91. *See also* Recharging

Charkey, A., 96n

Charlip, S., 91n

Chemical reactions, in production of electricity, 248. *See also* Batteries; Fuel cells

Chicago, Ill., 26–27; Transit Authority, 135

Chinitz, Wallace, 116n

Chloride salts, 168

Chlorine-lithium cell, 101, 207, 209, 217, 243, 251, 264

"Chopper," in DC motors, 199, 205, 224, 260

Chrysler Corporation, 127, 129, 139–40, 142, 146–47, 222, 255

Cincinnati, Ohio, 17–19, 21, 26–27

Cirpios, G., 301n

Citroen, 124, 139

Civil Aeronautics Board, 287

Clark, J. M., Jr., 124n

Clark, Ronald H., 150n

Clean Air Acts, 288, 292

Coal, 51, 149–50

Coal-gas, as turbine fuel, 146

Cobalt additive, for lead-acid batteries, 223

Cobalt-lead battery, 88

Coke, as fuel, 149–50

Cole, David, 115n

Combustion, in automobile engines, 12, 51, 252–54. *See also* External combustion engines; Internal combustion engines

Commutator (DC) motors, 191, 196–98, 259

COMSAT, 291

Comstock and Westcott, 168

Comuta electric car, 223–24

Condensation problem, in steam engines, 161–63, 256

Condenser, in steam engines, 169–74, 257

"Conjugate" electric motors, 191

Constant pressure reciprocating ICE, 121–22, 252

Control problem, in electric motors, 190, 197–201, 259–60

Coordinating Research Council, 292

Copper, 100, 170, 202, 250, 260, 285

Cornell "Urbmobile," 208

Corrosion, in steam engines, 168

Costs, in automotive transport, 4–7, 268–79, 284; battery systems, 93, 105–8, 250, 272–79; diesel engines, 274–79; of emission control, 130–32, 254; ECEs, 274–79; electric motors, 194, 200–205, 260; electric vehicles, 208–15, 261–62, 274–79; fuel, 125, 134–36, 253, 276–77, 284; fuel cells, 229, 303, 305, 308–9; gas turbines, 146–47, 179, 255, 274–79; hybrid systems, 229, 237, 239, 274–79; ICEs, 272–79; steam engines, 149, 158, 179; Stirling engines, 182–84, 274–79

Counterflow engine, 162–63

Crankcase blow-by, 12, 15, 126, 253; ventilation of, 126, 131

Cruising, in driving cycle, 16–17, 36–37, 44, 48, 210–11, 245

Culture, and automotive transport, 5

Curtiss-Wright, 114–15, 252

Daimler-Benz, 139, 240, 263

Dalin, G. A., 91n, 96n, 210n

Datsun, 159

Davis, J. P., 176n

Dawes, Bailey P., 186n

Dawson, David, 147n, 179n, 184n

DC motors, 190–205, 224, 241, 258–60, 263

Deafness, 25

Deceleration, 16, 36–39, 48, 210–12, 245. *See also* Braking

Deformable solid bodies, as energy source, 109

De Laval turbine wheel, 138

Delta electric car, 224

Denver, Colo., 20

Despic, A. R., 296n

de Tomaso Automobili (Italy), 226

"Detonation," 124n, 125

Detroit, Mich., 284

"Detroit" horsepower, 58

DeWitt, T. W., 103n

Dey, A. N., 100n

Diesel, Rudolph, 60

Diesel-electric hybrid, 240, 263

Diesel engines, 116, 249, 288; characteristics of, 43, 52, 59–64, 247, 265–67,

269; compared with other engines, 141, 149, 156–57, 183–84; costs, 274–79; emissions from, 3*n*, 16–17, 20, 272; noise levels of, 28–31
Diesel fuel, 4, 51, 61–62, 64, 125, 146, 257
Differential turbine, 145–46
Direct conversion of energy, 51, 75–85, 247
Discharge rate, of batteries, 211–12, 215, 233–234
Doble, Abner, 148, 150–51, 154*n*, 162, 169
Doble, Warren, 150
Doble steam engine, 148, 150–52, 165–67, 256–57
Dooley, J. L., 151*n*
Dotto, G. A., 118, 186*n*
Dowtherm A and E, 173–74
Doyle, E. F., 153*n*, 241*n*
Dravnieks, A., 301*n*
Drazic, D. M., 296*n*
"Driving cycle," 35–36; California, 35–36, 154, 245; and electric-car power costs, 210–13; energy requirements for, 245–47; and hybrid power systems, 230–31, 242; Pittsburgh, 37–39, 49–50, 212, 233–36, 245–46, 262
Drumm (nickel-zinc) cell, 93, 228
Dual-ratio carburetor, 125–26
Dunbell, John, 137
DuPont Corporation, 126
Duprey, R. L., 15*n*, 21*n*
Duty cycle. *See* "Driving cycle"
"Dynamic" braking, 48*n*

E-Engines, 51, 78, 247. *See also* Internal combustion engines
Eagle-Pitcher, 93
ECEs. *See* External combustion engines
Edison, Thomas A., 88; Edison effect, 85
Effluent charges (taxes), 289
EG&G, Inc., 291
Eisenberg, M., 100*n*
Electric batteries, 262–63. *See also* Batteries
Electric Fuel Propulsion, Inc., 88, 223
Electric motors: for bicycles, 226, 228; for trucks, 222, 224–25, 228–29
Electric propulsion systems for automobiles, 43, 48*n*, 190–205, 246, 258–61, 270; batteries in, 198, 206–18, 260–63; braking in, 201, 210–13, 240, 260; costs of, 194, 200–205, 260; effect on economy of, 285; emissions from, 206, 273; hybrid, 113, 189, 228, 240, 251, 258, 263; solid state circuitry in, 197, 199–201, 204–5, 223–26, 241, 259–60. *See also* AC induction motors; DC motors; Electric vehicles

Electric Storage Battery Co., 94, 99–100, 222
Electric utilities: night rate reductions, 290–91; power plants, 69, 206, 285
Electric vehicles, 198–99, 206–29, 240, 242, 260–62, 274–79, 290–91; GM research on, 92, 196, 201, 224–25, 229, 304
Electrochemical conversion of energy, 75–85. *See also* Batteries; Fuel cells
Electrochemical energy storage, 86–108, 249–51
Electrochemistry, 248–49
Electrochimica Corporation, 99–100
Electrodes: air, 93–98; in fuel cells, 76–77, 79–84, 248–49
Electrolytes: molten, 93, 100–105, 210, 250–51; organic, 93, 98–100, 250
Electronic control systems, 190, 199–201, 260; solid state circuitry, 197, 199–201, 204–5, 223–26, 241, 259–60
Electrovair (automobile), 92, 196, 224
Electrovan, 224–25, 229, 304
Eliason, R., 101*n*
Elliott, W. E., 97*n*
"Elliptocline" engine, 156–58
Emissions, automotive, 15–31, 267–68, 272–73; control of, 126–28, 130–32, 254, 278–79, 286–92; from diesel engines, 62; from ECEs, 268; difficulty of enforcing standards of, 132, 254, 288; from electric engines, 206, 273; from gas turbines, 146–47, 255, 272; government standards re, 6, 25, 129–33, 253, 279, 288, 292 (*see also under* California); from hybrid systems, 237, 273; from ICEs, 12–20, 26, 28–31, 113–15, 122–36, 252–54, 272, 288; from Rankine-cycle steam engines, 257; from rotary air engines, 188; from Stirling engines, 184, 258, 273. *See also* Air pollution; Carbon monoxide; Hydrocarbons, in automotive emissions; Nitrogen compounds
Energy conversion, for automotive transport, 51–85, 210–17, 230, 247–49; direct, 51, 75–85, 247; electrochemical (*see* Batteries; Fuel cells); requirements for, 35–50, 245–47
Energy Conversion Ltd., 94, 307
Energy storage, 86–112, 230–35, 249–51. *See also* Batteries
Engines, automotive, ideal, 265. *See also* Diesel engines; ECEs; Electric propulsion systems; Gas turbines; Hybrid energy systems; ICEs; Rankine-cycle engines; Rotary engines; Steam propulsion systems; Stirling-cycle engines

Environmental Protection Agency, 16, 292
Ericsson cycle, 71–72, 180, 248
ESB, Inc., 222–23
Ethanol, as fuel, 134–35
Eutectics, in thermal storage, 108, 168
Evaporative losses, 13–15, 128–31, 253–54
Exhaust products. *See* Emissions, automotive; Residuals
Expansion ratio, 163–64, 166
External combustion engines, 84, 237, 242, 247, 285; costs, 274–79, emissions from, 268; in hybrid systems, 108, 230, 235, 262–63; need for dynamic industry for promotion of, 291; noncondensing, 180–89, 257–58; thermal buffering in, 241. *See also* Rankine-cycle engines

Fairchild Semi-Conductor, 201
Faraday disk. *See* Homopolar motor
Faraday unit, defined, 78n
Farbstein, Leonard, 288
Federal Communications Commission, 287
Federal Power Commission, 214, 287
Federal safety legislation, 287
Federal standards for emissions control, 129–33, 253, 279, 288, 292
Fee, G. G., 300n
Ferrell, T. A., 209n
Fiat (Italy), 130, 139, 142, 201, 217
Findl, E., 297n
Fischer, A. K., 104n
Fischman, Leonard L., 135n
Fisher, Joseph L., 135n
Fluid circulation system, in steam engines, 169–73
Flynn, G., Jr., 168n
Flywheels, as energy source, 109–13, 211n, 234–37, 251, 262–63
Ford Motor Co., 128, 130, 139, 141–42, 147, 154, 195, 254–55; sodium-sulfur battery developed by, 103–4, 209–10, 213, 251, 261
Formaldehyde, 147
Forziati, A., 301n
"Free" turbine systems, 143–46
Freeway construction, 7
Freon, as fuel, 159–60, 171–72, 174, 257
Frictional losses, 41, 246
Frysinger, Galen R., 83n, 243, 308n
Fuel, for automotive transport, 4, 51, 133–36, 247, 254, 266–67; costs, 125, 134–36, 253, 276–77, 284; in diesel engines, 4, 61–62, 64, 125, 146, 257; in gas turbines, 66–67, 146; in ICEs, 122, 235 (*see also* Gasoline; Petroleum); in Otto-cycle engines, 55, 57, 266–68. *See also* Air-fuel ratio

Fuel cells, 51, 76–85, 210, 248–49, 295–302; costs, 229, 303, 305, 308–9; in hybrid power systems; 216, 237, 241–44, 262–64; present status of development of, 302–11; in tractors and motorbikes, 228; in trucks, 222, 224–25, 229. *See also* Batteries
Fuel injection, 123, 127–28, 131, 247, 252–54

Gaddy, L. D., Jr., 300n
Garrett Corporation, 177
Gas, compressed, as energy source, 109–10, 180–89
Gas, natural, 135, 146, 210, 254, 298, 305, 308, 311
Gas turbines, 51, 52n, 62–68, 120, 247–249, 258; for automotive vehicles, 137–47, 255, 284; braking with, 255; combined with flywheel, 111; combined with Rankine-cycle engine, 177; costs, 146–47, 179, 255, 274–79; emissions from, 146–47, 255, 272; in hybrid systems, 111, 177, 237, 262–63; performance characteristics of, 43, 113, 265–69
Gaseous pollutants. *See* Emissions, automotive; Residuals
Gasoline, 4, 12–15, 21, 51, 150, 285, 288, 298–99, 305, 308; taxes on, 216
Gasoline engine. *See* Internal combustion engines
Geimer, R. G., 176n
General Electric, 93–94, 139, 201, 224, 226, 296, 298, 304, 307
General Motors' research projects; electric vehicles, 92, 196, 201, 224–25, 229, 304; fuel cell, 301; gas turbines, 139, 141–42, 147, 255; hybrid systems, 237, 239, 263; ICEs, 126–28; lithium batteries, 99, 101, 207, 209, 243, 251; steam engines, 149, 163, 168, 256; Wankel engines, 114, 116, 252; zinc-air cell, 94
General Steam Corporation, 158, 256
General Telephone and Electronics, 90, 93, 98, 135
George, J. H. B., 205n, 222n–225n, 227n
"Getaway" capability, 40
Ghia S. P. A. (Italy), 225–26
Gibbs, R. A., 156–58
Gillis, E. A., 300n
Glass, W., 296n
Glicksman, R., 88n
Globe-Union Corporation, 97, 99
Goldsmith, J. R., 11n
Golf carts, 87, 219, 226, 249
Gould-National Batteries, Inc., 225
Government contracts, need for, 291

Government regulation, 6; of automotive emissions, 25, 129–33, 253, 279, 288, 292 (*see also* California); of automotive safety, 287
Gross national product (GNP), and automotive transport, 4, 284–85
Gulf General Atomic, 94–97, 209
Gulton Industries, 99–100, 201, 219–22
Gyrobus, 111

Harris, R. L., 152n
Harvey, R., 171n
Hass, G. C., 35n
Heat energy storage. *See* Thermal energy storage
Heat engines, 108, 251. *See also* Q-engines; Steam propulsion systems
Heath, C. E., 301n
Heinkel-Hirth (Germany), 139
Helium, 180–81, 184, 248, 258
Henny Kilowatt car, 198–99, 225, 227, 229
Henschel bus, 150, 256
Heredy, L. A., 98n
Herman "cam" engine, 158
Hesson, J. C., 168n
Hexafluoride, 257
Hexafluorobenzene (CP-28), 171–73
Highway construction, 4–5, 7
Hills, A. W. D., 167n
Hodgson, J. N., 176n
Hoess, J. A., 85n, 109n, 110n, 140n, 141n, 146n, 179n, 184n
Hoffman, G. A., 41n, 202–3, 208n, 235n
Holzwarth, Hans H., 138–39
Homopolar motor (Faraday disk), 190–91, 193–96, 259
Horsepower, computation of, 57–58
Hosick, Thomas, 156–58
"Hot-air" engines, 180, 186–89
"Hot-soak" losses from carburetor, 128–29, 253
Housing and Urban Development, Department of, 237
Huebner, George, Jr., 140n, 141n
Hulbert, R. E., 152n
Hybrid energy systems, 220–21, 230–44, 266–67, 270, 284–85; batteries in, 84, 113, 231, 233–37, 242–44, 262–64, 285; Brayton-cycle adaptations as, 121, 177, 258; costs, 229, 237, 239, 274–79; ECEs in, 108, 230, 235, 262–63; electric engines in, 113, 189, 228, 240, 251, 258, 263; emissions from, 237, 273; fuel cells in, 216, 237, 262–64; gas turbines in, 111, 177, 237, 262–63; ICEs in, 230, 235–41, 243, 262–63, 273; steam engines in, 156–58, 263

Hydrazine, as fuel, 134, 228, 241, 254, 299–300, 302–3, 306–8
Hydrocarbon fuel cells, 82–83, 216, 222, 243, 298–99, 302–3, 310
Hydrocarbons, in automotive emissions, 12–21, 24–31, 115, 147, 154, 184, 239, 252, 255; government standards re, 129–30, 253, 288; reduction of, 122, 125–30, 135, 268, 272–73
Hydrocarbons, as fuel, 51, 61–62, 257. *See also* Gasoline
Hydrogen: in automotive emissions, 12; as fuel, 51, 135–36, 254; as working fluid in Stirling engine, 180–81, 184, 239, 248, 258
Hydrogen fuel cells, 210, 224, 229, 243, 295–98, 302–8

ICEs. *See* Internal combustion engines
Idling, in driving cycle, 16, 36–37, 245
Illinois Institute of Technology Research Institute, 301
Inspection, of emission control equipment, 132, 254, 288
Institut Français de Pétrole, 124
Inter-Industry Emission Control (IIEC) Program, 129–30, 254
Internal combustion engines, 3, 51–68, 148, 151, 246–48; air-fuel ratio in, 12–13, 55–57, 123–25, 127–28, 253; costs, 272–79; effect of replacement on economy, 285; emissions from, 12–31, 113–15, 122–36, 252–54, 272, 288; in hybrid systems, 230, 235–41, 243, 262–63, 273; new developments in, 113–36, 252–54; performance comparisons of, 188–89, 197, 242, 265–67
"Internalization," defined, 286–87
International Harvester Co., 139, 255
International Research and Technology Corporation, 188
Interstate Commerce Commission, 287
Ionics, Inc., 296
"IR drop," 79
Iron, 168, 190, 202; in energy cells, 88–90, 94, 98, 105, 249–50
Isobutene, 21
Isuzu (Japan), 116–17

Jaderquist, E., 148n
Janz, G. J., 168n
Japan, 228
Jasinski, Raymond, 87n
Johns Hopkins Applied Physics Laboratory, 111, 251
Johnson, D. E., 158
Jones, Charles, 115
Jones, J. H., 20n

Joule, James, 62–63
Juda, W., 299n
Judge, Arthur W., 145n
Junker (Germany), 139

Kal-Pac Engineering Ltd. (Canada), 117–19, 252
Kalb, W. J., 123n
Kauertz, Eugene, 116; engine, 118, 252
Keen, Charles F., 152, 155, 163
Kennedy, J., 101n
Kerosene, 51, 146, 150, 152, 239, 257, 298
Kikin, B. M., 176n
Kinetic wheel generator, 111
Kinetics, Inc., 159, 256
Kirk, Robert, 147n, 179n, 184n
Kitrilakis, S. S., 153n, 241n
Klein, M., 297n
Kline, L. U., 110n
Kneese, Allen V., 11n
"Knocking," defined, 124n
Kordesch, K. V., 296n
Kovacik, V. P., 161n, 176n, 177n
Kreeger, A. H., 176n
Krypton-85, 268
Kummer, J. T., 103n

Labor unions, 284
Laher Spring Electric Car Corporation, 219
Laithwaite, E. R., 198n
Landaw, S. A., 11n
Landsberg, Hans H., 135n
Lansing-Bagnall (U.K.), 201, 229
Larsen, R. I., 21n, 22n
Lathrop, M., 152
Lauck, F., 51n
Laumeister, B. R., 224n
Lawson, L. J., 240n
Lead-acid cell, 76–79, 86–90, 213–17, 249, 303; costs, 209, 272–75; in electric vehicles, 93, 206–7, 219, 222–27, 229, 239, 249, 261–62; in hybrid systems, 237, 240, 263
Lead additives, 13, 125, 127–28, 131, 253, 267–68, 285
Lead chloride, 168
Lead Wedge electric car, 223–24
Lear Motors Corp., 156, 163, 240, 256, 263
Leesona-Moos Laboratories, 94, 96, 296
Lemale, Charles, 138
Liebhafsky, H. A., 81, 83n, 298n
Linear Alpha, Inc., 201, 225
Linn, W. Leon, 118, 186n
Lithium-chlorine cell, 101, 207, 209, 217, 243, 251, 264
Lithium compounds, 93, 96–105, 168, 176, 250–51, 285; costs, 273

Lithium-nickel fluoride battery, 222, 225
Litz, L. M., 296n
Livingston Electronics, 99
Lockheed Missiles and Space Co., 99–100, 111
Locomobile steamer, 148
"Logmotor," 198
Long Island Railroad, 290
Los Angeles, Calif., 15–21, 24, 28–29, 35–36, 129, 132, 254
Lozier, G. S., 88n
LP (liquid propane) gas, 134–35
Lubrication problem, in steam engines, 165, 174
Luchter, S., 75n, 177n, 178
Lyall, A. E., 91n, 100n
Lynn, D. A., 22n

Macosko, R. P., 176n
Magnesium, 51, 93, 168, 250
Magnuson, Warren, 291
Maintenance costs, 4, 275, 277–78, 284
Mallory (P.R.) & Co., 100, 118–21, 252, 258; engine developed by, 118–21, 186, 239
Mapham, N., 200n, 204n
Marathon Oil Co., 130
March, J. W., 196n
Marco, S. M., 110n
Markette electric car, 227
Marks, A. T., 85n
Mars electric cars, 223
Maschinen Fabrik Oerlikon, 111
Mass transit, 7; minicars for, 237–39
McCormick, J. E., 177n
McCulloch Corporation, 150–51, 155, 166; steam car, 256
McDonald, G. D., 97n
McDonnell Douglas Astropower Laboratory, 92, 96
McKenna, Richard P., 188n
McKenzie, D. E., 98n
Mechanical energy storage, 109–12, 249, 251
Mechanical Technology, Inc., 121–22, 177, 252
Meijer, R. J., 181n
Mercedes-Benz, 116, 252
Mercer, Austin, 116
Mercury, 160–61, 176–77, 195
Metal cells, for energy storage, 86–108, 250. *See also* Batteries
Meteorological conditions, and air quality, 21
Methane, 135, 146
Methanol, 134–35
Metrodynamics Corporation, 226

Meyer, W. A. P., 37n
Mills, Edwin S., 289n
Miner, S. S., 157n
Minicar system, 237–39, 263, 290
Minto, Wallace, 159
MIPB (mono-isopropyl biphenyl), 176–77, 257
Mitsubishi Heavy Industries (Japan), 130
MMM (3M) Co., 177
Mobil Oil Co., 130, 154
Molten electrolytes, 93, 100–105, 210, 250–51
Molten salts, 100–105, 108–9, 166–68, 210, 241, 243, 250–51, 257, 263–64
Monochlorobenzene (CP–27), 177–78
Monsanto Research Corporation, 300, 306
Moore, M. R., 167n
Morehouse, C. K., 88n, 90n
Morgan, D. T., 153n, 241n
Morse, Richard S., 156n
Moss, S. A., 138
Motor coaches. *See* Buses
Motorbike, electric, 228
Mrava, G. L., 176n
Mufflers (afterburners), 6, 127, 131, 253–54
Mungenast, J., 200n, 204n
Myers, P. S., 51n

National Aeronautics and Space Administration, 100, 136, 197, 311
National Air Pollution Control Administration, 16, 129, 154
National Union Electric Corporation, 198, 225
Natural gas, as fuel, 135, 146, 210, 254, 298, 305, 308, 311
Navy, U.S., 149, 151
New Haven Railroad, 149
New Jersey, 132
New York City, 17, 20, 291
Nickel-cadmium cell, 90–93, 206, 217, 222, 224, 250
Nickel-iron cell, 88–90, 249–50
Nickel-lithium fluoride cell, 100, 222, 225
Nickel-zinc (Drumm) cell, 93, 228
Nissan Motor Co., 130, 159
Nitrogen compounds, in engine emissions, 13–31, 115, 133, 147, 154, 184, 239, 255, 258; government standards re, 129–30, 253, 288; reduction of, 122, 126–30, 135, 268, 272–73
Noise pollution, 6, 8, 25–31, 113, 268
North American Aviation, 98, 202
North Star Electric Co., 225
NSU Motor Werke (Germany), 114, 116, 252
Nuclear power, 51, 67, 85, 268, 291

Octane number, defined, 125n
Oerlikon gyrobus, 111, 251
Oil. *See* Petroleum
Oklahoma, 284
Olefins, 21–22, 134
Opel Kadett, 239, 263
Organic compounds, as fuel in steam engines, 171–74, 176–77, 248, 257, 285
Organic electrolytes, 93, 98–100, 250
Oster, E. A., 296n
Otto, Karl, 52
Otto-cycle engines, 3, 16, 42, 52–59, 114–15, 125, 156, 178, 247, 266–69
Oxidants, in automotive emissions, 21
Oxygen, as exhaust product, 12–15

Pacific Lighting Corporation, 135
Palladium, in afterburners, 127
Paraffins, 21, 61, 134
Pargo-Columbia Car Corporation, 219
Parking, for electric minicars, 290
Parsons, Sir Charles, 137–38
Pattison, J. N., 35n
Paxton, Douglas, 158; engine, 150–51
Paxve, Inc., 159
Peattie, C. G., 295n
Peltier effect, 84n
Penn Central Railroad, 290
Pennsylvania, University of, 237, 263
Penny, Noel, 140n
Percival, W. H., 168n
Performance, comparison of propulsion systems, 265–68
Peterson, Karl, 170
Petroleum: industry, 5, 292; products, 4–5, 21–24, 134–35, 284–85. *See also* Gasoline
Peugeot, 123
Phase multiplication, in electric motors, 198
Philadelphia, Pa., 26–27, 239
Philips (N.V.) Research Laboratories, 180–81, 185, 258
Photochemical reactivity of petroleum fuels, 21–24
Photoelectric switching, 197
Pittsburgh driving cycle, 37–39, 49–50, 212, 233–36, 245–46, 262
Plastic bodies, for electric cars, 208, 261
Platinum, 83–84, 127, 301, 311
Plattner, J. W., 296n
Polarization, defined, 79–80
Pollution, air. *See* Air pollution
Pooler, F., Jr., 23n
Porsche (Germany), 116
Positive crankcase ventilation (PCV), 126, 131
Potassium, 102, 176

Power needs: in automobile engines, 12, 57–58, 245–46; in electric motors, 201–2, 210–17. *See also* Energy conversion

Powers, J. P., 101n

Pratt & Whitney (United Aircraft), 296, 307, 311

"Pressure ratio," 163

Pritchard, E., 156n, 236n; steam engine built by, 155–56, 163, 166, 256

Private enterprise, in automotive industry, 5–6

Propane, as fuel, 51, 134–35, 254, 298, 302, 311

Propylene carbonate, 250

Public interest in automotive transport, 3–11

Public utilities, 287–88

Puchy, C. G., 240n

Pulse control modulation (PCM), in AC motors, 199–200

Pure Oil Corporation, 124

Q(heat)-engines, 51, 63n, 68–75, 84–85, 247–48. *See also* Rankine-cycle engines; Stirling-cycle engines

Rabenhorst, David W., 111–12, 251

Ragone, D. V., 95n

Rail transit, 7, 20, 149–50, 290

Rankine-cycle (steam) engines, 43, 51, 71–75, 246, 248–49, 256–57, 266–69, 272; application to automotive vehicles, 148–79, 284–85; costs, 274–79

Rao, M. L. B., 100n

Rapid transit, rail, 20

Rateau, A., 139; compressor, 138

Rawcliffe, G. H., 197

"Reactor" exhaust manifold, 126–28

Recharging, in energy storage, 86, 249, 251, 261. *See also* Charging

Recht, H. L., 98n

Reciprocating engine, 121–23, 247. *See also* Internal combustion engines

Redding, T. E., 177n

Regeneration of batteries, in hybrid systems, 233, 235

Regenerative braking, in electric engines, 201, 210–13, 240, 260

Regenerator, in Stirling-cycle engines, 181

Regulation of automotive externalities, 6, 25, 129–33, 253, 279, 286–92

Reluctance switching, in electric motors, 197

Renault (France), 116–17, 124, 217; electrified, 198–99, 223, 225

Renner, Roy, 179n, 188n, 238n, 278n

Rental: of batteries, 207, 209–10, 261; of ICE-electric hybrid, 237

Research, on emissions control, 122–36

Research and development, need for public sponsorship of, 289–92

Reserve energy storage, in hybrid systems, 231–33

Residuals: disposal of, 4–6; from nuclear power plants, 268. *See also* Emissions, automotive

Resistance: air, 41, 246; rolling, 41–43, 246

"Revolving block" engine, 116

Revolvolator Co., 219

"Rhombic" drive, in Stirling-cycle engine, 181, 258

Rice, R. H., 139

Ridker, R. G., 11n

Rightmire, R. A., 101n

Rolfsmeyer, Melvin, 116

"Rolling resistance," 41–43, 246

Rolls-Royce, 116, 252

Rollsock seal, in Stirling-cycle engines, 182

Rotary engines, 113–22, 163, 186–99, 239, 252, 258

Rover Co., Ltd., 139–40, 142, 146–47, 255

Rowan Controller Co., 225–26

Rubidium, 168, 176

St. Louis, Mo., 26–27

Salts, molten, 100–105, 108–9, 166–68, 210, 241, 243, 251, 257, 263–64

San Diego Gas & Electric Co., 135

San Francisco, Calif., 26–27; bus project in, 290

Sanders, N., 85n

Sarasota vapor engine, 159

Saturates, in petroleum fuels, 21–22

SCRs (silicon-controlled rectifiers), 197, 199–200, 204, 224–25

Sealed Power Corporation, 123

Seebeck effect, 84n

Seiger, H. N., 91n, 100n

Sentinel truck, 149–50, 256

Shair, R. C., 100n

Shaw, M., 99n

Sherbondy turbocharger, 138

Shimotake, H., 104n, 105n, 168n

Shipps, P. R., 95n

Shuldiner, O. W., 20n

Silicon-controlled rectifiers (SCRs), 197, 199–200, 204, 224–25

Silicon steel, 190, 202

Silver: cadmium cell, 93, 217, 250; salts, 168; zinc cell, 91–93, 95, 209–10, 217, 224, 227–28, 250, 261, 303

Simmons, E. L., 84n, 301n

Singman, D., 301n

Skrotski, B. G. A., 143n

Slabiak, W., 240*n*
SLI lead-acid cell, 88–89, 215, 249
Small Business Administration, 291
Smith, F. Beckley, Jr., 37*n*
Smith, Richard, 170; steam engine developed by, 163, 170
Smith's Electric Vehicles Ltd., 229
Smog, 21–24, 129, 132, 254
"Smokejack," 137
Smyth, J. R., 88*n*
Snoke, D. R., 176*n*
Sodium, 93, 96, 98, 102–4, 250, 261, 285
Sodium-sulfur battery, 103–4, 209–10, 213–14, 223, 251, 261
Solid state circuitry, in electric motors, 197, 199–201, 204–5, 223–26, 241, 259–60
Soluble fuel cells, 299–302
Space program: electric-motor research in, 197; Rankine-cycle systems for, 161, 176–77
Spark plugs with a precombustion chamber, 125–26
Speed: effect on emissions, 16, 18–20; in energy conversion equations, 36–39, 245–46
Speed control circuitry, in electric cars, 210, 260
Springs, as energy source, 109
"Squirrel-cage" AC motors, 190–205, 240–41, 258–60, 263
Stall torque, in electric motors, 202–4
Standard Oil Co. of Ohio, 101, 130
Stanford Research Institute, 21
Stanley Engineering (U.K.), 229
Stanley steamer, 148, 159, 256
Starkey, W. L., 110*n*
Starkman, E. S., 134*n*
StaRRcar, 208, 219
Stationary lead-acid cells, 88–89
Steam Engine Systems Corporation, 156, 166–67
Steam Power Systems, 156
Steam propulsion systems, 75, 247–48, 256–57; binary vapor cycle in, 160–61, 177; in buses, 149–50, 156, 256–57, 290; condenser in, 169–74, 257; costs, 149, 158, 179; Doble, 148, 150–52, 165–67, 256–57; in hybrid systems, 156–58, 263; operational characteristics of, 48*n*, 161–69; Pritchard, 155–56, 163, 166, 256; in rail transit, 149–50; thermal efficiency in, 149, 151, 159–61, 164, 170–75, 177–78; in trucks, 149–50, 156, 256–57; use of organic compounds as fuel in, 171–74, 176–77, 248, 257, 285. *See also* External combustion engines; Rankine-cycle engines

Steel, 5, 168–70, 190, 202, 261, 285
Steinhagen, N. K., 15*n*, 126*n*, 127*n*
Stelber Industries, 226–28
Stern, A. C., 11*n*, 127*n*
Sternlicht, B., 178*n*, 179
Stir-lec (electric car), 224, 239, 263
Stirling, Robert, 180
Stirling cycle, 71, 180, 248, 268
Stirling-cycle engines, 51, 181–86, 248–49, 257–58, 270; costs, 182–84, 274–79; emissions from, 184, 239, 258, 273; GM research re, 168, 263
Stolze, F., 138
Storage cells, 249–50; commercially available, 87–93; under development, 93–105. *See also* Batteries
Storto, E., 300*n*
Stratified charge systems, 123–26, 131, 253
Stratton, L. J., 205*n*, 222*n*
Strumpf, Johannes, 162, 256
Sturm, W. E., 227*n*
Subsidies, in automotive transportation, 288–89
Sulfur battery, 103–4, 209–10, 213–14, 223, 251, 261
Sulfur oxides, as exhaust product, 13, 17, 20, 147, 272–73
Sulkes, M., 91*n*
Sun Oil Co., 130
Supercharging, defined, 125
Surd, D. J., 84*n*, 301*n*
Sweeney, M. P., 35*n*
Swinkels, D. A. J., 101*n*
Switching, electronic. *See* Solid state circuitry
Synchronous electric motors, 191, 193, 197
Szego, G. C., 176*n*, 300*n*, 305*n*

Tailpipe. *See* Mufflers
Tanks, military, 152, 240
Tarmy, B. L., 301*n*
Taxes, on automotive fuel, 216, 262, 288–89
Taylor, Theodore B., 186*n*
Technology, new, need for public sponsorship of, 289–92
Technology Research Institute, Illinois Institute of, 301
Texaco, 253; work done on TCP engine by, 124–25
Texas, 284
Texas Instruments, 93
Thermal battery, 234, 237, 241, 251, 262–63
Thermal efficiency: in battery–fuel cell hybrid systems, 243–44, 249; in diesel engines, 61–62; in gas turbines, 66, 138–39,

248; in noncondensing gas cycle external combustion engines, 180-81; in Otto-cycle engines, 55-57, 247; in Rankine cycle, 72-74, 256-57; in steam engines, 149, 151, 159-61, 164, 170-75, 177-78

Thermal (heat) energy storage, 108-9, 166-68, 241, 249. *See also* Thermal battery

Thermal Kinetics Corporation, 153

Thermal reactors, effect of lead on, 127

Thermionic devices, 51, 84-85, 153, 249

Thermo Electron Corporation, 153-55, 163, 170, 241, 256-57, 263

Thermoelectric devices, 51, 84-85, 249

Thermoplastics, in electric vehicles, 208

Thiophene (CP-34), 171-73, 241, 257, 263, 269

Thomson, Elihu, 139

Thornton, Richard D., 197, 205

Thyristors (silicon-controlled rectifiers), 197, 199-200, 204, 224-25

Thyssen (Germany), 138

Tires, effect on automotive performance, 41-43

Toluene, 172-74, 285

Toni, J. E. A., 97n

Torque, defined, 246

Toshiba (Japan), 201

Toyo Kogyo Ltd. (Japan), 116, 130, 252

Tractor: electric, 228, 240; steam, 151-52, 256

Transistors, in electric motors, 199-200, 204

Transmission, mechanical, 42, 75, 240, 246, 278

Transportation, Department of, 237, 289-90

Travelair (steam airplane), 149

Trip-making behavior, 7-10

Trochoidal rotor engines, 116-17

Trucks, 3-4, 17, 40; electric, 222-25, 228-29, 304; gas turbines for, 141, 147, 255; hybrid systems for, 239-41, 243; steam-powered, 149-50, 156, 256-57; Stirling-cycle engines for, 183

Tschudi Engine Corporation, 116, 252

Tschudi, T., 116n, 118

Tson, M., 168n

Turbines: in buses and trucks, 141, 147, 255; in rail transit, 290. *See also* Gas turbines; Vapor turbines

Turboalternator, 239-40

Turbocharger, Sherbondy, 138

Turbocompound engine, 120. *See also* Gas turbines

Turbo-electric engine, 240, 263

Tustin, A., 197n

Uniflow steam engine, 162-63, 256

Union Carbide Corporation, 228-29, 296, 300, 304, 306

Unions, labor, 284

United Aircraft, 296, 307, 311

United Kingdom, electric-car research in, 228-29

Urban areas: automotive emissions in, 15-21; driving cycles, 242 (*see also* California driving cycle; Pittsburgh driving cycle); effect of automobiles on, 3-11

"Urbmobile," 208

Uyehara, O. A., 51n

Van Witteveen, R. A. J. O., 180n

Vapor: binary, cycle in steam engines, 160-61, 177; engines, 158-59; generator, 169-70, 257; losses, 13-15, 128-31, 253-54; turbines (Rankine-cycle), 156, 175-79, 240, 263

Variable cutoff, in steam engines, 164-65, 256

Velocity ratio, in steam turbines, 175

Vinal, G. W., 86n, 90n

Virmel engine, 117, 121, 252

Volkswagen, 123, 201, 212-13, 217, 253, 261

Volvo, 142

W & E Vehicles (U.K.), 229

Walker, M. L., Jr., 36n, 50n

Walton, J. N., 150n

Wankel, Felix, 114, 252; rotary engine, 113-17, 120, 122, 239, 252

Warren, G. B., 121n

Washington, D.C., 26-27

Waste products. *See* Emissions, automotive; Residuals

"Water rate" and thermal efficiency, 159-61, 256

Weaver, R., 101n

Weber, N., 103n

Weight, of electric engines, 201-4, 260-61

Werner, C. F., 160n

Werth, J., 101n

West Penn Power Co., 226-27

Westinghouse Research Laboratories, 98, 105, 227, 296

White steamer, 148, 256

Whittaker Corporation, 99

Whittle, Frank, 139

Wilcox, H., 196n

Williams, F. C., 198

Williams Brothers steam engines, 154-55, 158, 163, 165-66, 169, 256

Williams Research Corporation, 146

Witzky, J. E., 124n

Wolozin, Harold, 289n
Wood, Gar, 227
Wood, Gar, Jr., 226–27
Wynveen, R. A., 301n

Xylene, 285

Yardney Electric Corporation, 90, 92–94, 96, 227–28, 250

Yuba Manufacturing Co., 151–52, 256

Zaromb Research Corporation, 98
Zeislar, F. L., 195n
Zimmer, C. E., 21n, 22n
Zinc, in energy cells, 91–96, 209–10, 217, 224, 227–28, 250, 261, 285, 303
Zinc-chloride, 168
Zoning, effects of, 7–8

THE JOHNS HOPKINS UNIVERSITY PRESS

This book was composed in Press Roman text with
Univers Bold display by the Jones Composition Company.
It was printed on 60-lb. Sebago by Universal Lithographers, Inc.
and bound in Joanna Arrestox cloth by L. H. Jenkins, Inc.